Related Titles

Hippler, R., Kersten, H., Schmidt, M., Schoenbach, K. H. (eds.)

Low Temperature Plasmas
Fundamentals, Technologies and Techniques

2008
ISBN 978-3-527-40673-9

d'Agostino, R., Favia, P., Kawai, Y., Ikegami, H., Sato, N., Arefi-Khonsari, F. (eds.)

Advanced Plasma Technology

2008
ISBN 978-3-527-40591-6

Ostrikov, K., Xu, S.

Plasma-Aided Nanofabrication
From Plasma Sources to Nanoassembly

2007
ISBN 978-3-527-40633-3

Smirnov, B. M.

Plasma Processes and Plasma Kinetics
580 Worked-Out Problems for Science and Technology

2007
ISBN 978-3-527-40681-4

Smirnov, B. M.

Principles of Statistical Physics
Distributions, Structures, Phenomena, Kinetics of Atomic Systems

2006
ISBN 978-3-527-40613-5

d'Agostino, R., Favia, P., Oehr, C., Wertheimer, M. R. (eds.)

Plasma Processes and Polymers
16th International Symposium on Plasma Chemistry Taormina/Italy June 22-27, 2003

2005
ISBN 978-3-527-40487-2

Boris M. Smirnov

Cluster Processes in Gases and Plasmas

WILEY-VCH Verlag GmbH & Co. KGaA

The Author

Prof. Boris Smirnov
Institute for High Temperatures
Russian Academy of Sciences
Moscow, Russian Federation
bmsmirnov@gmail.com

Cover Picture

Peter Hesse, Berlin, Germany

Library of Congress Card No.: applied for

British Library Cataloguing-in-Publication Data
A catalogue record for this book is available from the British Library.

Bibliographic information published by the Deutsche Nationalbibliothek
The Deutsche Nationalbibliothek lists this publication in the Deutsche Nationalbibliografie; detailed bibliographic data are available on the Internet at http://dnb.d-nb.de.

Typesetting le-tex publishing services GmbH, Leipzig
Printing and Binding Strauss GmbH, Mörlenbach

Printed in the Federal Republic of Germany
Printed on acid-free paper

ISBN 978-3-527-40943-3

Contents

Cluster Processes in Gases and Plasmas. Boris M. Smirnov

ISBN: 978-3-527-40943-3

Preface

As a system of identical bound atoms or molecules, a cluster is analogous to various types of particles such as Aitken particles, aerosols, dust, mist in gases, solid and liquid particles in suspensions and emulsions, colloids in colloid solutions, grains in solid solutions, and so forth. In all these systems a condensed phase is present in the form of individual particles inside some matter. Cluster concepts may be extended to particle structures such as fractal aggregates, chain aggregates, and fractal fibers. These concepts with respect to clusters and small particles in gases and plasmas are the subject of this book.

The goal of the book is to analyze some properties of clusters ranging in size from tens to billions of atoms and processes involving clusters and cluster structures. Being guided by simple methods for this analysis, we apply mostly analytic theoretical methods and conduct our analysis in the form of individual problems, so that each problem answers a certain question with respect to cluster parameters or cluster processes. These problems may be useful both for students studying the physics of clusters or their applications and for professionals who seek an explanation for a certain cluster question and the method for solving a particular cluster problem. To connect these problems with the contemporary state of cluster physics, some results of original investigations are presented.

Cluster Processes in Gases and Plasmas. Boris M. Smirnov

ISBN: 978-3-527-40943-3

1
Introduction

Small macroscopic particles and clusters as systems of a finite number of bound atoms or molecules have been investigated since the nineteenth century. In this respect, one recalls some results from that century, among which is the conclusion by Michael Faraday [1, 2] that the radiation of candle flames results from the emission of soot particles. Another example relates to the generation of fine gold particles located in colloidal suspension [3]. Then one can govern by the final particle with the addition of salt, whereas adding micellar surfactant molecules (amphilic molecules) will prevent gold particles from sticking. This method was developed and became widespread in the twentieth century. In addition to this, Ostwald investigations [4, 5] explained the character of the growth of small particles as the result of atom evaporation from small particles and their attachment to large particles, later called "Ostwald ripening". One more example relates to so-called Aitken particles, which are small particles that scatter sunlight in the upper atmosphere [6, 7]. All these results became the basis for the contemporary understanding of the behavior of small particles and clusters and also provided the methods for their generation and detection.

Some cluster concepts arise from the study of aerosols, dust and mist particles in the Earth's atmosphere, domains and grains in solid solution, solid and liquid particles in suspensions and emulsions, islands and films on surfaces, and colloids in liquid solution due to the behavior of these objects and processes in gases and plasmas involving small particles and clusters. In particular, such concepts of cluster growth as coagulation and coalescence emerged from the concepts of "blood coagulation" and "coalescence of bones" in physiology. Therefore, the contemporary analysis of the properties of clusters and their behavior in gases and plasmas will be based on the results of previous investigations for analogous objects.

As specific physical objects, clusters arose in the eighties after the discovery of magic numbers, that is, numbers of atoms (or molecules) at which solid clusters have a heightened stability compared to that at neighboring sizes. Magic numbers correspond to complete atomic structures for clusters as systems of bound atoms or molecules, and the values of magic numbers depend on the character of interaction between cluster atoms. Cluster parameters as a function of the number of atoms have extrema at the magic numbers of atoms. For example, a cluster with a magic number of atoms has a higher binding energy of atoms and ionization potential

Cluster Processes in Gases and Plasmas. Boris M. Smirnov

ISBN: 978-3-527-40943-3

than clusters with neighboring numbers of atoms. As physical objects, clusters occupy an intermediate position between atomic particles (atoms and molecules) on the one hand and macroscopic atomic systems (solids and liquids) on the other.

The goal of this book is the analysis of certain properties of clusters ranging in size from tens to billions of atoms as well as the analysis of processes involving clusters and cluster structures. These systems include cluster sources and generators where clusters are formed and grow under nonequilibrium conditions in gases or plasmas, and also the complex plasma that is an ionized gas containing particles or clusters. The structures formed by solid clusters or solid particles such as fractal aggregates, chain aggregates, and fractal fibers are studied in the book. Guided largely by simple methods based on analytic theoretical methods, we perform our analysis using individual problems, each of which considers a certain aspect of cluster behavior or cluster processes. The results of these problems will allow us to draw certain conclusions and make certain evaluations regarding clusters and complex plasmas. The problems may be useful both for students studying the physics of clusters or their applications as well as for professionals who seek an explanation for a particular cluster question.

The problems under consideration in this book are related to two types of cluster applications. First, understanding cluster processes allows us to analyze some natural and laboratory processes and phenomena. For example, in this manner one can optimize combustion processes from the standpoint of the formation of soot and solid combustion products. Second, an understanding of cluster behavior, together with a new experimental technique that allows one to analyze nanoparticles, will lead to the development of new branches of nanotechnology, for example, production of nanometer porous or sandwichlike films. We will give a more detailed list of cluster applications in the conclusions.

The issues raised in this book are represented in various books of cluster physics such as [8–22]. Moreover, the material in [17] serves as a basis for this book.

Part I Cluster Properties and Cluster Processes

2
Fundamentals of Large Clusters

A cluster is, by definition, a system of a finite number of identical bound atoms, and in this context clusters possess an intermediate position between atoms and molecules, on the one hand, and bulk solids and liquids, on the other. In principle, the definition of a cluster may be broader and also include systems of identical molecules or systems of repelling particles. In this book we will use clusters consisting of identical bound atoms; such clusters are almost located in a vacuum or in gases, that is, these clusters are almost free from interaction with the environment.

Clusters may be considered a specific physical object due to magic numbers, which are numbers of atoms for solid clusters located in a gas at which the population of clusters as a function of the number of cluster atoms has local minima [23–30], and various cluster parameters have extrema at these numbers of atoms as a function of an atomic number. Correspondingly, these cluster parameters are nonmonotonic functions of their size. But if some cluster properties or processes involving clusters are determined by other means and a nonmonotonic dependence of their parameters on cluster size is not of principle, one can simplify this dependence for these cluster parameters taking it to be monotonic. In this case we have a cluster model in which a cluster is a piece of a bulk atomic system.

The simplest model of this type that will be used throughout the book is the liquid drop model for clusters. Within the framework of this model, a cluster is modeled by a drop that is cut off from the bulk liquid, and the atomic density inside the cluster and a macroscopic liquid are identical. Evidently, this model is valid more or less for liquid clusters and its applications for solid clusters corresponds to an average of cluster parameters over their size. Therefore, the size dependence for cluster parameters is monotonic, and the liquid drop model is useful for cluster properties that are not strongly size selective.

Clusters have a developed surface, that is, a number of surface atoms in clusters may be a significant part of the total atoms in a cluster. Therefore, some cluster properties and processes are determined by interactions involving surface atoms, and the behavior of surface atoms influences cluster properties, including the stability of charge clusters and processes of atom evaporation from the cluster surface, as well cluster growth as a result of atom attachment to the cluster surface.

Cluster Processes in Gases and Plasmas. Boris M. Smirnov

ISBN: 978-3-527-40943-3

2.1
Models for Large Clusters and Processes with Their Participation

Problem 2.1

Determine the diffusion cross section in a collision between a large spherical cluster and an atomic particle within the framework of the hard sphere model.

The cross section of scattering of atomic particles (electrons, ions, atoms) is determined by the interaction potential between colliding particles. If an atomic particle is scattered on a large cluster, it interacts with cluster surface atoms, and a region of strong interaction is of the order of atomic size near the cluster surface. For a large cluster this size is small compared to the cluster radius r_o, and this leads to the hard sphere model for interactions between an atomic particle and a cluster [17, 31] with the following model potential:

$$U(r) = \begin{pmatrix} 0, & r > r_o \\ \infty, & r \leq r_o \end{pmatrix} . \tag{2.1}$$

In the case of an elastic scattering in the collision of an atomic particle with a cluster, this interaction potential leads to the following relation between the scattering angle ϑ and the impact parameter ρ of collision in the classical character of motion (Figure 2.1):

$$\frac{\rho}{r_o} = \sin\frac{\pi - \vartheta}{2} = \cos\frac{\vartheta}{2} .$$

This gives for the differential cross section of scattering

$$d\sigma = 2\pi\rho d\rho = \pi r_o^2 d\cos\vartheta . \tag{2.2}$$

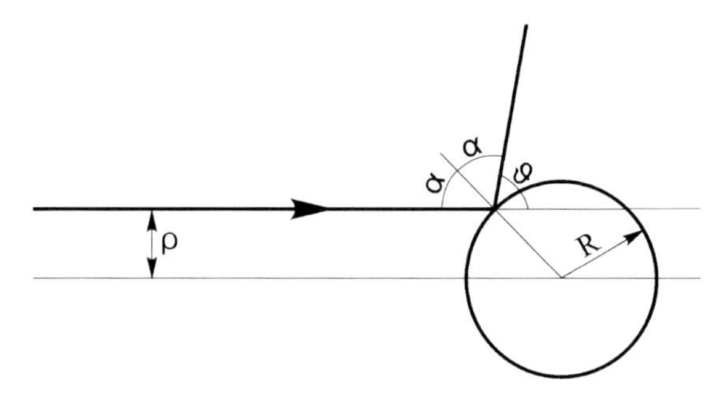

Fig. 2.1 Elastic scattering of an atomic particle on a cluster within the framework of the hard sphere model for collision. The following relations follow from this figure $\rho = R\sin\alpha$, $\vartheta = \pi - 2\alpha$, where ρ is the impact parameter of collision, R is the hard sphere radius, ϑ is the scattering angle.

From this it follows for the diffusion (transport) cross section for elastic scattering of an atomic particle on a cluster within the framework of the hard sphere model

$$\sigma^* = \int (1 - \cos\vartheta) d\sigma = \pi r_o^2 . \tag{2.3}$$

Problem 2.2

Determine the cross section of absorption of an atomic particle by a large spherical cluster within the framework of the hard sphere model.

We assume that each contact of an atomic particle with the cluster surface leads to particle attachment, so that the cross section σ_{abs} of particle absorption is equal to the cross section of their contact, that is,

$$\sigma_{abs} = \int_{\rho=0}^{\rho=r_o} d\sigma = \pi r_o^2 . \tag{2.4}$$

This is also the cross section of charge transfer in the collision of an atomic particle with a neutral cluster if each contact between them leads to charge transfer for a cluster.

Problem 2.3

Derive the relation between the number of cluster atoms $n \gg 1$ and the cluster radius within the framework of the liquid drop model.

Using the liquid drop model, we assume a cluster to be similar to a liquid drop whose density coincides with that of the corresponding bulk liquid. We define the Wigner–Seitz radius r_W [32, 33] such that the volume of a sphere of this radius is equal to the volume of a bulk system per atom. Then the radius r_n of a cluster containing n atoms is equal within the liquid drop model to

$$n = \left(\frac{r_n}{r_W}\right)^3 \quad \text{or} \quad r_n = r_W n^{1/3} . \tag{2.5}$$

On the other hand, the number of cluster atoms n is

$$n = \frac{4\pi\rho r_n^3}{3m_a} ,$$

where m_a is the atomic mass and ρ is the liquid density. From this it follows

$$r_W = \left(\frac{3m_a}{4\pi\rho}\right)^{1/3} \quad \text{or} \quad r_W = \frac{r_n}{n^{1/3}} . \tag{2.6}$$

The Wigner–Seitz radius is the fundamental cluster parameter. Tables 2.1 and 2.2 contain the values of the Wigner–Seitz radius for some metals and semiconductors [17, 34]. The mass number density p is taken at the melting point: in abesence this information p is taken at room temperature.

Table 2.1 Parameters of some metals and semiconductors and clusters of these elements. T_m and T_b are, respectively, the melting and boiling points for bulk materials, ρ is the density of a liquid at the melting point (in the absence of this value, the material density at room temperature is given), r_W is the Wigner–Seitz radius, and k_o [17] is the reduced rate constant at $T = 1000$ K.

Element	T_m, K	T_b, K	ρ, g/cm^3	r_W, Å	k_o, 10^{-11} cm^3/s
Li	454	1615	0.512	1.71	17
Be	1560	2744	1.69	1.28	7.9
Na	371	1156	0.927	2.14	14
Mg	923	1363	1.584	1.82	9.8
Al	933	2730	2.375	1.65	7.6
K	336	1032	0.828	2.65	16
Ca	1115	1757	1.378	2.26	12
Sc	1814	3103	2.80	1.85	7.4
Ti	1941	3560	4.11	1.66	5.8
V	2183	3680	5.5	1.54	4.8
Cr	2180	2944	6.3	1.48	4.4
Fe	1812	3023	6.98	1.47	4.2
Co	2750	5017	7.75	1.44	3.9
Ni	1728	3100	7.81	1.44	3.9
Cu	1358	2835	8.02	1.47	4.2
Zn	693	1180	6.57	1.58	4.5
Ga	303	2680	6.08	1.65	4.8
Ge	1211	3106	5.32	1.63	4.5
Rb	312	961	1.46	2.85	13
Sr	1050	1655	6.98	1.71	4.5
Zr	1128	4650	5.8	1.84	5.1
Nb	2750	5100	8.58	1.68	4.3
Mo	2886	4912	9.33	1.60	3.8
Rh	3237	3698	10.7	1.56	3.8
Pd	1828	3236	10.4	1.60	3.5
Ag	1235	2435	9.32	1.66	3.8
Cd	594	1040	8.00	1.77	4.3
In	430	2353	7.02	1.86	4.7
Sn	505	2875	6.99	1.89	4.7
Sb	904	1860	6.53	1.95	5.0
Cs	301	944	1.843	3.06	12
Ba	1000	1913	3.34	2.54	7.9
La	1191	3737	5.94	2.10	5.4
Hf	2510	4876	12	1.81	3.5
Ta	3290	5731	15	1.68	3.0
W	3695	5830	17.6	1.61	3.6
Re	3454	5880	18.9	1.58	2.7
Os	3100	5300	20	1.56	2.6
Ir	2819	4700	19	1.59	2.6
Pt	2041	4098	19.8	1.57	2.6
Au	1337	3129	17.3	1.65	2.8
Hg	334	630	13.6	1.80	3.3
Tl	577	1746	11.2	1.93	3.8
Pb	600	2022	10.7	1.97	3.9
Bi	544	1837	10.0	2.02	4.1
Th	2023	5061	11.7	1.99	3.8
U	1408	4404	17.3	1.77	2.9
Pu	913	3500	16.7	1.70	2.9

Table 2.2 Lanthanide parameters. T_m and T_b are, respectively, the melting and boiling points for bulk lanthanides, ρ is the density of a liquid at the melting point (in the absence of this value, the material density at room temperature is given), r_W is the Wigner–Seitz radius, and k_o is the reduced rate constant at $T = 1000$ K [34].

Lanthanide	T_m, K	T_b, K	ρ, g/cm^3	r_W, Å	k_o, 10^{-11} cm^3/s
Ce	1072	3669	6.55	1.97	5.1
Pr	1204	3785	6.5	1.99	5.1
Nd	1294	3341	7.0	1.96	4.9
Pm	1441	3000	7.2	1.99	4.8
Sm	1350	2064	7.5	1.98	4.7
Eu	1065	1870	5.13	2.23	6.1
Gd	1586	3539	7.4	1.95	4.8
Tb	1629	3396	7.65	2.04	4.7
Dy	1685	2835	8.37	2.05	4.4
Ho	1747	2968	8.34	2.01	4.4
Er	1802	3136	8.86	2.00	4.3
Tm	1818	2220	8.56	2.27	4.4
Yb	1097	1466	6.21	2.03	5.4
Lu	1936	3668	9.3	2.02	4.2

Problem 2.4

Determine the rate constant of atom attachment to a cluster surface within the framework of the liquid drop model from a surrounding vapor. Consider the atom attachment process as a result of an atom–cluster contact.

The cluster M_n as a system of n bound atoms M is considered on the basis of the liquid drop model as a bulk liquid drop of spherical shape whose density coincides with the density of the bulk liquid. Then the radius r_n is expressed through a number of cluster atoms n by the relation (2.5). The process of atom–cluster collision with atom attachment to the cluster surface is characterized by the cross section $\sigma_n = \pi r_n^2 = \pi r_W^2 n^{2/3}$, and the rate of atomic attachment, that is, the total atomic flux to the cluster surface, is equal to [17, 35, 36]

$$\nu_n = N v \sigma_n = N k_o n^{2/3} \,, \quad \text{where } k_o = \sqrt{\frac{8T}{\pi m_a}} \pi r_W^2 \,. \tag{2.7}$$

Here v is the average atomic velocity, T is the gaseous temperature, and m_a is the atomic mass. Tables 2.1 and 2.2 give the values of the reduced rate constant k_o for some metals or semiconductors at a temperature of 1000 K.

Problem 2.5

Obtain the first two expansion terms for the total cluster energy E over a small parameter $1/n^{1/3}$ (n is the number of cluster atoms) for a large cluster at room temperature.

Introducing the binding energy ε_o per atom for a macroscopic system of atoms, we have for the cluster energy E, accounting for the surface energy E_{sur},

$$E = -\varepsilon_o n + E_{sur} ,$$

where E_{sur} is the surface energy, which for a liquid drop is proportional to the surface area. We therefore represent the surface energy in the form

$$E_{sur} = An^{2/3} , \tag{2.8}$$

where A is the specific surface energy. Correspondingly, the total cluster energy is given in the form of expansion over a small parameter $n^{-1/3}$ [37]:

$$E = -\varepsilon_o n + An^{2/3} . \tag{2.9}$$

In particular, from this it follows for the atom binding energy $\varepsilon_n = -dE_n/dn$ in a large liquid cluster

$$\varepsilon_n = -\frac{dE_n}{dn} = \varepsilon_o - \frac{\Delta\varepsilon}{n^{1/3}} , \quad \Delta\varepsilon = \frac{2}{3}A . \tag{2.10}$$

One can connect the specific surface energy A for a liquid drop model with the specific surface tension γ of the liquid, because, according to the definition of the surface tension, the drop surface energy is

$$E_{sur} = 4\pi r_o^2 \gamma .$$

Comparing the expressions for the surface energy of a cluster, we have [17]

$$A = 4\pi r_W^2 \gamma . \tag{2.11}$$

Table 2.3 contains the values of the energy parameters ε_o and A for some metals and semiconductors [17]. Next, the atom binding energy of a bulk system ε_o may be found from the temperature dependence for the saturated vapor pressure p_{sat}, which has the form [38, 39]

$$p_{sat}(T) = p_o \exp\left(-\frac{\varepsilon_o}{T}\right) ,$$

and the parameters of this formula are given for liquids in Table 2.3.

Table 2.3 Energetic parameters of clusters of some elements. ε_o is the binding energy of bulk per atom, A is the specific surface energy, and p_o is a preexponent for the saturated vapor pressure over a macroscopic plane surface [17, 34].

Element	ε_o, eV	A, eV	p_o, 10^5 atm	Element	ε_o, eV	A, eV	p_o, 10^5 atm
Li	1.16	0.99	1.3	Rh	5.42	3.8	7.7
Be	3.12	1.4	23	Pd	3.67	2.9	6.0
Na	1.08	0.73	0.63	Ag	2.87	2.0	15
Mg	1.44	1.4	1.1	Cd	1.06	1.4	14
Al	3.09	2.0	11	In	2.38	1.5	0.17
K	0.91	0.62	0.37	Sn	3.10	1.6	0.24
Ca	1.67	1.4	0.72	Sb	1.5	1.04	0.03
Sc	3.57	–	8	Cs	0.78	0.51	0.24
Ti	4.82	3.2	300	Ba	1.71	1.4	0.17
V	5.1	3.7	150	Ta	8.1	4.7	250
Cr	3.79	2.4	30	W	8.59	4.7	230
Mn	2.44	–	2.0	Re	7.62	5.3	42
Fe	3.83	3.0	11	Os	7.94	4.7	230
Co	4.10	3.1	35	Ir	6.5	4.5	130
Ni	4.13	2.9	7	Pt	5.4	3.6	40
Cu	3.40	2.2	15	Au	3.63	2.5	12
Zn	1.22	1.5	1.6	Hg	0.62	1.23	7.7
Ga	2.76	1.5	2.0	Tl	1.78	1.3	2.0
Rb	0.82	0.54	0.28	Pb	1.95	1.4	1.0
Sr	1.5	1.3	0.32	Bi	1.92	1.2	50
Zr	6.12	3.8	52	Th	5.6	–	5
Nb	7.35	4.5	360	U	4.95	3.8	5
Mo	6.3	4.5	59	Pu	3.5	–	1

Problem 2.6

Defining a short-range interaction between atoms of a cluster as a pairwise interaction between atoms – nearest neighbors, express the cluster energy at zero temperature through the total number of bonds between nearest neighbors.

Taking the pairwise interaction potential between atoms and nearest neighbors, as is realized in a diatomic molecule, we take into account that this interaction potential has a minimum at an equilibrium distance R_e between atoms of a diatomic molecule, where the interaction potential is $U(R_e) = -D$, that is, D is the breaking energy per bond for this interaction potential. If cluster atoms form a structure with Q bonds between nearest neighbors, the total energy E of cluster atoms is equal to

$$E = -QD\,.$$

The optimal structure of a system of an infinite number of cluster atoms at zero temperature is the close packed structure where each internal atom has 12 nearest neighbors. This gives, for the parameter ε_o in formula (2.9) that characterized the cluster energy in the limit of an infinite number of atoms,

$$\varepsilon_o = 6D\,, \tag{2.12}$$

and the surface cluster energy E_{sur} in the case of a short-range interaction is defined as

$$E_{sur} = 6nD + E(n)\,. \tag{2.13}$$

Here $E(n)$ is the total energy of a cluster consisting of n atoms with a given structure.

On the other hand, if in a cluster of n atoms there are n_k atoms with k nearest neighbors, we have for the total cluster energy

$$E(n) = -D\sum_k \frac{n_k}{2}\,, \quad n = \sum_k n_k\,, \tag{2.14}$$

and we take into account that one bond relates to two atoms. This equation gives for the cluster surface energy [17, 40]

$$E_{sur} = \sum_k \left(6 - \frac{k}{2}\right) n_k\,. \tag{2.15}$$

As is clear, the addition of an atom with k nearest neighbors to a cluster increases the cluster surface energy by the value $(6-k)D$.

Note that according to the principles of thermodynamics [41–43], if a cluster consisting of a certain number of atoms has a different surface shape, the most stable shape relates to the minimum surface energy E_{sur} or the minimum specific surface energy A.

Problem 2.7

Analyze the optimal parameters of metal clusters on the basis of a jellium model. Determine the quantum numbers of electrons.

According to the jellium model, which corresponds to the mean field approximation, the positive charge is distributed over some space, and electrons are located in the self-consistent field of a smearing positive charge and other electrons. According to the Wigner concept [32, 33], the correlation between electrons is of principle, and the correlation interaction chooses an optimal dimension of the total system or an optimal distance between cores and electrons. Evidently, for the liquid state the number density of electrons is characterized by the Wigner–Seitz radius (2.6), and the radius of a large metal cluster is given by formula (2.5).

Within the framework of the jellium model, the potential well for electrons exists in a restricted region $r < r_o$, where r is the distance from the cluster center and

r_o is the cluster radius. Evidently, the simplest form of the potential $U(r)$ of a self-consistent field that acts on electrons within the framework of the jellium model is

$$U(r) = \begin{cases} -U_o\,, & r \le r_o \\ 0\,, & r > r_o \end{cases}.$$

We also give the potential of a self-consistent field that is suitable for sodium [26]:

$$U(r) = \frac{U_o}{\exp[(r - r_o)/a] + 1}\,,$$

where for sodium $U_o = 5.93\,\text{eV}$ is the sum of the Fermi energy (3.23 eV) and the work function (2.7 eV for bulk sodium), and the radius of action of atomic forces is $a = 1.5a_o$. In any case, we have within the framework of the jellium model that electrons of a metal cluster are distributed in a spherical self-consistent field $U(r)$.

In this case we have that the energy of an individual electron ε_n in a metal cluster is determined by the Schrödinger equation for the wave function ψ_n of this electron:

$$-\frac{\hbar^2}{2m_e}\frac{d^2}{rdr^2}(r\psi_n) = [\varepsilon_n - U(r)]\psi_n\,.$$

One can see an analogy with an electron distribution in atoms, and we use this analogy. Indeed, each valence electron in a cluster is characterized by the following quantum numbers: n – the principal quantum number, l – the electron orbital momentum, m – the electron momentum projection onto a given direction, and σ – the spin projection onto a given direction. According to the Pauli principle, a certain combination of quantum numbers $nlm\sigma$ may relate only to one electron. Next, for this form $U(r)$ for the potential of a self-consistent cluster field the degeneration takes place for the quantum numbers $m\sigma$, and clusters similar to atoms have a shell electron structure. Note that in atoms the Coulomb field of a nucleus is the basic one in a self-consistent field acting on electrons, and this leads to the relation $n \ge l + 1$ for the electron quantum numbers. In the cluster case this condition is absent, and therefore the sequence of electron shells differs from those in the atomic case.

We now analyze the magic numbers within the framework of the jellium model. Experimentally, magic numbers are determined on the basis of cluster mass-spectrometry when clusters are ejected from a gas containing these clusters. The population of states for magic numbers is higher than that for neighboring numbers of cluster atoms, and therefore the intensity of the cluster flux has local maxima at magic numbers of atoms, as is shown in Figure 2.2a for small sodium clusters [26]. In another method for finding cluster magic numbers, we take into account that magic numbers correspond to the maximum binding energy of cluster atoms. Let us represent the energy of atoms $E(n)$ for a cluster consisting of n atoms as

$$E(n) = \overline{E(n)} + \Delta\varepsilon_n\,, \tag{2.16}$$

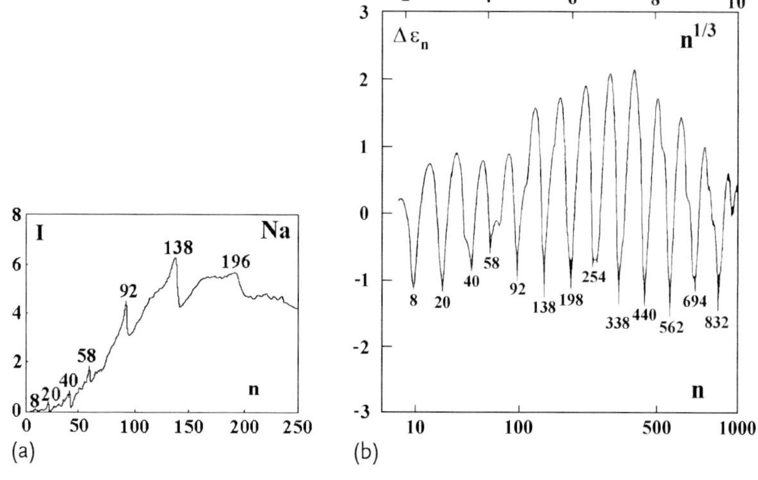

Fig. 2.2 Magic numbers of sodium clusters followed from mass-spectrum of sodium clusters formed in an expanded vapor [26] (a) dependence on the atom numbers n for an additional energy $\Delta\varepsilon_n$ of the last atom (formula (2.29)) compared with the average atom energy for this shell [44] (b). The indicated magic numbers correspond to complete atom shells for the jellium cluster model.

where the average value of atomic energy $\overline{E(n)}$ is given by (2.9), and $\Delta\varepsilon_n$ characterizes the binding energy of the n-th atom. Since the quantity $E(n)$ is negative, the quantity $\Delta\varepsilon_n$ has a minimum at magic atomic numbers, as is shown in Figure 2.2a [44].

Magic numbers of cluster atoms correspond to complete atomic shells. On the basis of experimental data and the jellium model one can find the sequence of filling of lower electron shells for clusters of alkali metals as follows: $1s^2 1p^6 1d^{10} 2s^2$ $1f^{14} 2p^6 1g^{18} 2d^{10} 1h^{22} 3s^2$. Here we use the usual notations for quantum numbers nl of electron shells, so that the first value is the principal electron quantum number and the second one is the electron orbital momentum, so that the values s, p, d, f, g, h correspond to $l = 0, 1, 2, 3, 4, 5$. The superscript indicates the number of electrons in a given shell. Thus, the magic numbers of this cluster that correspond to filled electron shells are 2, 8, 18, 20, 34, 40, 58, 68, 80, 82, and so on. Clusters with these numbers of atoms have a heightened stability. As is seen, the Pauli exclusion principle leads to a specific behavior of metallic clusters due to the exchange interaction between electrons. Note that the jellium model describes well clusters of alkali metals and is not suitable for other metal clusters.

Problem 2.8

Analyze the jellium model of metal clusters within the limits of a high-electron-number density.

Repulsion of electrons at high densities is determined by the exchange interaction between electrons due to the Pauli exclusion principle, while attraction is connected with the correlation interaction with other electrons and the Coulomb interactions with positively charged cores and other electrons. When the number density of electrons is high, many electrons are located in the same space point, and the jellium model for clusters becomes similar to the Thomson atom model. An electron subsystem in this limiting case is a degenerate electron gas whose parameters at low temperatures are determined by the Pauli exclusion principle, and the electron momentum is restricted by the Fermi momentum p_F that at zero temperature follows from the relation

$$n = 2 \int_{p \le p_F} \frac{d\mathbf{p} d\mathbf{r}}{(2\pi\hbar)^3} .$$

Here n is the total number of electrons in a given volume, the factor 2 accounts for two directions of the electron spin, and $d\mathbf{p}$ and $d\mathbf{r}$ are elements of the electron momentum and a volume. The above equation leads to a simple relation between the electron number density $N_e = n / \int d\mathbf{r}$ and the Fermi momentum p_F, which is the maximum electron momentum, or the Fermi energy $\varepsilon_F = p_F^2/2m_e$, which is the maximum electron energy, for this distribution (m_e is the electron mass):

$$p_F = (3\pi^2\hbar^3 N_e)^{1/3} , \quad \varepsilon_F = \frac{p_F^2}{2m_e} = \frac{(3\pi^2 N_e)^{2/3}\hbar^2}{2m_e} .$$

From this we have the following mean kinetic energy of electrons:

$$\overline{\varepsilon} = \overline{\frac{p^2}{2m_e}} = \frac{3p_F^2}{10m_e} = a N_e^{2/3} , \quad a = 2.87 \frac{\hbar^2}{m_e} .$$

The interaction potential of an electron with positive and negative charges of this plasma that includes also the exchange interaction between electrons is of the order $e^2 N_e^{1/3}$ and corresponds to electron attraction. On this basis, one can represent the average electron energy $\overline{\varepsilon}$ in the form

$$\overline{\varepsilon} = a N_e^{2/3} - b N_e^{1/3} , \qquad (2.17)$$

where both numerical coefficients a, b are of the order of atomic values.

Minimization of formula (2.17) gives for the optimal electron number density $N_e = (b/2a)^3$, which corresponds to the average energy per electron,

$$\overline{\varepsilon} = -\frac{b^2}{4a} , \quad |\overline{\varepsilon}| \sim \frac{m_e e^4}{\hbar^2} .$$

We obtain from this that both a typical distance between neighboring electrons and a typical electron binding energy are of the order of atomic values.

Table 2.4 Parameters of solid and liquid inert gases at the triple point. Here R_e is the distance between two atoms of the minimum interaction potential, $-D$, a is the distance between the nearest atoms in the crystal lattice, $\rho_o = \sqrt{2}/a^3$, ρ_s, ρ_l are the densities of the solid and liquid states at the triple point, T_m is the temperature at the triple point (the melting point), ΔH_{fus} is the fusion enthalpy, and ε_{sub} is the specific sublimation energy for the solid state at the triple point (the binding energy per atom) [31, 46].

Parameter	Ne	Ar	Kr	Xe	Average value
R_e, Å	3.091	3.756	4.011	4.366	
D, K	42.2	143	201	283	
a/R_e	1.028	1.000	0.996	0.993	1.004 ± 0.014
ρ_o, g/cm^3	1.604	1.770	3.049	3.704	
ρ_s/ρ_o	0.900	0.917	0.927	0.956	0.925 ± 0.020
ρ_l/ρ_o	0.777	0.801	0.801	0.830	0.802 ± 0.019
T_m/D	0.583	0.585	0.576	0.570	0.578 ± 0.006
$\Delta H_{fus}/D$	0.955	0.990	0.980	0.977	0.98 ± 0.01
ε_{sub}/D	6.0	6.5	6.4	6.4	6.3 ± 0.3

Problem 2.9

Analyze the parameters of liquid inert gases near the triple point within the framework of the model, characterizing the liquid state by the coordination number, which is the average number of nearest neighbors, with an identical distance from a test atom. Use experimental data for the liquid state of inert gases.

Cluster melting leads to changes in the cluster parameters, and the phase transition in clusters differs in some respects from that for bulk systems [45]. In Table 2.4, we give some parameters of liquid inert gases near the triple point and use below the data from this table to determine the coordination number of liquid inert gases for the coordination model. As a matter, within the framework of this model we consider the liquid state of a system of bound atoms as the crystal atomic structure with vacancies. The presence of these vacancies leads to a decrease in the density of the liquid state of this atomic system in comparison with the solid state; it also leads to a decrease in the specific sublimation energy in the liquid state.

Guided by the above model of a system of bound atoms with fixed knots, let us extract a volume V of a solid inert gas. This volume contains $\rho_s V/m_a$ atoms, where m_a is the atomic mass. Each atom of this volume has 12 nearest neighbors, that is, the total number of bonds in the extracted volume is equal to $6\rho_s V/m_a$. Because the density of the liquid state equals ρ_l, the number of vacancies in this volume is $(\rho_s - \rho_l)V/m_a$. The formation of each vacancy leads to a loss of 12 bonds, that is, the number of bonds in this volume and the liquid state is $6(2\rho_l - \rho_s)V/m_a$. We have from this [47]

$$q = 24 - \frac{12\rho_s}{\rho_l}, \tag{2.18}$$

for the average number q of nearest neighbors for the liquid state.

Table 2.5 The average number q of nearest neighbors in liquid inert gases at the triple point within the coordination model and the ratio k of the average number of neighboring atoms to the average number of vacancies for this model [47].

	Ne	Ar	Kr	Xe	Average value
ρ_l/ρ_s	0.863	0.874	0.864	0.868	0.865 ± 0.004
Formula (2.18), q	10.10	10.27	10.11	10.19	10.17 ± 0.08
Formula (2.18), k	5.74	5.94	5.35	5.63	5.66 ± 0.25
Formula (2.19), q	10.07	10.15	10.14	10.19	10.14 ± 0.04
Formula (2.19), k	5.22	5.49	5.45	5.56	5.43 ± 0.15

The other way to determine the average number of nearest neighbors in the liquid state at the melting point uses the energetic parameters of melting. Let us introduce the energy per atom ε that is spent on the atomization of the liquid state at the melting point. For the solid state this value is equal to $\varepsilon + \Delta H_{fus}$ at the melting point, where ΔH_{fus} is the specific fusion energy. Because each atom partakes in 12 bonds of the solid state, the number of nearest neighbors of atoms of the liquid state is given by

$$q = \frac{12}{1 + \frac{\Delta H_{fus}}{\varepsilon}} . \tag{2.19}$$

Table 2.5 contains the numbers of nearest neighbors according to this equation. For this calculation we replace the value $\varepsilon(T_m)$ with the evaporation energy $\varepsilon_{ev}(T_b)$ at the boiling point, which leads to an increase of q. As is seen from the Table 2.5 data, both methods give close values of numbers of nearest neighbors for liquid inert gases. Averaging over the various inert gases and the methods of its determination, we obtain the result $q = 10.15 \pm 0.06$. From this it follows that one vacancy of liquid inert gases at the melting point is related to 5.6 ± 0.2 atoms. Note that we use the rough model for average parameters of the liquid state, and this model does not describe the principal peculiarities of the melting process.

2.2 Stability of Charged Metal Clusters

Problem 2.10

Find the condition of stability of a small liquid spherical metal particle (the Rayleigh problem [48]).

Let us consider a strongly charged cluster within the framework of the liquid drop model as it was first made by Rayleigh [49]. Coulomb repulsion of charges tries to break this drop, while the surface tension tends to conserve the spherical drop shape. Under some relation between the parameters of these interactions, this drop becomes unstable and is broken down into parts. This problem is important for the stability of nuclei [50, 51] and clusters (see [52, 53]). We now consider this problem within the framework of the model of a charged liquid drop [48].

Under these conditions, the cluster surface tension retains its spherical form, while a Coulomb repulsion of charges of the cluster tends to change its form. Assuming the cluster charge is uniformly distributed over its surface, we represent the cluster potential energy in the form

$$U = \frac{Z^2 e^2}{2C} + \gamma S \,, \tag{2.20}$$

where Z is the cluster charge in units of electron charges, C is the cluster electric capacity, S is the area of the cluster surface, and γ is the surface tension. For a spherical cluster we have $C = r_o$, $S = 4\pi r_o^2$, where r_o is the cluster radius. Let us divide the cluster into two identical spherical clusters. Then the radius of each cluster is $r' = r_o/2^{1/3}$ for the liquid drop model, and the charge is $Z' = Z/2$. Using the expression for the potential energy of each cluster, we obtain the condition of the stability of the initial cluster in the form

$$\frac{Z^2 e^2}{2r} + 4\pi r^2 \gamma \le 2\left(\frac{Z'^2 e^2}{2r'} + 4\pi r'^2 \gamma\right) .$$

From this we obtain the criterion of instability of a charged liquid cluster drop

$$Z^2 \ge 16\pi \frac{r_o^3 \gamma}{e^2} \frac{2^{1/3} - 1}{2 - 2^{1/3}} = 0.35 \cdot 16\pi \frac{r_o^3 \gamma}{e^2} \,. \tag{2.21}$$

Problem 2.11

Find the condition of stability for a small liquid spherical metal particle with respect to small deformations. Take elastic properties of the liquid drop such that a deformation in one direction does not lead to deformations in perpendicular directions (the Poisson coefficient equals zero).

According to the results of the previous problem, the instability of a charged liquid drop develops when its fission into two equal parts is energetically profitable. In order to achieve this, it is necessary to overcome the energetic barrier. But under the critical conditions, the cluster drop is stable with respect to a small change in the cluster shape, because in this case the surface tension of the cluster drop exceeds the electric pressure due to a cluster charge. The subsequent charge increase leads to instability with respect to weak oscillations of a cluster surface when the pressure $p_t = 2\gamma/r_o$ due to the surface tension is attained by pressure under the action of the electric field of charges, which is $p_e = ZeE/(4\pi r^2) = Z^2 e^2/(4\pi r^4)$. This leads to the following condition of the drop instability:

$$Z^2 \ge 8\pi \frac{r_o^3 \gamma}{e^2} \,. \tag{2.22}$$

This is valid for a fragile drop material if the normal pressure with respect to the cluster surface leads to its destruction.

In the case of an elastic drop material, the equality $p_t = p_e$ does not lead to drop destruction because for an increasing drop radius the pressure due to surface tension exceeds the electrostatic one. On the other hand, the development of this instability proceeds together with a change in the cluster shape. Then the instability threshold corresponds to higher charges and is given by the following criterion:

$$Z^2 \geq 16\pi \frac{r_o^3 \gamma}{e^2} \,. \tag{2.23}$$

Criteria (2.22) and (2.23) for the cluster instability are stronger than criterion (2.21). Hence, if criterion (2.21) for the instability with respect to strong deformations is fulfilled and the instability criteria (2.22) and (2.23) as a result of small deformations are not valid, the cluster is found in a metastable state. Note that under the action of a Coulomb interaction the cluster takes an ellipsoidal shape that decreases the range of parameters in which the metastable state exists.

Problem 2.12

Based on experimental data, analyze the stability of multicharged cluster ions.

Applying the liquid drop model to charged clusters, one can see that the assumptions of this model are not valid for clusters. Within this model, we assume that a cluster charge is large and located on the cluster surface, and, when this charged cluster is divided into parts, the charge of each part is also large. In reality, a cluster charge is not large, and its parts resulting from cluster fission have a unit charge or close to one. Therefore, the above criteria (2.21)–(2.23) for the threshold of the Rayleigh instability are model ones, and the validity of the instability criterion follows from a comparison of these formulas with the experimental results.

Next we consider this problem in the following form. Let us introduce the critical number $n_{cr}(Z)$ of cluster atoms, so that if the number of atoms for a cluster with a charge Z exceeds this value, the cluster under consideration is stable. We represent the instability threshold in the form

$$n_{cr} = C \cdot \frac{Z^2 e^2}{16\pi r_W^3 \gamma} = C \frac{Z^2}{b} \,, \quad b = \frac{16\pi \gamma r_W^3}{e^2} = \frac{4AZ^2}{e^2} \,. \tag{2.24}$$

We introduce in this formula the parameter C as the proportionality parameter between the critical cluster size and its charge square; this parameter C in formula (2.24) is equal to 0.35, 0.5, and 1 if the instability threshold is given by criteria (2.21), (2.22), and (2.23), respectively; the specific surface cluster energy A is connected with the specific surface tension γ of a cluster material by formula (2.11), and the Wigner–Seitz radius r_W is given by formula (2.5). If the liquid drop model describes the stability of a multicharged cluster ion, the critical cluster size n_{cr} is proportional to the charge square Z^2, and parameter C lies between 0.35 and 1.

Table 2.6 The parameters of a liquid drop consisting of inert gas atoms and parameters of the pairwise interaction potential for identical inert gas atoms. Here A is the specific surface energy of a large liquid drop defined by formulas (2.9) and (2.11), r_W is the Wigner–Seitz radius given by formula (2.5), parameter b is defined by formula (2.24); R_e is the distance between two inert gas atoms where the interaction potential has a minimum, and its value at this point is $-D$.

Inert gas	A, meV	r_W, Å	b	R_e, Å	D, meV
Ne	15.3	1.86	0.0079	3.09	3.64
Ar	53	2.23	0.033	3.76	12.3
Kr	73	2.38	0.048	4.01	17.3
Xe	97	2.57	0.069	4.36	24.4

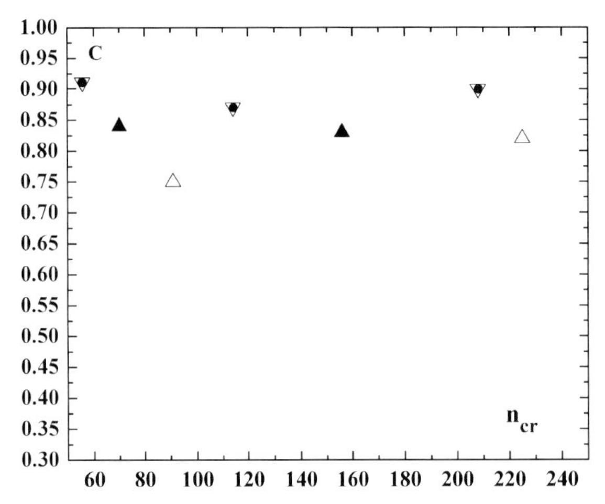

Fig. 2.3 The dependence of the parameter C defined by formula (2.24) on the basis of measurements [25, 54–57] of the critical (appearance) sizes for multicharged ions of inert gases. Filled triangles correspond to multicharge argon ions, open triangles relate to multicharge krypton ions, and inverted open triangles correspond to multicharge xenon ions.

We now use these equations for large liquid clusters consisting of inert gas atoms. The parameters of these equations as well as the parameters of the pairwise interaction potential for two identical atoms are given in Table 2.6; in addition, a comparison of formulas (2.24) with experimental data [25, 54–57] for multicharged ions of inert gases with $Z = 2$–4 is made in Figure 2.3. As is clear, the liquid drop model describes more or less the Rayleigh instability of these ions, so that the critical cluster size is proportional to the charge square, and the average value of the proportionality is $C = 0.85 \pm 0.05$ [36] on the basis of the above experimental data.

The value C obtained on the basis of experimental data [58] for sodium clusters is represented in Figure 2.4. As is clear, the value of this parameter is outside the

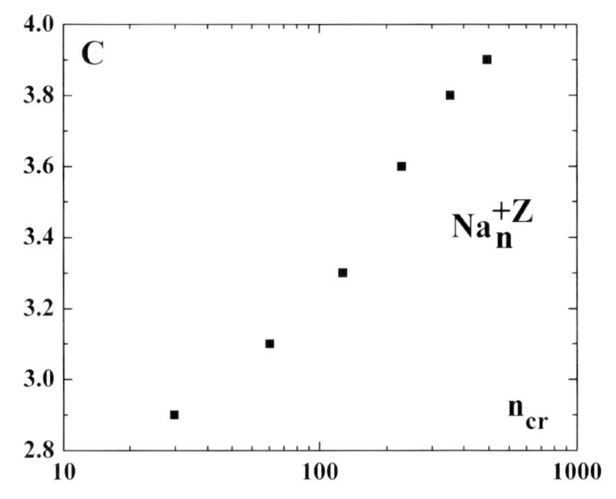

Fig. 2.4 The dependence on the critical (appearance) size for the parameter C defined by formula (2.24) for multicharged sodium ions on the basis of measurements [58].

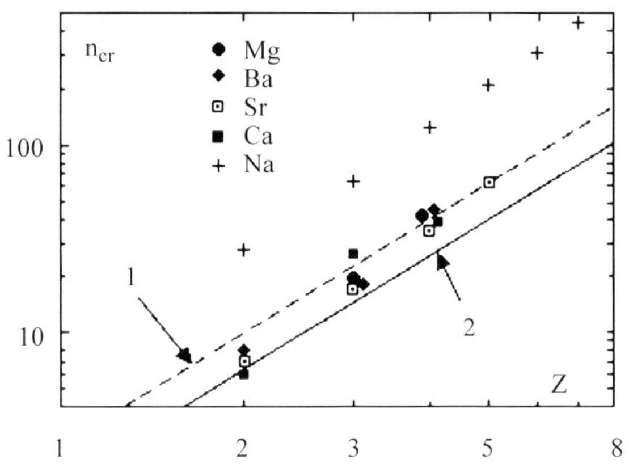

Fig. 2.5 The critical size for multicharge ions of sodium and earth alkali metals according to [58, 59, 61]. 1 – formula (2.23) for sodium multicharged ions, 2 – formula (2.23) for sodium multicharged ions of alkali earth metals.

limits of formulas (2.21)–(2.23). This means that the liquid drop model is not suitable for the sodium case. The same result follows from a comparison of the liquid drop model with experiments [59] for clusters of alkali earth metals (Figure 2.5), though the excess of the higher limit (2.23) for value C is less than this according to experiments [58, 60, 61] for the alkali metal clusters.

Problem 2.13

Analyze the features of decay of multicharge cluster ions when clusters consist of weakly bound atoms (molecules) or metal atoms.

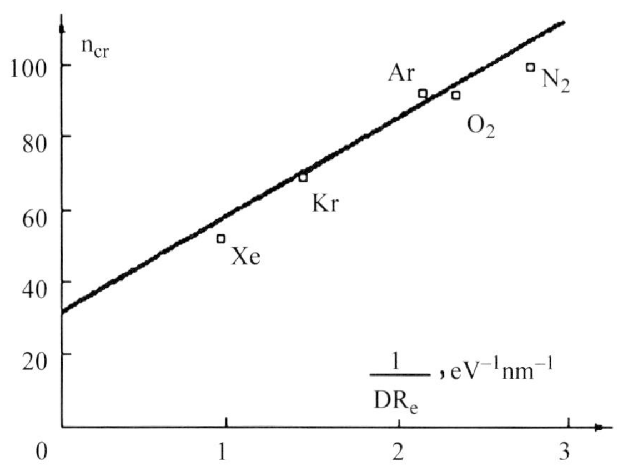

Fig. 2.6 The critical size of double charged ions of weakly bound atoms and molecules; D is the dissociation energy of a diatomic energy of this kind, R_e is the equilibrium distance between nuclei of such a diatomic molecule [63].

The above comparison between theoretical and experimental results shows that the charged liquid drop model works for charged clusters of inert gases or clusters consisting of weakly bound atoms or molecules, but it is not suitable for metal clusters. Evidently, the first conclusion is a consequence of a weak interaction between atoms, since interactions between nearest neighbors in these clusters dominate. Therefore, the interaction parameters in clusters may be expressed through the interaction potential between two atoms. In particular, according to criteria (2.21)–(2.23) we have

$$\frac{Z^2}{n_{cr}} \backsim \frac{A r_W}{e^2} \backsim \frac{D R_e}{e^2} ,$$

where D is the well depth for the interaction of two atoms and R_e is the equilibrium distance between atoms that corresponds to the minimum of the pairwise interaction potential. Figure 2.6 [63] checks this dependence for clusters consisting of weakly bound atoms or molecules for doubly charged cluster ions ($Z = 2$).

Note that criteria (2.21)–(2.23) for the appearance size of multicharge ions correspond to its decay in almost equal parts (symmetric decay). In the case of metal multicharge atoms antisymmetric decay with different sizes of fractions is preferable [61, 62]. We also emphasize that an appearance size for clusters consisting of weakly bound atoms is several times more than that for metal atoms. For example, the appearance size of doubly charged cluster ions of alkali metals is 20–25 [60] and is less compared to that of Figure 2.6 in the case of weakly bounded atoms. Therefore antisymmetric decay of multicharged clusters is typical for metal clusters.

2.3 Macroscopic Solid Particles with a Pairwise Interaction of Atoms

Problem 2.14

Construct a macroscopic structure of atoms with a short-range interaction between them by adding the planes of the direction $\{111\}$.

One can divide a pairwise interaction potential for a bulk system of many bound atoms into two parts – short-range and long-range interaction potentials. In the case of a short-range interaction between atoms in a bulk system, only nearest neighbors interact. Then, the atoms of a bulk system of bound atoms can be modeled by hard balls of identical radius. In particular, this takes place in condensed rare gases whose properties are mostly determined by the interaction of nearest neighbors. If a long-range interaction is added to a short-range one, these balls become soft, and they are compressed into a bulk system if a long-range interaction corresponds to the attraction of atoms at large distances between them.

At low temperatures a bulk condensed system with a pairwise interaction of atoms forms a crystal lattice of the close packed structure if a short-range interaction gives a significant contribution to the total binding energy of the atoms of this bulk system. There are two structures of close packing in which atoms form a face-centered cubic (fcc) lattice or a hexagonal lattice. Each internal atom of these structures has 12 nearest neighbors. Below we consider such bulk systems of atoms.

Atoms with a short-range interaction form crystals of a close packed structure at zero temperature (Figure 2.7), where atoms are modeled by balls of a diameter $a = R_e$. Each ball of this structure has six nearest neighbors. The next layer is added in such a way that balls of a new layer are placed in hollows between atom-

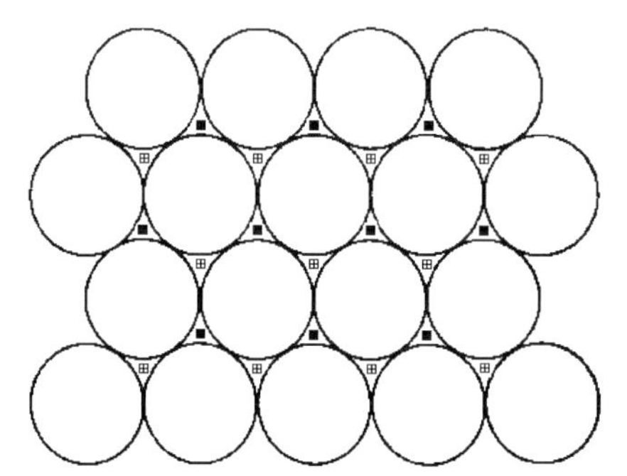

Fig. 2.7 Growth of close packed macroscopic structures by addition of planes 111. The close packed structure of a lattice with atoms modeled by balls of a diameter a located in planes of symmetry $\{111\}$. Dark squares mark locations of centers for atoms of the lower plane which is placed at a distance $a\sqrt{2/3}$ from the crossing plane. Positions of atoms of the upper plane are marked by dark squares for the hexagonal structure and by crossed squares for the fcc structure.

balls of the previous layer. Because the distance between nearest atom neighbors is a, the distance between these planes is equal to $b = a\sqrt{3}/2$.

The following plane is constructed in the same way by locating the balls of a new layer in the hollows of the previous plane. There are two methods to locate the balls of this plane with respect to those of the first one. If the projections of atom-balls of the third plane coincide with those of the first plane, we obtain a hexagonal lattice. In the other case, the fcc structure is realized. Let us denote the first plane of this structure as A and the second one as B. Then the sequence of planes $ABABAB$ corresponds to the hexagonal lattice, and the combination $ABCABCABC$ corresponds to the fcc-lattice.

Problem 2.15

Determine the symmetry of fcc and hexagonal structures.

An infinite system of bound atoms forms a crystal lattice at low temperatures, and we consider the lattice for a short-range interaction of atoms when the distance between nearest neighbors is a. Modeling atoms of the crystal lattice by balls, we note the translational symmetry of this lattice, which means the conservation of this lattice as a result of a shift by a lattice constant. It is convenient to describe the crystal lattice as a Bravais lattice. Within the framework of the Bravais lattice, we give the coordinates of atom centers in the form [64]

$$\mathbf{r} = n_1\mathbf{a}_1 + n_2\mathbf{a}_2 + n_3\mathbf{a}_3 , \tag{2.25}$$

where n_1, n_2, n_3 are integers and the vectors $\mathbf{a}_1, \mathbf{a}_2, \mathbf{a}_3$ form the basis of this lattice. The basis vectors of a given lattice may be expressed through the unit vectors $\mathbf{i}, \mathbf{j}, \mathbf{k}$ directed along the x-, y-, and z-axes. Table 2.7 gives the parameters of the lattices for a short-range interaction of atoms, that is, for the fcc and hexagonal lattices.

Let us introduce the notations for directions of crystal planes on which the centers of bound atoms are placed. These planes are described by Miller indices – x, y, z coordinates of the vector that is perpendicular to this plane and passes through the origin. As the Miller indices are used the minimal whole components of this vector. The Miller indices are denoted by (m_1, m_2, m_3), and the value $\overline{m}_1$ is taken instead of $-m_1$. In particular, if the vector $\mathbf{b} = m_1\mathbf{i} + m_2\mathbf{j} + m_3\mathbf{k}$ is perpendicular to a given plane, the Miller indices of this plane are m_1, m_2, m_3.

Table 2.7 The basis vectors for the face-centered cubic (*fcc*) and hexagonal (*hex*) lattices; *a* is the lattice constant.

Lattice	$\mathbf{a}_1/a$	$\mathbf{a}_2/a$	$\mathbf{a}_3/a$
fcc	$(\mathbf{j}+\mathbf{k})/2$	$(\mathbf{i}+\mathbf{k})/2$	$(\mathbf{i}+\mathbf{j})/2$
fcc	$\mathbf{i}$	$\mathbf{i}/2+\mathbf{j}\sqrt{3}/2$	$\mathbf{j}\sqrt{\frac{1}{3}}+\mathbf{k}\sqrt{\frac{2}{3}}$
hex	$\mathbf{i}$	$\mathbf{i}/2+\mathbf{j}\sqrt{3}/2$	$\mathbf{j}\sqrt{\frac{1}{6}}\left[1+(-1)^{n_3}\right]+\mathbf{k}\sqrt{\frac{2}{3}}$

As follows from the Table 2.6 data, the fcc lattice is preserved as a result of the following transformations:

$$x \longleftrightarrow -x\,,\quad y \longleftrightarrow -y\,,\quad z \longleftrightarrow -z\,,\quad x \longleftrightarrow y \longleftrightarrow z\,. \tag{2.26}$$

The fcc lattice has the cube symmetry $\mathbf{O_h}$ [48]. If we have the plane of Figure 2.7 as the basis of the hexagonal lattice, this lattice is preserved under transformations

$$z \leftrightarrow -z\,;\quad \Phi \rightarrow \Phi \pm \pi/3\,. \tag{2.27}$$

Here the frame of reference is based on a plane {111} in the Miller notations, where each atom-ball has six nearest neighbors on this plane. The z-axis is directed perpendicular to this plane, and Φ is the polar angle for the polar z-axis. As is seen, the symmetry of the hexagonal lattice is lower than that for the fcc lattice.

Problem 2.16

Determine the number of possible plane facets of a bulk particle of an fcc structure with a pairwise interaction of atoms. Find the specific binding energy of surface atoms for a short-range interaction of atoms.

The fcc structure, due to a high symmetry, has three types of planes that can be occupied by the centers of atom-balls. Within the Miller notation of a plane through the coordinates of a line passing through the origin and perpendicular to this plane, these planes are {100}, {110}, and {111} and are represented in Figure 2.8. Because of the symmetry (2.26) for the fcc structure, there are several planes of one type. They are transformed into each other as a result of operations (2.26). This gives 6 different planes of the {100} type, 12 planes of the {110} type, and 8 planes of the {111} type. Thus, the maximum number of plane facets of a crystalline particle of an fcc structure is 26.

We now calculate the number of nearest neighbors for atoms located on the surface of each of the above planes (Figure 2.8). For the short-range interaction potential between atoms, the binding energy for each atom of a {100}-plane has 4 nearest neighbors of this plane and 4 nearest neighbors of the previous one, that is, an atom of a surface {100}-plane has 8 nearest neighbors. An atom of a surface {111}-plane has 6 nearest neighbors of this plane and 3 nearest neighbors of the previous one, that is, an atom ball of a surface {111}-plane has 9 nearest neighbors. Next, there are two nearest neighbors among surface atoms of a {110}-plane, 4 nearest neighbors from atoms of the previous plane, and 1 nearest neighbor from the penultimate plane, that is, an atom of a surface {110}-plane has 7 nearest neighbors. This shows that geometric figures with surface facets of directions {111} and {100} are energetically favorable for bulk crystalline particles of an fcc structure with a pairwise atomic interaction.

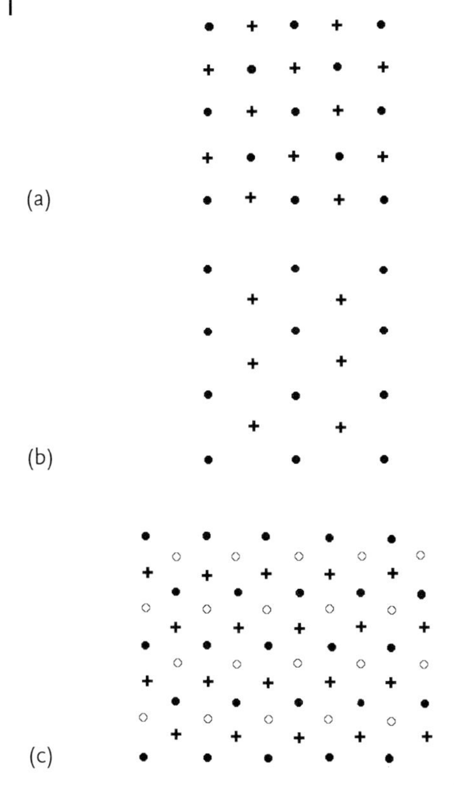

Fig. 2.8 The structures of fcc-planes. Positions of atom centers of the surface layer are indicated by dark circles, centers of atoms of the previous layer are marked by crosses, and centers of atoms of the penultimate layer are shown by open circles. (a) $\{100\}$-plane; the distance between neighboring lines of atoms of the plane is a, the distance between neighboring planes is $a\sqrt{2}/2$; (b) $\{110\}$-plane; the distance between neighboring lines of atoms of the plane is $a\sqrt{2}$, the distance between neighboring planes is $a/\sqrt{2}$; (c) $\{111\}$-plane; the distance between neighboring lines of atoms of the plane is $a\sqrt{3}/2$, the distance between neighboring planes is $a\sqrt{2/3}$.

Problem 2.17

Construct a regular bulk particle by cutting the identical pyramids near the vertices of a cube with directions $\{100\}$ by planes of directions $\{111\}$ [65]. Analyze the shapes of symmetric figures in the limiting cases.

Let us take a cube of length $2L$, choose a frame of reference with a $\{100\}$ plane direction, and cut off regular triangular pyramids near each cube vertex by intersecting planes of direction $\{111\}$. As a result, we obtain families of various polyhedra. A figure thus formed has 14 facets from which 6 have the direction $\{100\}$ and 8 have the direction of the intersecting plane $\{111\}$.

Introduce a parameter β such that βL is the length of the crossing edges of the pyramids. This parameter ranges from 0 to 2. At $\beta < 1$ the facets of the figure include six octagons and eight regular triangles, and $\beta = 1$ corresponds to

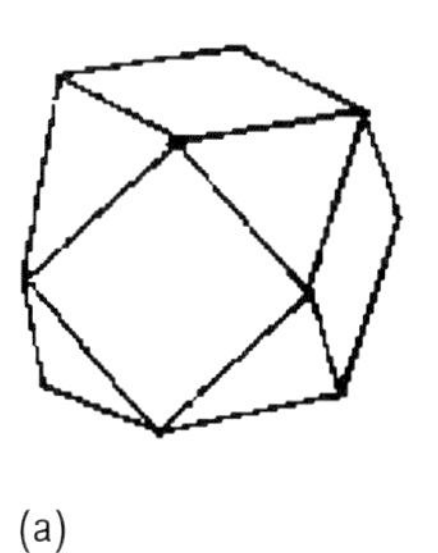
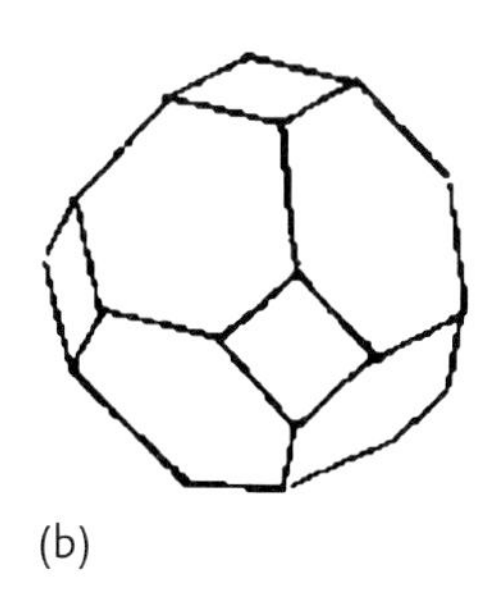
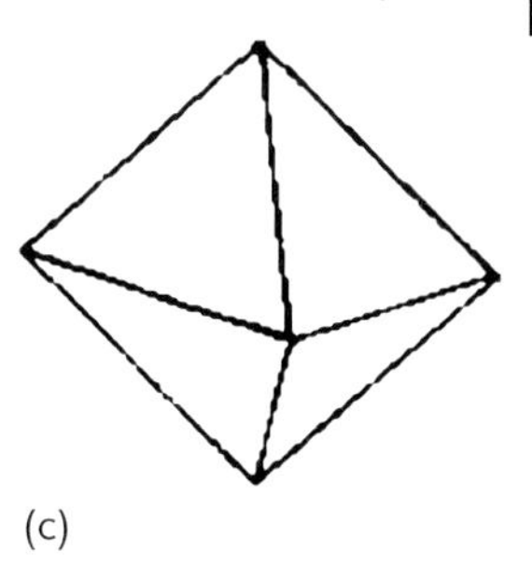

(a) (b) (c)

Fig. 2.9 Regular structures constructed from a cube by cutting off identical regular pyramids near cube vertices: (a) cuboctahedron $\beta = 1$, (b) regular truncated octahedron $\beta = 1.5$, (c) octahedron $\beta = 2$

a cuboctahedron whose surface consists of six squares and eight regular triangles (Figure 2.9a). At $\beta > 1$ the surface of a formed figure includes six squares and eight hexagons, and these hexagons are regular at $\beta = 1.5$ (Figure 2.9b). We call this figure a regular truncated octahedron. In the case $\beta = 2$, an octahedron is formed (Figure 2.9c). It has six vertices, which are the centers of the surface squares of the initial cube, and the octahedron surface consists of eight regular triangles. The surface of a truncated octahedron consists of six squares and eight hexagons, and the squares are characterized by the Miller indices (100), (010), (001), $(\bar{1}00)$, $(0\bar{1}0)$, $(00\bar{1})$ ($\bar{1}$ means -1). The sum of these directions is denoted as $\{100\}$, that is, it is accepted that squares have the directions $\{100\}$. In the same manner, the regular hexagons of the surface of a truncated octahedron have the direction $\{111\}$. All the facets of the octahedron are transformed into each other as a result of the symmetry transformations (2.26) for the fcc structure. Below we consider these figures for the analysis of fcc clusters.

Problem 2.18

Analyze the shell structure of a symmetric cluster of a close packed structure that is cut off from the bulk fcc or hexagonal crystal.

A symmetric fcc cluster that is cut off from a crystal lattice of an fcc structure satisfies the symmetry (2.26) that preserves the lattice as a result of indicated transformations. The origin of the frame of reference may be placed either in the center of an elemental cell or at an atom of the lattice. Thus, there are two types of fcc clusters – with and without a central atom.

Taking as our frame of reference a cluster with $\{100\}$ planes for an fcc cluster that is cut off from the fcc crystal lattice, we have the following coordinates for the nearest neighbors of an atom with coordinates xyz:

$$\begin{aligned} &x, \; y \pm a/\sqrt{2}, \; z \pm a/\sqrt{2}; \\ &x \pm a/\sqrt{2}, \; y, \; z \pm a/\sqrt{2}; \\ &x \pm a/\sqrt{2}, \; y \pm a/\sqrt{2}, \; z, \end{aligned} \tag{2.28}$$

where a is the distance between nearest neighbors. It is convenient to introduce reduced values for atomic coordinates, expressing them in units $a/\sqrt{2}$. Then the coordinates $z\,x\,y$ of each atom are whole numbers. We call a cluster shell a system of atoms whose positions are transformed into one another as a result of transformations (2.26). Thus, the coordinates of atoms of one shell differ by the sign of one or several coordinates and by the transposition of coordinates $z\,x\,y$. As is seen, the maximum number of atoms in one shell is equal to $3 \cdot 2 \cdot 2 \cdot 2 = 48$.

In the case of a symmetric hexagonal cluster that satisfies the symmetry (2.27), we take a plane $\{111\}$ as the basis of the hexagonal lattice, and each atom of this plane has six nearest neighbors. Take as a cluster shell the system of atoms located at the same distance from the central atom. Because the symmetry of the hexagonal lattice is lower than that for the fcc lattice (the cube symmetry $\mathbf{O_h}$), the number of atoms of one shell of the hexagonal cluster is lower than that of the fcc cluster. In accordance with the symmetry (2.27) of the hexagonal lattice, the maximum number of atoms in one shell is equal to $2 \cdot 6 = 12$ in this case.

Problem 2.19

Evaluate the specific binding energy of atoms of the crystal lattice with a pairwise interaction of atoms.

One can express the specific energetic parameters of a condensed system of atoms with a pairwise interaction between atoms $U(R)$ as a function of the distance between atoms R through the parameters of this interaction in a diatomic molecule consisting of these atoms. Take the equilibrium distance between the atoms in a diatomic molecule R_e and the depth of the potential well D as the basic parameters of the interaction potential, which are

$$U'(R_e) = 0\,; \quad U(R_e) = -D\,.$$

The total binding energy of bound atoms E_b in a symmetric system with a pairwise interaction of atoms is

$$E_b = -\frac{1}{2}\sum_k n_k\, U(ka)\,, \tag{2.29}$$

where a is the distance between nearest neighbors and n_k is the number of bonds of length ka. The factor $1/2$ accounts for each bond corresponding to two atoms. Below we use this formula to determine the energetic parameters of systems of bound atoms.

Problem 2.20

Determine the specific binding energy of atoms in an fcc crystal and the distance between nearest neighbors for the Lennard-Jones interaction potential of atoms.

Table 2.8 Parameters of shells of fcc structure (the shell zxy is given by reduced coordinates of its atom, $r^2 = k^2 a^2$ is the square of the distance from the center, a is the distance between nearest neighbors, and n_k is the number of shell atoms) [68].

Shell	r^2/a^2	n_k	Shell	r^2/a^2	n_k
011	1	12	044	16	12
002	2	6	334	17	24
112	3	24	035	17	24
022	4	12	006	18	6
013	5	24	244	18	24
222	6	8	116	19	24
123	7	48	235	19	48
004	8	6	026	20	24
114	9	24	145	21	48
033	9	12	226	22	24
024	10	24	136	23	48
233	11	24	444	24	8
224	12	24	055	25	12
015	13	24	017	25	24
134	13	48	345	25	48
125	15	48	046	26	24

The Lennard-Jones potential of interaction of two atoms has the following form [66, 67]:

$$U(R) = D \cdot \left[\left(\frac{R_e}{R} \right)^{12} - 2 \cdot \left(\frac{R_e}{R} \right)^{6} \right] . \tag{2.30}$$

If we take a test atom of an fcc crystal lattice as the origin, we can distribute the positions of other atoms in this reference frame over shells such that the atoms of certain shells are located at the same distance from the test atom. Table 2.8 gives the number of atoms located on such shells [68].

On the basis of formula (2.29) we have for the crystal sublimation energy ε_{sub}, which is the total binding energy of the crystal per atom,

$$\varepsilon_{\text{sub}} = \frac{E_b}{n} = -\frac{1}{2} \sum_k n_k U(ka) ,$$

where the notations are the same as in Table 2.8, that is, n_k is the number of lattice atoms located at a distance ka from the origin, and summation is made over bonds of one test atom located at the origin. We take the number of bonds of a given length for a crystal of the fcc structure n_k from Table 2.8, so that this formula has the form

$$\varepsilon_{\text{sub}} = -\frac{1}{2}\sum_k n_k U(ka) = -\frac{D}{2}\sum_k n_k\left(\frac{R_e}{ka}\right)^{12} + D\sum_k n_k\left(\frac{R_e}{ka}\right)^6 . \quad (2.31)$$

Using the data of Table 2.8, we calculate each of these sums. As a result, we obtain the sublimation energy in the form [69]

$$\frac{\varepsilon_{\text{sub}}}{D} = -\frac{C_1}{2}\left(\frac{R_e}{a}\right)^{12} + C_2\left(\frac{R_e}{a}\right)^6 , \quad (2.32)$$

where $C_1 = \sum_k (n_k/k^{12})$, $C_2 = \sum_k (n_k/k^6)$. Taking a finite number of terms of this equation for $k \le k_o$, we replace summation by integration for $k > k_o$, accounting for the fact that the average number density of atoms for a crystal of a close packed structure is $\sqrt{2}/a^3$. Then we have for this part of the sums

$$\Delta C_1 = \frac{1}{2}\sum_{k\ge k_o} n_k(ka)^{-12} = \frac{1}{2}\int_{k_o}^{\infty}\frac{\sqrt{2}}{a^3}\frac{4\pi r^2 dr}{r^{12}} = \frac{4\pi\sqrt{2}}{9r_o^9} ,$$

$$\Delta C_2 = \sum_{k\ge k_o} n_k(ka)^{-6} = \int_{k_o}^{\infty}\frac{\sqrt{2}}{a^3}\frac{4\pi r^2 dr}{r^6} = \frac{4\pi\sqrt{2}}{3r_o^3} ,$$

where $r_o = ak_o$. Finally, using all the terms of Table 2.8, we get $C_1 = 12.131$, $C_2 = 14.454 \pm 0.002$. The error is determined by the choice of the lower limit of integration, which is taken such that r^2/a^2 lies between 26 and 27 (in expressions for $\Delta C_1, \Delta C_2$ we take k_o such that $r^2/a^2 = 26$).

The lattice constant a in formula (2.32) can vary. Optimizing the sublimation energy (2.32) in order to obtain the maximum binding energy of crystal atoms, we get

$$a = 0.971 R_e , \quad \varepsilon_{\text{sub}} = \frac{C_2^2 D}{2C_1} = 8.61 D . \quad (2.33)$$

Note that the binding energy per atom due to the interaction of nearest neighbors is $6D$, that is, the interaction of nearest neighbors gives a contribution of approximately 70% to the energy of the Lennard-Jones crystal with the fcc lattice.

Problem 2.21

Separate the binding energy of atoms of the Lennard-Jones crystal of the fcc-structure in the energy of interaction of nearest neighbors, the energy of interaction of nonnearest neighbors, and the strain energy. Evaluate each of these terms.

Let us represent the specific binding energy in the form [70]

$$\varepsilon_{\text{sub}} = \varepsilon_{\text{nn}}(R_e) + \varepsilon_{\text{nnn}}(R_e) + \varepsilon_{\text{str}} , \quad (2.34)$$

where $\varepsilon_{nn}(R_e)$ and $\varepsilon_{nnn}(R_e)$ are the interaction energies between nearest neighbors and nonnearest neighbors, respectively, and the strain energy is given by

$$\varepsilon_{str} = \varepsilon_{sub}(a) - \varepsilon_{sub}(R_e) , \tag{2.35}$$

where a is the optimal distance between nearest neighbors, which corresponds to the maximum binding energy of the crystal. For the Lennard-Jones crystal (i.e., for the crystal with the Lennard-Jones interaction potential of atoms) we have

$$\begin{aligned} \varepsilon_{nn}(R_e) &= 6D , \\ \varepsilon_{nnn}(R_e) &= \left(C_2 - \frac{C_1}{2} - 6 \right) D = 2.39D , \\ \varepsilon_{str} &= \frac{(C_2 - C_1)^2}{2C_1} D = 0.22D . \end{aligned} \tag{2.36}$$

The first term is determined by a short-range interaction of atoms in a cluster, and the second term corresponds to a long-range interaction. The third term is responsible for deviation in the distance between nearest neighbors from the equilibrium one.

Problem 2.22

Determine the specific binding energy of atoms in a crystal of hexagonal structure and the distance between nearest neighbors for the Lennard-Jones interaction potential of atoms.

The symmetry of the hexagonal structure is given by formula (2.27) if the hexagonal crystal lattice is constructed on the basis of a plane $\{111\}$. This symmetry corresponds to the conservation of this atomic system if atoms are reflected with respect to the initial xy plane of the $\{111\}$ symmetry or are rotated around the z axis by an angle of $\pi/3$. Take an atom as the origin and construct atomic shells around it so that atoms of one shell can exchange their positions as a result of the above transformation. Numbering the layers of this structure, we obtain the positions of atoms of the hexagonal structure represented in Table 2.9 [68]. Note that the hexagonal structure has a lower symmetry than the fcc one, so that the average number of atoms of one shell of the hexagonal structure is smaller than that of the fcc structure.

Applying formula (2.31) to the hexagonal Lennard-Jones crystal, we obtain for its parameters

$$C_1 = 12.232 ; \quad C_2 = 14.454 \pm 0.002 .$$

As is seen, within the limits of the accuracy used, the energetic parameters of the Lennard-Jones crystal coincide for the fcc and hexagonal structures. Though a more accurate calculation shows the advantage of the hexagonal structure, this fact has no practical significance.

Table 2.9 Parameters of shells of the hexagonal structure ($r^2 = k^2 a^2$ is the square of the distance from the center, where a is the distance between nearest neighbors and n_k is the number of shell atoms). Positions of layers for a given shell are indicated.

Layer	r^2/a^2	n_k	Layer	r^2/a^2	n_k
0	1	6	1	9	6
1	1	6	2	29/3	12
1	2	6	1	10	12
2	8/3	2	3	31/3	12
0	3	6	4	32/3	2
1	3	12	1	11	12
2	11/3	12	3	34/3	6
0	4	6	2	35/3	12
1	5	12	4	35/3	12
2	17/3	12	0	12	6
1	6	6	3	37/3	12
3	19/3	6	0	13	12
2	20/3	12	1	13	12
0	7	12	4	41/3	12
1	7	12	3	43/3	6
3	22/3	6	2	44/3	12
3	25/3	12	4	44/3	12
0	9	6	1	15	12

Problem 2.23

Determine the specific binding energy of the fcc crystal for the Morse pairwise interaction potential between atoms.

The Morse interaction potential of two atoms has the form [71]

$$U(R) = D\left[e^{2\alpha(R-R_e)} - 2e^{\alpha(R-R_e)}\right] . \tag{2.37}$$

It has a minimum at $R = R_e$, and near the minimum we have for the interaction potential

$$U(R) = -D + \alpha^2 (R - R_e)^2 .$$

According to formula (2.29) by analogy with relation (2.31), we have for the specific binding energy of atoms of the fcc crystal

$$\varepsilon_{\text{sub}} = D\left[e^{\alpha R_e} F(\alpha a) - \frac{1}{2} e^{2\alpha R_e} F(2\alpha a)\right] , \tag{2.38}$$

where according to (2.29) and (2.37) we have

$$F(\alpha a) = \sum_k n_k \exp(-\alpha r_k) . \tag{2.39}$$

Here r_k is the distance between the central and kth atoms, n_k is the number of atoms of this shell, and a is the distance between nearest neighbors of the lattice. Values of the function $F(\alpha a)$, which are calculated on the basis of the data in Table 2.8 for the fcc structure, are given in Table 2.10. Table 2.10 also contains an effective number of atoms $F(\alpha a)\exp(\alpha a)$ that partake in this interaction, and the derivative of this function that is given by

$$F'(\alpha a) = \frac{dF}{d(\alpha a)} = -\sum_k (n_k r_k / a) \exp(-\alpha r_k) . \tag{2.40}$$

Note that the value $F(\alpha a)\exp(\alpha a)$ tends to 12, that is, to the number of nearest neighbors for large αa.

Formula (2.38) allows us to connect the equilibrium distance between atoms of the diatomic R_e with the distance between nearest neighbors of the crystal a. Because the crystal binding energy has a maximum at this distance between nearest neighbors, we have

$$\exp(\alpha R_e) \equiv f(\alpha a) = \frac{F'(\alpha a)}{F'(2\alpha a)} , \tag{2.41}$$

so that the optimal specific binding energy per atom for this crystal is given by [72, 73]

$$\frac{\varepsilon_{sub}}{D} = f(\alpha a) F(\alpha a) - \frac{1}{2} f^2(\alpha a) F(2\alpha a) .$$

These parameters for the lattice of the fcc structure are given in Table 2.10 [72, 73].

Problem 2.24

Determine the parameter α of the Morse interaction potential at which the specific binding energy of the fcc crystal becomes identical for the Lennard-Jones and Morse interaction potentials of atoms. Find components of this energy for the Morse interaction potential.

This value of α leads to the specific binding energy $\varepsilon_{sub} = 8.61$ that corresponds to $\alpha a = 4.17$ for the Morse interaction potential. According to the equations of the previous problem, this corresponds to $\alpha R_e = 4.46$, that is, $a = 0.935 R_e$. As is seen, the crystal compression for the Morse interaction potential is more than in the case of the Lennard-Jones one.

Let us represent the specific binding energy of crystal atoms according to formula (2.34) in the form of the sum of the interaction energies between nearest neighbors $\varepsilon_{nn}(R_e)$, nonnearest neighbors $\varepsilon_{nnn}(R_e)$, and the strain energy $\varepsilon_{str} =$

Table 2.10 Parameters of fcc lattice with Morse interaction potential between atoms (Morse crystal).

αa	αR_e	$F(\alpha a)$	$-dF(\alpha a)/d(\alpha a)$	$F(\alpha a)\exp(\alpha a)$	ε_{sub}/D
1	2.77	34.3 ± 0.3	104 ± 1	93 ± 1	50.8
2	3.03	3.87	6.47 ± 0.01	28.6	21.3
3	3.57	0.910	1.17	18.3	12.2
4	4.31	0.274	0.311	14.9	8.94
5	5.17	0.0911	0.0975	13.5	7.52
6	6.10	0.0318	0.0330	12.8	6.84
7	7.06	0.0114	0.0116	12.5	6.47
8	8.03	0.00412	0.00417	12.3	6.29
9	9.02	0.001503	0.001514	12.2	6.18
10	10.01	$5.499 \cdot 10^{-4}$	$5.523 \cdot 10^{-4}$	12.1	6.11

$\varepsilon_{sub}(a) - \varepsilon_{sub}(R_e)$. Then we obtain for each term of the sum (2.34) in the case of the Morse crystal

$$\varepsilon_{nn}(R_e) = 6D\,, \quad \varepsilon_{nnn}(R_e) = 2.136D\,, \quad \varepsilon_{str} = 0.474D\,.$$

Thus for the Morse crystal the contributions to the total binding energy from the interaction between nonnearest neighbors and from the strain energy are equal to 25% and 5.5%, respectively. For the Lennard-Jones crystal these values follow from formula (2.36) and are equal to 28% and 2.6%. From this and the character of crystal compression one can conclude that crystal parameters depend on the form of the interaction potential between atoms.

2.4 Macroscopic Solid Surfaces

Specific properties of clusters and small particles are determined by a relatively large area of their surface. Because many cluster atoms are found on the surface or near the surface, the structure of the cluster surface is of importance for the cluster parameters. The cluster surface is similar to that of the corresponding bulk system, only edge and vertex atoms do not influence the surface properties of bulk systems. In addition, since the number of surface atoms of a bulk system is small compared to the total number of atoms, one can separate surface properties from other ones for bulk. Therefore, it is convenient to analyze the surface properties of clusters on the basis of those for bulk systems.

The basic energetic parameter of the bulk surface is the surface energy of bulk. Below we consider this value for crystalline particles with a pairwise atomic interaction. On the one hand, the surface energy of a bulk crystalline particle is small compared to the total binding energy of the atoms. Hence, the surface energy can be separated from the total energy for a bulk particle. On the other hand, the surface bulk energy consisting of a given number of atoms depends on the shape of the bulk surface. Hence, the surface energy allows us to optimize the shape of bulk particles depending on the interaction potential between atoms. An analysis of the surface energy of bulk crystalline particles is given in the following problems.

Problem 2.25

Construct a regular geometric figure for a crystalline particle taking a cube of side length $2L$ as a basis and cutting off from this cube regular pyramids whose vertices coincide with the cube vertices, and a side length is $L\beta$ ($\beta \le 2$) [65], as is described in Problem 2.17. When a cluster has the shape of this figure with interactions between nearest neighbors in this cluster, determine the surface energy of this cluster in the limit $L \gg a$, where a is the distance between nearest neighbors, and find the optimal figure that is characterized by the minimal specific surface energy.

The facets of formed figures have directions {100} and {111}, so that facets located on squares of the initial cube have directions {100} and cutting surfaces have directions {111}. We consider separately the cases $1 > \beta > 0$ and $2 > \beta > 1$. In the first case, facets consist of six octagons and eight triangles; in the second case, facets are six squares and eight trapeziums. In the limit $\beta = 2$, the geometric figure is an octahedron, so that squares are transformed into points, and trapeziums are converted into equilateral triangles. At $\beta = 1.5$ we have a regular truncated octahedron, and at $\beta = 0$ this figure is a cube.

In considering the case $1 > \beta > 0$, we have for the area of separate facets

$$s_{100} = 4L^2 - 2\beta^2 L^2 = 4L^2\left(1 - \frac{\beta^2}{2}\right), \qquad s_{111} = (L\beta)^2\sqrt{3}/2 .$$

In evaluating the total surface energy of this cluster for a short-range interaction of atoms, when only nearest neighbors partake in the interaction, we take into account that the formation of a {100} facet for a macroscopic system leads to a loss of four bonds per surface atom, and the area per atom is equal to a^2 (a is the distance between nearest neighbors). Thus we obtain for the specific surface energy of this facet $\varepsilon_{\text{sur}}^{100} = 2/a^2$ in energy units of one bond breaking. In the same manner, accounting for the fact that the formation of surface {111} leads to a loss of three bonds per surface atom, and the area per atom is equal to $a^2\sqrt{3}/2$, we obtain for the specific surface energy of this facet $\varepsilon_{\text{sur}}^{111} = \sqrt{3}/a^2$, where we take into account that each broken bond relates to two atoms. The surface area for facets of a given direction is $s_{100} = 4L^2(1 - \beta^2/2)$, and $s_{111} = (L\beta)^2\sqrt{3}/2$. As a result, we have for the surface cluster energy, which is expressed in units of breaking of one bond,

$$E_{\text{sur}} = 6s_{100}\varepsilon_{\text{sur}}^{100} + 8s_{111}\varepsilon_{\text{sur}}^{111} = 12\frac{L^2}{a^2}(4 - \beta^2), \quad 1 \ge \beta \ge 0 .$$

The volume of this cluster is

$$V = 8L^3 - 8V_o = 8L^3 - \frac{4}{3}(L\beta)^3, \quad 1 \ge \beta \ge 0 ,$$

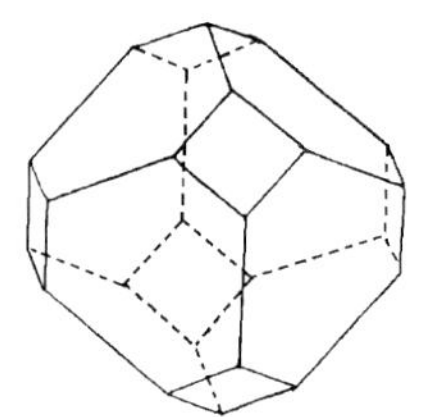

Fig. 2.10 A truncated octahedron.

where $V_o = (L\beta)^3/6$ is the volume of an individual pyramid, because its height is $h = L\beta/\sqrt{3}$, and the area of its basis is $(L\beta)^2\sqrt{3}/2$. Because the number density of atoms for a close packed structure is $\sqrt{2}/a^3$, the number of cluster atoms in this case is $V\sqrt{2}/a^3$. From this we have for the specific surface energy A defined by formula (2.8)

$$E_{\rm sur} = An^{2/3}\,, \quad n^{2/3} = 2^{1/3}\frac{V^{2/3}}{a^2}A = \frac{3}{2^{1/3}}\frac{4-\beta^2}{(1-\beta^3/6)^{2/3}}\,, \quad 1 \geq \beta \geq 0\,. \tag{2.42}$$

This equation gives $A = 6 \cdot 2^{2/3} = 9.52$ for a cubic cluster of a close packed structure and $A = 9 \cdot (18/25)^{1/3} = 8.07$ for a cuboctahedral cluster with a short-range interaction of atoms.

Let us use the same equations in the case $2 \geq \beta \geq 1$. We have the area of cluster facets $s_{100} = 2(2-\beta)^2L^2$, $s_{111} = (3\beta - \beta^2 - 3/2)L^2\sqrt{3}$, and the surface energy is equal in this case to

$$E_{\rm sur} = 6s_{100}\varepsilon_{\rm sur}^{100} + 8s_{111}\varepsilon_{\rm sur}^{111} = 12L^2(5-2\beta)/a^2\,, \quad 2 \geq \beta \geq 1\,.$$

Next, the cluster volume in this case is

$$V = (8\beta^3/3 - 12\beta^2 + 12\beta + 4)L^3\,, \quad 2 \geq \beta \geq 1\,.$$

From this we have by analogy with formula (2.42) for the specific surface energy

$$A = \frac{6(5-2\beta)}{(4\beta^3/3 - 6\beta^2 + 6\beta + 2)^{2/3}}\,, \quad 2 \geq \beta \geq 1\,. \tag{2.43}$$

This function has a minimum of $A = 2^{1/3} \cdot 6 = 7.56$ at $\beta = 1.5$, which corresponds to a regular truncated octahedron. This proves that the regular truncated octahedron given in Figure 2.10 is the optimal geometrical figure for short-range atomic interactions. In addition, for $\beta = 1$ according to formula (2.43) we get $A = 9 \cdot (18/25)^{1/3}$, as we obtained earlier, and in the octahedron case $\beta = 2$ we obtain $A = 3 \cdot 18^{1/3} = 7.86$. Figure 2.11 contains the dependence $A(\beta)$ for crystallites under consideration.

Problem 2.26

Compare three definitions of cluster sphericity for symmetric clusters with the symmetry center. In the first case, the sphericity coefficient is defined as

$$\xi = \frac{r_2 - r_1}{\overline{r}}\,, \tag{2.44}$$

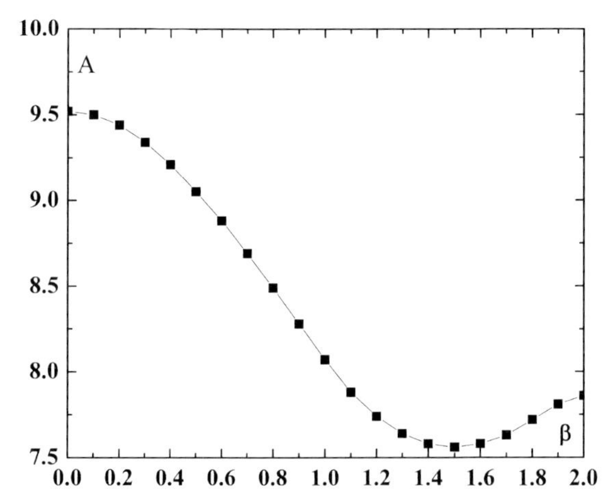

Fig. 2.11 The specific surface energy for a fcc cluster has the shape of the geometric figure of Problem 2.25. The values of A are given by formulas (2.42) and (2.43).

where r_1, r_2 are the distances from the center for the nearest and farthest surface atoms, and for the mean cluster radius we take $\overline{r} = (r_1 + r_2)/2$. In the second case, the sphericity coefficient of a geometric figure is introduced as

$$\psi = 1 - \frac{4\pi r^2}{S} = 1 - (36\pi)^{1/3}\frac{V^{2/3}}{S}, \tag{2.45}$$

where S is the surface area of this figure and r is the radius of a ball whose volume equals the figure's volume. In the third case, the sphericity coefficient of a symmetric cluster is taken as

$$\eta = \frac{\Delta r}{\overline{r}}, \quad \overline{r} = \sum_i r_i, \quad \Delta r^2 = \sum_i (r_i - \overline{r})^2 = \overline{r^2} - \overline{r}^2, \tag{2.46}$$

and r_i is the distance of a surface atom from the cluster center. Apply these equations to large crystallites of the cubic shape.

All the definitions give a zero sphericity coefficient for the spherical cluster surface and are lower the closer the cluster surface is to the spherical shape. The first definition (2.44) of the spherical coefficient is a rough one and gives for the cube shape

$$\xi = \frac{2(\sqrt{3}-1)}{\sqrt{3}+1} = 0.54 .$$

The second definition (2.45) gives for the cube cluster shape ($S = 6a^2$, $V = a^3$, where a is a length of the cube side)

$$\psi = 1 - \left(\frac{\pi}{6}\right)^{1/3} = 0.19 .$$

In evaluating the spherical coefficient (2.46), when a cluster is a symmetric geometric figure consisting of identical facets, we divide each facet into identical rectangular triangles, and the line from the figure center to the facet center is perpendicular to the facet. We denote by h a line length between the centers of the figure and the facet and by a a leg length for a rectangular triangle that joins the centers of the facet and its side. Then we have for the rectangular triangle area S the average distance $\overline{r}$ between the centers of the figure and the points of the facets and the average distance square $\overline{r^2}$ according to the following equations:

$$S = \frac{a^2}{2} \int_0^{\varphi_o} \frac{d\varphi}{\cos^2 \varphi} \,,$$

$$\overline{r} S = \int_0^{\varphi_o} \frac{d\varphi}{3} \left[\left(h^2 + \frac{a^2}{\cos^2 \varphi} \right)^{3/2} - h^3 \right] ,$$

$$\overline{r^2} S = \int_0^{\varphi_o} \frac{d\varphi}{4} \left(\frac{2h^2 a^2}{\cos^2 \varphi} + \frac{a^4}{\cos^4 \varphi} \right) .$$

One can reduce these formulas to the form

$$S = \frac{a^2 \tan \varphi_o}{2} \,,$$

$$\overline{r} = \frac{2}{3a^2 \tan \varphi_o} \int_0^{\varphi_o} \frac{dx}{1+x^2} [(h^2 + a^2 + a^2 x^2)^{3/2} - h^3] \,, \tag{2.47}$$

$$\overline{r^2} = h^2 + \frac{a^2}{2} + \frac{a^2}{6} \tan^2 \varphi_o \,.$$

Let us apply these equations to the cube case where $h = a, \varphi_o = \pi/4$. Then extracting a part of the cube surface that is a rectangular triangle and occupies 1/8 of the quadratic facet. This gives $S = a^2/8, \overline{r} = 1.281a, \overline{r^2} = 5a^2/3$. From this it follows that $\delta r^2 = \overline{r^2} - (\overline{r})^2 = 0.0257$, $\delta r = 0.160$, and $\eta = \delta r/\overline{r} = 0.125$, that is, $\eta < \psi \ll \xi$.

Evidently, the definition of the spherical coefficient in the form of formula (2.44) does not account for the space atom distribution on a facet and gives a large value for this quantity. Therefore below we reject this definition of the spherical coefficient and will base our analysis mostly on definitions (2.45) and (2.46), which account for the distribution based on distances from the cluster center and are suitable for clusters with incomplete atomic shells.

Problem 2.27

Determine the sphericity coefficient (2.45) of a regular geometric figure as is obtained from a cube with a side length $2L$ by cutting off the regular pyramids whose vertices are the cube vertices [65].

In accordance with the operation of Problem 2.25, we take a cube with a side length $2L$ as a basis and cut off eight pyramids whose vertices are cube vertices; a side length of these pyramids is $L\beta$, where $0 \le \beta \le 2$. We find the surface area and the volume of the geometric figure obtained and then the sphericity coefficient on the basis of formula (2.45).

We find firstly the range $0 \le \beta \le 1$ when the surface of this figure consists of six octagons of directions {100} of the area $s_{100} = 4L^2 - 2(L\beta)^2$, and eight triangles of the area $s_{111} = L^2\beta^2\sqrt{3}/2$, and the total area of the figure surface S and its volume V are equal to

$$S = 6s_{100} + 8s_{111} = 4L^2(6 - 3\beta^2 + \beta^2\sqrt{3}) ,$$
$$V = 8L^3\left(1 - \frac{\beta^3}{6}\right) , \quad 0 \le \beta \le 1 .$$

Formula (2.45) gives for the spherical coefficient

$$\psi = 1 - (36\pi)^{1/3}\frac{\left(1 - \frac{\beta^3}{6}\right)^{2/3}}{6 - 3\beta^2 + \beta^2\sqrt{3}} , \quad 0 \le \beta \le 1 . \tag{2.48}$$

In the limit $\beta = 0$ when this figure is a cube, formula (2.48) gives

$$\psi = 1 - \left(\frac{\pi}{6}\right)^{1/3} = 0.194$$

in accordance with the above result. In another limit, $\beta = 1$, when this figure is transformed into a cuboctahedron, according to formula (2.48) we have

$$\psi = 1 - \frac{(25\pi)^{1/3}}{3 + \sqrt{3}} = 0.095 .$$

In another range of the parameter β, when $1 \le \beta \le 2$, the surface of the figure under consideration consists of six squares of direction {100} of the area $s_{100} = 4L^2 - 2(L\beta)^2$ and eight trapeziums of the area $s_{111} = L^2\beta^2\sqrt{3}/2$. The total area of the figure surface S and its volume V are equal to

$$S = 6s_{100} + 8s_{111} = 4L^2[3(2 - \beta)^2 + \sqrt{3}(6\beta - 2\beta^2 - 3)] ,$$
$$V = \left(\frac{8}{3}\beta^3 - 12\beta^2 + 12\beta + 4\right)L^3 , \quad 1 \le \beta \le 2 .$$

From this we have on the basis of formula (2.45) for the spherical coefficient

$$\psi = 1 - (9\pi)^{1/3}\frac{\left(\frac{2\beta^3}{3} - 3\beta^2 + 3\beta + 1\right)^{2/3}}{3(2 - \beta)^2 + \sqrt{3}(6\beta - 2\beta^2 - 3)} , \quad 1 \le \beta \le 2 . \tag{2.49}$$

In the limit $\beta = 1$, when this figure is a cuboctahedron, formula (2.49) gives

$$\psi = 1 - \frac{(25\pi)^{1/3}}{3 + \sqrt{3}} = 0.095$$

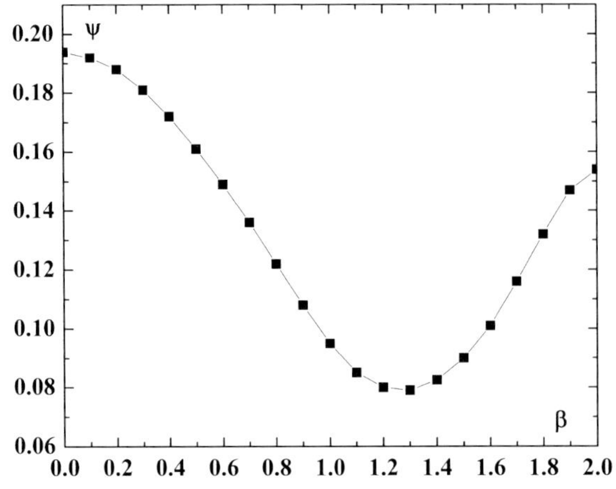

Fig. 2.12 Defined by (2.45) the spherical coefficient $\psi(\beta)$ for the geometric figure obtained in Problem 2.25 by cutting off identical regular pyramids from a cube near its vertices. The values of this spherical coefficient are given by formulas (2.48) and (2.49).

in accordance with the above result. In the case $\beta = 2$, this figure is transformed into an octahedron, and from (2.49) we have

$$\psi = 1 - \frac{(\pi)^{1/3}}{\sqrt{3}} = 0.154\,,$$

as we obtained above. Figure 2.12 gives the dependence $\psi(\beta)$ for the crystallite under consideration.

Problem 2.28

Determine the specific energy of the surfaces {100} and {111} of an fcc structure for the Lennard-Jones interaction potential of atoms.

Let us take an infinite fcc crystal and divide it into two parts by a plane {100} or {111}. The specific surface energy corresponds to the interaction potential of these parts per unit area. Let us denote the interaction energy of a test surface atom with all the atoms of the kth layer by ε_k (this atom is located in the zeroth layer). Then the specific surface energy is given by

$$\varepsilon_{\text{sur}} = \sum_{k=1}^{\infty} k \frac{\varepsilon_k}{s}\,, \tag{2.50}$$

where s is the area per atom, and we take into account that each bond relates to two atoms. Thus, for the Lennard-Jones interaction potential, the specific surface energy is equal to

$$\varepsilon_{\text{sur}} = \frac{1}{s}\left[B_6 \frac{R_e^6}{a^6} - B_{12} \frac{R_e^{12}}{2a^{12}}\right]\,, \tag{2.51}$$

Table 2.11 Distances of nearest atoms for a test atom of {100} surface of an fcc structure.

r_{ik}^2/a^2	First layer	Second layer	Third layer	Fourth layer
1	4	–	–	–
2	–	1	–	–
3	8	4	–	–
4	–	4	–	–
5	4	–	4	–
6	–	4	–	–
7	8	8	8	–
8	–	–	–	1
9	8	–	4	4
10	–	4	–	4
11	–	–	8	–
12	–	8	–	4

where the dissociation energy of the diatomic is taken as $D = 1$, a is the distance between nearest neighbors, R_e is the equilibrium distance for the diatomic, s is the surface area per atom, and

$$B_6 = \sum_{i,k} k n_{ik} \frac{a^6}{r_{ik}^6} , \quad B_{12} = \sum_{i,k} k n_{ik} \frac{a^{12}}{r_{ik}^{12}} . \tag{2.52}$$

Here r_{ik} is the distance from a test atom to the ith atom of the kth layer, and n_{ik} is the number of such atoms.

First we consider the surface {100}. The area per atom of this surface is equal to $s = a^2$, and the distance between neighboring layers is $a/\sqrt{2}$. Table 2.11 contains numbers of atoms [17] whose distance from a test atom is $r_{ik}^2 \leq 12$, and atoms are located over layers. We take the sum in formula (2.52) up to $r_{ik}^2 = 12$, and for $r_{ik}^2 > 12$ we replace the sum by integrating both for atoms of the first four layers and for atoms of subsequent layers. Correspondingly, sums (2.52) are represented in the form of three terms:

$$B_6^{100} = 5.35 + 10a^6 \int_{r_o}^{\infty} \frac{2\pi\rho d\rho dz}{(\rho^2 + z^2)^3} + \sum_{k=5}^{\infty} k a^4 \int_{ka/\sqrt{2}}^{\infty} \frac{2\pi\rho d\rho}{(\rho^2 + k^2 a^2)^3} = 5.39 ,$$

where we take $r_o^2 = 13$. In the same way we obtain $B_{12}^{100} = 4.06$.

Let us use the same method to determine the parameters of the surface {111}. Table 2.12 lists the positions of the nearest atoms for a test atom of this surface [17]. In this case we get

$$B_6^{111} = 4.116 + \frac{10\pi}{\sqrt{3}} \cdot \frac{a^4}{r_o^4} + \frac{3\sqrt{3}\pi}{8k_o^2} = 4.44 , \quad B_{12}^{111} = 3.07 ,$$

where $k_o = 5$.

Table 2.12 Distances of nearest atoms for a test atom of {111} surface of an fcc-structure.

r_{ik}^2/a^2	First layer	Second layer	Third layer	Fourth layer
1	3	–	–	–
2	3	–	–	–
3	6	3	–	–
4	–	3	–	–
5	6	6	–	–
6	3	–	1	–
7	6	6	6	–
8	–	3	–	–
9	3	6	6	–
10	6	–	–	–
11	6	3	6	3
12	–	6	3	3

Let us take the distance between the nearest neighbors of the surface as in the Lennard-Jones crystal $a = 0.975 R_e$. Then, taking into account that for the {100} surface the area per atom is $s = a^2$, and that for the {111} surface is $s = a^2\sqrt{3}/2$. Formula (2.51) gives for the specific surface energies of the indicated surfaces that $\varepsilon_{100} = 3.66/a^2$, $\varepsilon_{111} = 3.59/a^2$.

Problem 2.29

Determine the specific surface energy of crystalline particles that take the form of a truncated octahedron for the Lennard-Jones interaction potential of atoms. Find the optimal parameters of this particle.

The specific surface energy of such particles is given by formula (2.43), which has the form

$$A = \frac{a^2}{2L^2} \cdot \frac{\varepsilon_{100} S_{100} + \varepsilon_{111} S_{111}}{(4\beta^3/3 - 6\beta^2 + 6\beta + 2)^{2/3}},$$

where ε is the specific surface energy of the corresponding surface and $S_{100} = 12(2-\beta)^2 L^2$ and $S_{111} = 4\sqrt{3}(6\beta - 2\beta^2 - 3)L^2$ are the total areas of the facets of the corresponding directions. Using the results of the previous problem for the Lennard-Jones interaction potential, we transform this equation into the form

$$A = (50.5 - 13.2\beta - 2.9\beta^2) \cdot (4\beta^3/3 - 6\beta^2 + 6\beta + 2)^{-2/3}. \tag{2.53}$$

This formula has a minimum at $\beta = 1.3$, which corresponds to $A = 15.1$. For the cuboctahedron ($\beta = 1$) this formula gives $A = 15.4$; for the regular truncated octahedron ($\beta = 1.5$) we have $A = 15.2$, and for an octahedron ($\alpha = 2$) we get $A = 16.3$. It follows from this that, although the optimal figure differs from the regular truncated octahedron, the difference in the specific surface energies for these figures is relatively small.

Problem 2.30

In the case of the Morse interaction potential of atoms, determine the specific surface energies of crystalline particles of an fcc structure and find its optimal configuration as a function of the adjustable parameter α.

Using the Morse potential (2.37) of atomic interaction, we obtain for the specific surface energy

$$\varepsilon_{\text{sur}} = D\left[e^{\alpha R_e} G(\alpha a) - \frac{1}{2} e^{2\alpha R_e} G(2\alpha a)\right],$$

where according to formula (2.50) we have

$$G(\alpha a) = \sum_k k n_{ik} \exp(-\alpha r_{ik})/s\,. \tag{2.54}$$

Here k is the layer number, i is the number of an atom located in this layer, n_{ik} is a number of such atoms, r_{ik} is the distance from a test surface atom to this one, s is the surface area per atom in the layer of the corresponding direction, and below we take $D = 1$, that is, express the energies in units of breaking of one bond. The first terms of this series for the $\{100\}$ and $\{111\}$ surface planes have the form

$$G_{100} = 4e^{-\alpha a} + 2e^{-\alpha a\sqrt{2}} + 16e^{-\alpha a\sqrt{3}} + 8e^{-2\alpha a} + 16^{-2\alpha a\sqrt{5}},$$

$$G_{111} = 3e^{-\alpha a} + 3e^{-\alpha a\sqrt{2}} + 12e^{-\alpha a\sqrt{3}} + 6e^{-2\alpha a} + 18e^{-\alpha a\sqrt{5}},$$

and the first term of each expansion corresponds to a short-range interaction, that is, it accounts for the interaction between nearest neighbors only.

The results of numerical calculations are given in Table 2.10, where the above sum is restricted by $r_{ik}^2 \le 12$. In order to estimate the accuracy of neglecting more distant atoms, let us compare the contribution of farther atoms in the sum ΔG and the contribution of nearest neighbors from the first layer $G_{nn} = n_1 \exp(-\alpha a)$, where n_1 is the number of nearest neighbors of a test surface atom in the first layer. This value is $n_1 = 4$ for the $\{100\}$ plane and $n_1 = 3$ for the $\{111\}$ surface plane. We have in the case $\alpha r_o \gg 1$

$$\frac{\Delta G}{G_{nn}} = \frac{231}{\beta a n_1} \exp(-2.6\alpha a)\,,$$

for $r_o = \sqrt{13}$. In particular, for $\alpha a = 3$ and for the $\{111\}$ surface plane this ratio is 0.01. Because we are guided by $\alpha a \ge 3$, below we neglect ΔG.

From this one can find the specific energy per surface atom

$$g(\alpha a) = \exp(\alpha R_e) G(\alpha a) - \frac{1}{2} \exp(2\alpha R_e) G(2\alpha a)\,. \tag{2.55}$$

Table 2.13 Parameters of a bulk crystalline of an fcc structure for the Morse interaction potential between atoms [73].

αa	αR_e	G_{100}	G_{111}	g_{100}	g_{111}	A	β_{opt}
2	3.03	2.00	1.70	37.6	32.1	155	1.30
3	3.57	0.387	0.321	11.3	9.52	46.4	1.33
4	4.31	0.103	0.0832	5.53	4.57	22.4	1.35
5	5.17	0.0322	0.0255	3.64	2.94	14.5	1.39
6	6.10	0.0109	0.0^2850	2.85	2.28	11.3	1.41
7	7.06	0.0^2384	0.0^2296	2.48	1.95	9.69	1.43
8	8.03	0.0^2138	0.0^2106	2.25	1.74	8.71	1.45
10	10.01	0.0^3184	0.0^3139	2.08	1.59	7.96	1.47
12	12.00	0.0^4247	0.0^4186	2.02	1.52	7.66	1.49

In particular, in the case of a short-range interaction of atoms, the value g equals one half of the bonds between nearest neighbors outside the surface, that is, it is $g_{100} = 2$ for the {100} surface plane and $g_{111} = 3/2$ for the {111} surface plane. These values correspond to asymptotic values of g in the limit of large αa. Values of the function $g(\alpha a)$ are given in Table 2.13. Note that the connection of the distance between nearest neighbors a and the equilibrium distance of the diatomic R_e are determined in accordance with the Table 2.6 data.

Problem 2.31

Determine the specific surface energy for the optimal figure of a truncated cube and Morse interaction potential between atoms.

Using formula (2.42) for the surface energy, one can determine it for any figure whose surface consists of {100} and {111} plane facets. On the basis of formula (2.42), using definition (2.8) of the specific surface energy $E_{sur} = An^{2/3}$ we obtain for this quantity

$$A = \frac{2\left[3(2-\beta)g_{100} + 6(6\beta - 2\beta^2 - 3)g_{111}\right]}{(4\beta^3/3 - 6\beta^2 + 6\beta + 2)^{2/3}},$$

where g_{100} and g_{111} are the specific surface energies per surface atom for the corresponding plane of facets. In the case of a short-range interaction of atoms, when we account for the interaction between nearest neighbors only, these parameters are $g_{100} = 2$ and $g_{111} = 3/2$, and the specific surface energy has a minimum at $\beta = 3/2$, where this value is $A = 2^{1/3} \cdot 6$. This takes place at large values of the parameter α of the Morse interaction potential.

At arbitrary values of the parameter α the optimal value β differs from 3/2. Below we assume this value β_{opt} to be close to 3/2, so that one can expand the expression for A over a small parameter $\beta - 3/2$. Then the specific surface energy takes the following form:

$$A = 2^{1/3} \cdot 6 \cdot \left[g + \frac{\Delta g}{2} + \frac{g}{4}(\beta - 3/2)^2 + 1.5\Delta g(\beta - 3/2) \right],$$

where $g = g_{100}/4 + g_{111}/3$, $\Delta g = g_{111}/3 - g_{100}/4$.

From this formula it follows for the optimal value of β and the minimum value of the specific surface energy

$$\beta_{\text{opt}} - 3/2 = -3\Delta g/g\,, \quad A = 6 \cdot 2^{1/3} \cdot [g + \Delta g/2 - 9\Delta g^2/(4g)]\,.$$

Using the values of g_{100} and g_{111} from Table 2.13, we find values of this equation that are given in Table 2.13. Note that $\beta_{\text{opt}} < 1.5$, that is, the optimal crystalline figure differs from a regular truncated octahedron such that the general edges of squares and hexagons are longer than the general edges of two hexagons.

Problem 2.32

Show the validity of the Wulff criterion [74] for the optimal configuration of the bulk systems under consideration when bound atoms form a truncated octahedron and for a short-range interaction of atoms. According to the Wulff criterion, the optimal crystalline figure with plane facets satisfies the relation

$$\frac{\varepsilon^i_{\text{sur}}}{R_i} = \frac{\varepsilon^j_{\text{sur}}}{R_j}\,, \tag{2.56}$$

where $\varepsilon^i_{\text{sur}}$ is the specific surface energy for the ith facet and R_i is the distance from the center to this facet.

Let us find the optimal form among the bulk figures under consideration of crystalline particles and a short-range interaction potential between atoms on the basis of the Wulff criterion. We have that the distance from the cube center to a surface square equals L, and the distance to surface hexagons is equal to $L\sqrt{3}(1 - \beta/3)$. From this we obtain on the basis of formula (2.56) for the optimal value of β that

$$\beta_{\text{opt}} = 3 \cdot \left(1 - \frac{\varepsilon_{111}}{\varepsilon_{100}\sqrt{3}}\right)\,. \tag{2.57}$$

In the case of a short-range interaction ($\varepsilon_{111} = \sqrt{3}, \varepsilon_{100} = 2$) this formula gives $\beta = 1.5$, in accordance with results from the optimization of the relation (2.43). In the case of the Lennard-Jones interaction potential of atoms ($\varepsilon_{111} = 3.59D/a^2, \varepsilon_{100} = 3.66D/a^2$), formula (2.57) gives that the optimal structure of a particle corresponds to $\beta = 1.3$. This coincides with the minimum of the specific surface energy (2.53). Thus, the Wulff criterion is valid for both interaction potentials.

In the case of the Morse interaction potential (2.57) takes the form

$$\beta = 3 - 2\frac{g_{111}}{g_{100}} .$$

This equation gives optimal values of the parameter β for the Morse interaction potential that coincide with the data of Table 2.13 within an accuracy of about 1%.

Problem 2.33

Determine the specific surface energy of a spherical liquid particle with a short-range interaction between its atoms.

Assume that an internal atom of a liquid drop has k nearest neighbors on average, and the distance between nearest neighbors is a. Because the optimal structure of the surface corresponds to the {111} plane of the close packed structure, the surface energy per unit area is equal to $(k/2 - l/2)\,D/s$, where l is a number of nearest neighbors for a surface atom, D is the well depth for the pair interaction potential of two nearest neighbors, and $s = a^2\sqrt{3}/2$ is the surface area per particle. Then the surface energy of a liquid particle is $(D = 1)$

$$E_{\mathrm{sur}} = \frac{4\pi}{\sqrt{3}}\frac{r^2}{a^2}(k - l) .$$

Accounting for the variation of the particle density as a result of their melting, we have for a number of atoms of a liquid drop

$$n = \frac{4\pi}{3}r^3\frac{\sqrt{2}}{a^3}\frac{\rho_{\mathrm{l}}}{\rho_{\mathrm{s}}} ,$$

where $\sqrt{2}/a^3$ is the number density of atoms in a solid particle of a close packed structure, ρ_{s} and ρ_{l} are the densities of a bulk system of this sort for the solid and liquid state, respectively, at the melting point.

Thus, the specific surface energy of a liquid drop for this model is given by

$$A = \frac{E_{\mathrm{sur}}}{n^{2/3}} = 3^{1/6}(2\pi)^{1/3}(k - l)\frac{\rho_{\mathrm{s}}}{\rho_{\mathrm{l}}} .$$

Taking the parameters of the liquid drop in accordance with the data of Table 2.5, we have $k = 10.16$, $l = 9k/12$, $\rho_{\mathrm{l}}/\rho_{\mathrm{s}} = 0.865$. This leads to the specific surface energy $A = 4.13$, so that the total binding energy of atoms of the drop E and the specific binding energy of surface atoms $\Delta\varepsilon = dE/dn$ are equal, respectively, to

$$E = 5.08n - 4.13n^{2/3} , \quad \Delta\varepsilon = 5.08 - \frac{6.2}{n^{1/3}} .$$

2.5 Thermodynamics of Large Liquid Clusters in Parent Vapor

Problem 2.34

Derive the expression for the free enthalpy of a large liquid drop located in a parent vapor.

When the vapor pressure exceeds the saturated vapor pressure p_{sat}, growth is possible for a liquid cluster located in a vapor. Therefore, the measure of cluster growth is the supersaturation degree given by

$$S = \frac{p}{p_{sat}(T)} = \frac{N}{N_{sat}(T)}, \tag{2.58}$$

where p is the vapor pressure, N is the number density of free atoms, and $N_{sat}(T)$ is the number density of atoms at the saturated vapor pressure and at an indicated temperature T. The temperature dependence for the saturated vapor pressure has the form

$$p_{sat}(T) = p_o \exp\left(-\frac{\varepsilon_o}{T}\right), \tag{2.59}$$

with ε_o the atom binding energy in a cluster.

If a nucleating vapor is an admixture to a buffer gas, we deal with the partial pressure of a vapor of a given sort. Because as the parameters of the vapor state are the pressure p and temperature T, the thermodynamic parameter is the free enthalpy G. We denote by G_n the free enthalpy of a liquid cluster (liquid drop) consisting of n atoms. The cluster chemical potential that is defined as $\mu_n = G_{n+1} - G_n$ characterizes an equilibrium of a liquid cluster with a parent gas. The equilibrium of this drop with the surrounding vapor is determined by equality of the chemical potentials of the drop μ_n and a surrounding parent vapor, and we determine the free enthalpy G_n of the drop consisting of n atoms.

Let us represent the free enthalpy of a large liquid cluster by the sum of two terms that account for the space and surface energies in accordance with formula (2.9):

$$G_n = (\varepsilon_g - \varepsilon_o)n + An^{2/3}, \tag{2.60}$$

where a cluster radius is $r_o = r_W n^{1/3}$ according to formula (2.5) (r_W is the Wigner–Seitz radius), ε_o is the binding energy of internal cluster atoms, ε_g is the energy of an atom in the surrounding gas, and A is the specific surface energy. The free enthalpy is taken with respect to the enthalpy of a parent gas, that is, the free enthalpy of the gas is zero. Below we find the equilibrium conditions for the system of a drop and its vapor at a given temperature and pressure.

Using relations (2.58) and (2.59), that gives

$$-\ln S = \frac{\varepsilon_o}{T} + \text{const},$$

and substituting this into formula (2.60), we obtain for the free enthalpy of a cluster

$$G_n = An^{2/3} - n(T \ln S + \text{const}) .$$

The value of const will be found from the equilibrium condition $\mu_n = 0$ for a plane surface, which corresponds to the limit $n \to \infty$ and, according to the definition of the saturated vapor pressure, has the form $s = 1$. This condition gives const $= 0$, so that the free enthalpy of a large liquid cluster modeled by the liquid drop model has the form

$$G_n = -nT \ln S + An^{2/3} . \tag{2.61}$$

In terms of the cluster radius r_o that is connected by formula (2.5) with a number n of cluster atoms, this equation has the form

$$G_n = -\frac{r_o^3}{r_W^3} T \ln S + 4\pi r_o^2 \gamma . \tag{2.62}$$

Problem 2.35

Find the critical size of a liquid cluster (or the critical radius) that corresponds to the minimum stability of a liquid cluster (liquid drop).

The critical cluster size n_{cr} is given by the relation

$$\frac{\partial G_n}{\partial n} = 0 ,$$

and is determined by the equation

$$n_{cr} = \left(\frac{2A}{3T \ln S} \right)^3 . \tag{2.63}$$

The corresponding critical cluster radius is equal to

$$r_{cr} = n_{cr} r_W = \frac{2\gamma m_a}{\rho T \ln S} , \tag{2.64}$$

where the Wigner–Seitz radius r_W is given by formula (2.6).

The critical cluster size is the principal parameter that characterizes an equilibrium between liquid clusters and a parent vapor. The free enthalpy of a cluster is, at the critical cluster size, equal to

$$G(n_{cr}) = \frac{n_{cr}}{2} T \ln S = \frac{A}{3} n_{cr}^{2/3} = \frac{A}{3} \left(\frac{2A}{3T \ln S} \right)^2 , \tag{2.65}$$

and characterizes the formation energy for a cluster of the critical radius. Since the probability of this event is $\sim \exp[-G(n_{cr})/T]$, the nucleation process proceeds long enough in a pure vapor, that is, a vapor can be found for a long time in a supersaturated state.

If we express the free enthalpy of a cluster through a cluster radius r_o according to formula (2.5), this quantity and its derivations are given by

$$G(r_o) = 4\pi\gamma\left(r_o^2 - \frac{2r_o^3}{3r_{cr}}\right); \quad \frac{dG(r_{cr})}{dr_o} = 0, \quad \frac{d^2G(r_{cr})}{dr_o^2} = -16\pi\gamma\, r_{cr}, \tag{2.66}$$

where the surface tension γ is connected with the cluster-specific surface energy A by the relation (2.11) $A = 4\pi r_W^2\gamma$.

It is convenient to introduce a temperature for a current gas pressure, so that at the temperature T_* the current gas pressure coincides with the saturated vapor pressure, that is,

$$p = p_o \exp\left(-\frac{\varepsilon_o}{T_*}\right).$$

Then (2.61) gives

$$G_n = An^{2/3} + n\varepsilon_o\frac{T - T_*}{T_*}. \tag{2.67}$$

From this we have for the critical cluster size n_{cr}, expressing it through the critical temperature T_*,

$$n_{cr} = \left[\frac{2AT_*}{3\varepsilon_o(T_* - T)}\right]^3. \tag{2.68}$$

Problem 2.36

Determine the critical chemical potential of a cluster located in the parent vapor by modeling this cluster as a liquid drop.

By definition, the cluster chemical potential is

$$\mu_n = G_{n+1} - G_n$$

or, for large cluster sizes,

$$\mu_n = \frac{\partial G_n}{\partial n}.$$

The chemical potential characterizes the equilibrium of a liquid cluster with a parent gas, and in the case of cluster equilibrium with a surrounding vapor their chemical potentials are equal. Because the chemical potential of a parent gas is zero, the equilibrium requires for the cluster $\mu_n = 0$, and this equilibrium is realized at the critical point.

Using formula (2.61) for the free enthalpy of a liquid drop, we have for the cluster chemical potential within the framework of the liquid drop model

$$\mu_n = G_{n+1} - G_n = \frac{2A}{3}\left(n^{-1/3} - n_{\rm cr}^{-1/3}\right) , \tag{2.69}$$

and near the critical size ($|n - n_{\rm cr}| \ll n_{\rm cr}$) the chemical potential of the liquid drop has the form [75]

$$\mu_n = \frac{2A}{9n_{\rm cr}^{4/3}}(n_{\rm cr} - n) . \tag{2.70}$$

Problem 2.37

Determine the critical size for a charged particle located in the parent vapor.

Let us calculate the part of the free enthalpy ΔG of a charged cluster that is given by

$$\Delta G = \int_0^\infty \frac{\varepsilon E^2}{8\pi} d\mathbf{R}$$

and accounts for cluster interactions with an external electric field. Here $\mathbf{R}$ is a point coordinate for the frame of reference where the origin is located at the cluster center and is the point of ion location, $\mathbf{E}$ is the electric field strength, and ε is the dielectric constant of matter, which we assume to be unity outside the particle and a constant inside it. Then the electric field strength is equal to $e/(\varepsilon R^2)$ inside the cluster and e/R^2 outside it. From this it follows for the electric part of the free enthalpy of a cluster [76]

$$\Delta G = \frac{e^2}{R}\left(1 - \frac{1}{\varepsilon}\right) + \text{const} ,$$

where const does not depend on the cluster radius. Taking this term into account, we obtain for the free enthalpy of the particle, instead of formula (2.62),

$$G_n = 4\pi r_o^2 \gamma - \frac{r_o^3}{r_W^3} T \ln S + \frac{e^2}{2r_o}\left(1 - \frac{1}{\varepsilon}\right) + \text{const} .$$

The equilibrium condition between the cluster and a surrounding vapor corresponds to the free enthalpy maximum, that is, to the condition $dG/dr = 0$, which leads to the following equation for the critical radius:

$$8\pi r_{\rm cr}^2 \gamma - \frac{3r_{\rm cr}^2}{r_W^3} T \ln S - \frac{e^2}{2r_{\rm cr}}\left(1 - \frac{1}{\varepsilon}\right) = 0 . \tag{2.71}$$

Problem 2.38

Determine the supersaturation degree when the growth of a charged particle in the parent gas takes place for any particle size.

Let us rewrite formula (2.71) in the form

$$\ln S = \frac{2\gamma m_a}{r_{cr}\rho T}\left[1 - \frac{e^2}{16\pi r_{cr}^3 \gamma}\left(1 - \frac{1}{\varepsilon}\right)\right],$$

and this expression has its maximum at the following value of the critical radius:

$$r_{cr} = \left[\frac{e^2}{4\pi\gamma}\left(1 - \frac{1}{\varepsilon}\right)\right]^{1/4}.$$

This corresponds to the supersaturation degree (2.58)

$$\ln S_{max} = \frac{m_a \gamma}{r_{cr}\rho T} = \frac{e^2 r_W}{3 T r_{cr}},$$

where the Wigner–Seitz radius is given by formula (2.5), and the specific surface energy is defined by formula (2.8) and is connected with the surface tension γ by formula (2.11). Note that if a cluster is neutral, its growth is preferable when the cluster size exceeds the critical one, and the cluster evaporates when its size is less than the critical size. If the degree of supersaturation exceeds the value S_{max} for a charged cluster, the equation for the supersaturation degree s has no solution, that is, a cluster of any size grows.

3
Structures of Solid Clusters with Pairwise Atomic Interaction

Clusters as systems of bound atoms as well as bulk atomic systems may be found in the solid and liquid aggregate states. In the case of liquid clusters, the liquid drop model is an appropriate one for large clusters. Solid clusters exhibit certain structures where atoms are distributed over atomic shells as we had above for metal clusters within the framework of the jellium model. Below we consider the simplest case for a system of bound atoms with a pairwise interaction between atoms.

The principal cluster property that distinguishes a cluster as a physical object from macroscopic atomic systems is the magic numbers of solid clusters. Cluster parameters such as the cluster ionization potential, the cluster electron affinity, and the atom binding energy are characterized by higher values for magic numbers of atoms than for atom numbers higher or lower by one. The magic cluster numbers correspond to complete cluster structures. Taking a cluster to be constructed from separate shells, layers, or blocks, one can find these magic numbers.

It is clear that the crystal structure of very large clusters coincides with that of bulk crystals. When we cut these large clusters off macroscopic crystals, we call them crystallites when their surfaces are complete, that is, if these clusters consist of magic numbers of atoms. Crystallites are cluster structures with large numbers of atoms. But because the ratio of the number of surface atoms to the total number of atoms for small clusters is large compared with that in large clusters, small clusters may have another structure. Then an increase in cluster size may lead to a change in the optimal structure in some size range that is determined by the character of the atomic interaction inside the clusters. Competition among cluster structures is a specific property of solid clusters

In the case of a pairwise interaction between atoms, energetic cluster parameters can be expressed through parameters of the interaction potential of two isolated atoms. In considering such clusters we are guided by clusters of inert gases where the interaction potential between neighboring atoms in a system of bound atoms is small compared with a typical electron energy. Because of the weakness of this interaction, one can neglect three body and many body interactions between atoms. This simplifies the problem and allows one to ascertain the role of short-range and long-range interactions for cluster properties.

A cluster as a system of a finite number of bound atoms possesses a position intermediate between atoms and bulk solids. Atoms as a system of bound elec-

Cluster Processes in Gases and Plasmas. Boris M. Smirnov

ISBN: 978-3-527-40943-3

trons have a shell structure for electrons, and solids containing an infinite number of atoms form a crystal lattice at low temperatures. Clusters have a shell structure, and cluster shells are formed by atoms or electrons for dielectric and metal clusters, respectively. We restrict our consideration to clusters with a pairwise interaction of atoms. When all shells of a given atom are complete, this cluster has a magic number of atoms that corresponds to extrema for various cluster parameters as a function of the number of atoms. Because the number of cluster shells is infinite, a cluster is characterized by an infinite number of magic numbers at zero temperature.

One can consider such clusters to be cut off from a bulk crystal. There are 230 groups of symmetry for crystalline lattices [64], and hence clusters can have any one of these depending on the character of interaction inside the system. Below we restrict ourselves to two lattice structures that correspond to a pairwise interaction potential between atoms when a short-range interaction is important for a system of many bound atoms. The face centered cubic(fcc) and hexagonal crystal lattices are realized in this case, and they correspond to a close-packed structure. Then each internal atom of the lattice has 12 nearest neighbors, and this is the maximum possible number of nearest neighbors in such structures. Below we consider clusters of close-packed structures and analyze the energetic parameters of these clusters.

Magic numbers of atoms correspond to complete shells or layers and are characterized by extrema of various cluster parameters as a function of the number of atoms. For example, the binding energy of a surface atom in a cluster consisting of a magic number of atoms exceeds that for clusters whose numbers of atoms are more or less by one than the magic number. The same conclusion holds true for the ionization potential and electron affinity of clusters. A cluster consisting of a magic number of atoms has maximum stability. Note that in contrast to macroscopic particles, edge and vertex atoms contribute to the cluster energy and other additive parameters.

Bulk particles and clusters of a given crystalline structure can form different geometric figures. The optimal cluster shell for a cluster of a given size depends on the parameters of a pairwise interaction of atoms. The energetic parameters of a cluster with incomplete shells are sensitive to the filling of certain shells or layers. The optimal cluster structure results from competition among some cluster shapes even when only one crystalline structure is realized. The competition among cluster shapes for close packed structures is the subject of this analysis.

3.1 Clusters of Close-Packed Structures

Problem 3.1

Formulate a method for constructing clusters of a fcc structure with a short-range interaction of atoms by means of successive addition of atoms to a cluster core [31, 68]. Define the magic numbers of an assembled cluster.

Within the framework of this method, we construct an fcc cluster as a result of adding new shells or their parts to a completed core. Because of the cluster symmetry, one can assemble a cluster by location of atoms on cluster shells, so the coordinates of one atom of a shell can be used for the description of this shell. Below we take for this goal an atom of a given shell with positive values of coordinates and $z \leq x \leq y$.

A short-range interaction between cluster atoms means that one can restrict the interaction to only between nearest neighbors. Then the cluster binding energy is equal to the number of bonds between nearest neighbors. Expressing distances in units $a/\sqrt{2}$, where a is the distance between nearest neighbors in the fcc crystal lattice, we have the following xyz coordinates of nearest neighbors for an internal lattice atom:

$$x \pm 1, y \pm 1, z\,; \quad x \pm 1, y, z \pm 1\,; \quad x, y \pm 1, z \pm 1\,. \tag{3.1}$$

From this one can formulate the method for assembling an fcc cluster with a short-range interaction of atoms. We first take a filled cluster core such that each atom of this core transfers in positions occupied by other atoms as a result of transformations (2.26). In the second stage we add new atoms step by step to this core, so that the additional atoms have the maximum number of nearest neighbors. As a result, we obtain an atomic configuration with the maximum number of bonds between nearest neighbors of cluster atoms. Because a growing cluster has an almost spherical form, the number of simultaneously filled shells is restricted for clusters that are not large. In reality, after the addition of one or several atoms of one shell to a core, the addition of atoms of another shell is energetically profitable, and atoms of a new shell are nearest neighbors of these.

Then the cluster growth proceeds by the addition of new blocks consisting of atoms of different shells. Tables 3.1 and 3.2 give the sequence of filling of fcc clusters with and without a central atom. Comparison of these data for cluster structures with and without a central atom allows us to choose the energetically optimal fcc structure of clusters. As follows from the data in Tables 3.1 and 3.2, the growth of clusters that are not small proceeds by the addition of blocks that are elements of plane facets. Magic numbers of clusters correspond to the filling of individual blocks.

Thus, though the above method of cluster assembly is not universal, it allows us to choose optimal atomic configurations among almost symmetrical ones that have a symmetric core. We take two symmetric cores, with and without a central atom, and construct optimal cluster configurations for any number of cluster atoms.

Problem 3.2

Analyze the optimal construction of an fcc cluster consisting of 116 atoms within the frame of reference with and without a central atom.

The noncentered cluster of this size contains complete shells 122, 113, and 123, and its surface energy is 180 (Table 3.2). According to Table 3.1, such a centered

Table 3.1 The order of growth of face centered cubic (fcc) clusters with a central atom for short-range atomic interaction. The values in parentheses indicate the number of nearest neighbors for the filling shell [17, 77].

Filling shells	n	E_{sur}	Filling block
011	2–13	–	–
002(4)	13–19	42–54	–
112(3–5)+022(5)	19–55	54–114	110
013(4)	55–79	114–138	100
222(3)+123(4–6)	79–135	138–210	111
035(5)+004(4)+114(5)+024(6)	135–201	210–258	100
233(3–5)+224(5)+134(5–6)	201–297	258–354	111
015(4–6)+125(5–6)	297–369	354–402	100
044(5)+035(6)	369–405	402–414	110
006(4)+116(5)+026(6)	405–459	414–450	100
334(3–5)+244(5)+235(5–6)+ +145(5–6)+226(5)+136(6)	459–675	450–594	111
055(5)+046(6)	675–711	594–606	110
017(4–6)+127(5–6)+037(6)	711–807	606–654	100
008(4)+118(5)+028(6)	807–861	654–690	100
444(3)+345(4–6)+255(5)+336(5)+ +246(6)+156(5–6)+237(5–6)+147(6)	861–1157	690–858	111
066(5)+057(6)+228(5)+138(6)+048(6)	1157–1289	858–894	110
019(4–6)+129(5–6)+039(6)	1289–1385	894–942	100
455(3–5)+446(5)+356(5–6)+347(5–6)+ 266(5)+257(6)+338(5)+248(6)+ 158(6)+167(5–6)+239(5–6)+149(6)	1385–1865	942–1158	111
077(5)+068(6)+059(6)	1865–1925	1158–1170	110

cluster consists of a core of 79 atoms and blocks of seven atoms including atoms from 222 and 123 shells. Correcting this scheme, compose a large block that corresponds to two such blocks and three atoms from the 033 shell. The addition of one large block increases the number of cluster atoms by 17 and the cluster surface energy by 21. Thus, the centered fcc cluster of 116 atoms consists of a complete core of 79 atoms, two large blocks of 17 atoms, and three atoms above it (one atom from the 222 shell and two atoms from the 123 shell). The cluster surface energy is 186, that is, the noncentered cluster with this number of atoms is preferable.

Let us describe the optimal configuration of cluster atoms that corresponds to the complete noncentered structure in notations of the centered cluster, if we take into account that an atom with coordinates xyz in the coordinate frame of the noncentered cluster has coordinates $x+1$, yz in the frame of reference of the centered cluster. Then we obtain for the fcc cluster of 116 atoms that its optimal configuration of atoms in the centered frame of axes includes a core of 55 atoms, and above it

Table 3.2 The sequence of growth of fcc clusters without a central atom for a short-range interaction of atoms. The values in parentheses indicate the number of nearest neighbors for the filling shell [17].

Filling shells	n	E_{sur}	Filling block
001	1–6	–	–
111(3)	6–14	24–48	111
012(3–6)	14–38	48–84	110
003(4)	38–44	84–96	100
122(3–5)+113(5)+023(5–6)	44–116	96–180	110
014(4–6)	116–140	180–204	100
223(3–5)+133(5)+124(5–6)+034(5–6)	140–260	204–312	111
005(4)+115(5)+025(6)	260–314	312–348	100
016(4–6)	314–338	348–372	100
333(3)+234(4–6)+225(5)+ +144(5)+135(6)+126(5–6)	338–538	372–516	111
045(5–6)+036(6)	538–586	516–528	110
007(4)+117(5)+027(6)	586–640	528–564	100
018(4–6)	640–664	564–588	100
344(3–5)+335(5)+245(5–6)+236(5–6)+ +155(5)+146(6)++227(5)+137(6)	664–952	588–756	111
056(5–6)+047(6)	952–1000	756–768	110
128(5–6)+038(6)	1000–1072	768–792	100
009(4)+119(5)+029(6)	1072–1126	792–828	100
445(3–5)+355(4–6)+346(5–6)+256(5–6)+337(5)+ +247(6)+238(5–6)+166(5)+157(6)+148(6)	1126–1510	828–1020	111
067(5–6)+058(6)+229(5)+139(6)+049(6)	1510–1658	1020–1056	110
01,10(4–6)	1658–1682	1056–1080	100
00,11(4)+11,11(5)+02,11(6)	1682–1736	1080–1116	100
12,10(5–6)+03,10(6)	1736–1808	1116–1140	100

20 atoms from the 013 shell, 24 atoms from the 113 shell, four atoms from the 222 shell, four atoms from the 033 shell, one atom from the 004 shell, four atoms from the 114 shell, and four atoms from the 024 shell. As follows from Table 3.1, obtaining such an atomic configuration requires the displacement of many internal atoms, and hence such a configuration of atoms does not include centered clusters in our scheme of construction. This means that the schemes for assembling centered and noncentered clusters are different. These schemes assume that a cluster has an almost spherical shape, and this cluster can be constructed around a center. Thus, there are two schemes of construction of fcc clusters depending on the position of the center around which the cluster is assembled.

This example shows that the schemes to construct the optimal structure for a cluster with a short-range interaction between atoms are not universal. Intuitively

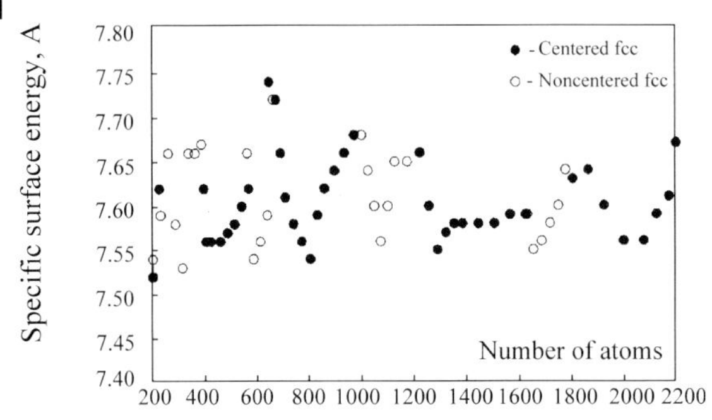

Fig. 3.1 The specific surface energy for optimal atom configurations of face centered cubic(fcc) clusters. Dark circles correspond to clusters with a central atom and open circles denote noncentered clusters.

we assume that the optimal atomic configuration "chooses" the maximum binding energy for an additive atom and use it as a principle for cluster assembly. In most cases this is fulfilled, and the total binding energy evaluated in these schemes is larger than that for some cluster figures. Nevertheless, one can find cases where these schemes give an error.

Problem 3.3

Determine the specific surface energy of fcc clusters with short-range atomic interactions as a function of their size.

The specific surface energy of a cluster is introduced on the basis of formula (2.9):

$$A(n) = \frac{\varepsilon_o n - E}{n^{2/3}} , \tag{3.2}$$

where $\varepsilon_o = 6$ for fcc clusters with a short-range interaction of atoms if we take the energy required for the breaking of one bond as an energetic unit. The optimal configuration of cluster atoms corresponds to the minimum values of A from two cluster structures with a central atom and without one. Figure 3.1 gives values of the function $A(n)$ that are obtained for magic numbers of cluster atoms on the basis of the data in Tables 3.1 and 3.2. Table 3.3 contains the values of the specific surface energies A for a range of cluster sizes $n = 1200$–1800 obtained on the basis of the data in Tables 3.1 and 3.2. The optimal configuration of cluster atoms is chosen from cluster structures with a central atom and without a central atom and is taken such that it is characterized by a lower value A. According to the data in Table 3.3, the average specific energy of clusters with a short-range atomic interaction is $A = 7.60 \pm 0.03$.

Problem 3.4

Determine the sphericity coefficient (2.46) for the fcc cluster of an optimal structure with a central atom and complete atomic shells for $n = 1157$.

Table 3.3 The specific surface energy A_1 and A_2 determined by (2.8) and on the basis of the data in Table 3.1 for A_1 and in Table 3.2 for A_2.

n	A_1	A_2	n	A_1	A_2	n	A_1	A_2
1197	–	7.64	1385	7.58	–	1654	–	7.55
1201	7.70	–	1414	–	7.72	1660	–	7.58
1222	–	7.66	1445	7.58	–	1678	–	7.64
1223	7.66	–	1462	–	7.73	1685	7.61	–
1245	7.62	–	1505	7.58	–	1696	–	7.68
1267	7.58	–	1510	–	7.75	1723	–	7.72
1270	–	7.67	1565	7.59	–	1744	–	7.73
1289	7.55	–	1570	–	7.66	1745	7.62	–
1318	–	7.69	1618	–	7.60	1756	–	7.72
1321	7.56	–	1625	7.60	–	1780	–	7.71
1366	–	7.70	1630	–	7.58	1805	7.63	–

Our analysis for this problem will be based on the data in Table 3.1 for optimal configurations of atoms in these clusters. Let us define atoms that belong to the cluster surface such that these atoms can form bonds added to the cluster in the course of its subsequent growth. According to the data in Table 3.1, for the fcc cluster of an optimal structure, its surface includes atomic shells starting from 055 and 046 that can form bonds with atoms of shells 066 and 057. Hence, the surface of a cluster with $n = 1157$ contains 482 atoms, and the distances from the cluster center for surface atoms vary from $a\sqrt{25} = 5.0a$ to $a\sqrt{33} \approx 5.7a$, and according to formula (2.44) we have for the sphericity coefficient $\xi \approx 0.28$, but this is a high value of the sphericity coefficient.

We now determine the average distance $\overline{r}$ and the average square distance $\overline{r^2}$ from the cluster center, and also the sphericity coefficient using a direct method:

$$\overline{r} = \frac{\sum_i n_i r_i}{\sum_i n_i}\,, \quad \overline{r^2} = \frac{\sum_i n_i r_i^2}{\sum_i n_i}\,, \quad \eta = \frac{\sqrt{\overline{r^2} - (\overline{r})^2}}{\overline{r}}\,,$$

where r_i is the distance of the ith atom from the cluster center and n_i is the number of such atoms. Under the conditions indicated we have

$$\frac{\overline{r}}{a} = 5.36\,, \quad \frac{\overline{r^2}}{a^2} = 28.805\,, \quad \Delta r = \sqrt{\overline{r^2} - (\overline{r})^2} = 0.28a\,, \quad \eta = \frac{\Delta r}{\overline{r}} = 0.053\,.$$

One can see that $\eta \ll \xi$.

Problem 3.5

On the basis of the data in Table 3.1, determine the relative number of surface atoms and the sphericity coefficient (2.44) for fcc clusters of an optimal structure

Table 3.4 Parameters of fcc clusters after filling a scheduled layer {111} constructed on the basis of the data in Table 3.1. n is the magic number of cluster atoms that correspond to the completion of a scheduled layer {111}, n' is the number of cluster atoms not included in the surface layer, ξ is the sphericity coefficient according to formula (2.44), and $E_{sur}/n^{2/3}$ is the specific surface energy (2.8) expressed as the breaking energy per bond.

n	n'	$(n-n'+1)/n$	ξ	$A = E_{sur}/n^{2/3}$
135	55	0.585	0.168	7.98
297	135	0.549	0.242	7.952
675	369	0.452	0.181	7.719
1157	675	0.416	0.139	7.785
1865	1157	0.379	0.154	7.643

with a central atom and complete atomic shells when the filling of a subsequent layer of the direction {111} is finished.

We define the surface layer, including in this layer the atoms that acquire new bonds when new atoms are added to the cluster in the course of its growth. Table 3.4 contains the magic numbers of cluster atoms n corresponding to the filling of blocks of direction {111}, and n' is the number of cluster atoms formed in the cluster core.

As follows from Table 3.4, the relative number of surface atoms decreases with cluster growth, but the sphericity parameter (2.44) is a nonmonotonic function of cluster size and cannot be a cluster characteristic. We also note that formula (2.8) for the cluster surface energy holds true for clusters consisting of the magic atom numbers that are included in Table 3.4. The specific cluster energy for these clusters with a short-range atomic interaction is $A = 7.82 \pm 0.15$, that is, the accuracy of formula (2.8) for clusters under consideration is approximately 2%. This accuracy characterizes the closeness of the cluster shell shape to the spherical one.

Problem 3.6

Calculate the number of atoms and the atom binding energy for the sequence of cuboctahedral clusters with a short-range interaction potential of atoms.

The cuboctahedron is a symmetric figure (Figure 2.9a) whose surface consists of six squares and eight equilateral triangles. The cuboctahedron contains 24 edges and 12 vertices. The distance between nearest neighbors for the cuboctahedral structure is fixed. Therefore, accepting the edge length to be ma, where a is the distance between nearest neighbors, we call m the figure number in the series. As an example of such a cluster structure, Figure 3.2 gives the cuboctahedral cluster consisting of 923 atoms.

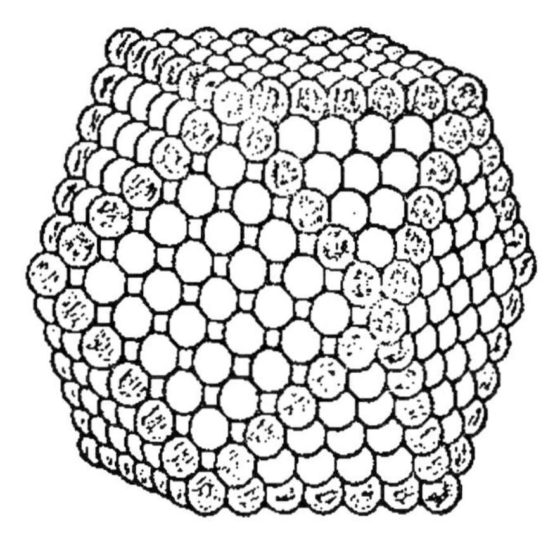

Fig. 3.2 Cuboctahedral cluster consisting of 923 atoms.

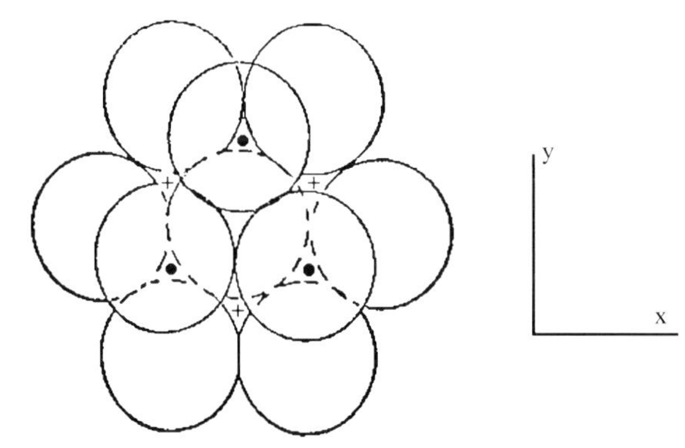

Fig. 3.3 Cuboctahedral cluster consisting of 13 atoms in the coordinate system of the plane {111}. Projections of the cuboctahedral cluster consisting of 13 atoms are represented in the coordinate system where xy is the plane {111}. Positions of three atoms of the upper layer and seven atoms of the central layer are given. Positions of centers of atoms of the lower layer are marked by crosses for the hexagonal structure and by circles for the fcc one.

Let us first consider a cluster consisting of 13 atoms (Figure 3.3). Take a plane {111} as the base of the cluster. This base consists of a regular hexagon with a central atom. The upper and lower layers contain three atoms that are located in hollows between atoms of the central layer so that each of these atoms has three nearest neighbors from the central layer and two nearest neighbors from its own layer. There are two possibilities for the relative location of atoms of the upper and lower layers, like the structures in Problem 3.1. If the projections of atoms of the upper and lower layers onto the central layer plane do not coincide, these atoms form a cuboctahedron – a figure of the fcc structure that has the symmetry (2.26). If these projections coincide, the figure formed, a hexahedron, has the hexagonal symmetry (2.27).

Cuboctahedral clusters have an fcc structure with a centered atom. Therefore, they can be included in the general scheme of construction of fcc clusters shown

Table 3.5 Parameters of cuboctahedral clusters with a short-range atomic interaction; m is the number of a cluster in this family, n is the number of cluster atoms, E_{sur} is the cluster surface energy, and E_o is the surface energy of the cluster of the optimal configuration of atoms among clusters of the fcc structure with this number of atoms and with a central atom (Table 3.1).

m	n	E_{sur}	E_o
1	13	42	42
2	55	114	114
3	147	222	220
4	309	366	362
5	561	546	535
6	923	762	729
7	1415	1014	963
8	2057	1302	1226

in Table 3.5. In our notation the cuboctahedral cluster with m closed shells includes atomic shells $\alpha\beta, \gamma$, where $\alpha, \beta, \gamma \leq m$ and $\alpha + \beta + \gamma \leq 2m$. Denote the number of vertices of a cuboctahedral cluster by $p = 12$, the number of edges by $q = 24$, the number of squares by $r = 6$, and the number of triangles by $s = 8$. Then we get for the number of surface atoms of a complete cuboctahedral cluster that

$$\Delta n = p + q(m-1) + r(m-1)^2 + s(m-1)(m-2)/2 = 10m^2 + 2\,.$$

From this the expression follows for the total number of atoms of the cuboctahedral cluster [65]:

$$n = 10m^3/3 + 5m^2 + 11m/3 + 1\,. \tag{3.3}$$

Let us determine the energy of a cluster with a cuboctahedral structure. Take into account that each internal atom has 12 nearest neighbors and that surface atoms have the following numbers of nearest neighbors: five for vertex atoms, seven for edge atoms, eight for square atoms, and nine for triangle atoms. Using the number of atoms of each unit, we obtain for the surface energy of the cuboctahedral cluster with m complete layers that

$$E_{sur} = 18m^2 + 18m + 6\,. \tag{3.4}$$

The total binding energy of atoms in the cuboctahedral cluster is given by [65]

$$E = 6n - E_{sur} = 20m^3 + 12m^2 + 4m\,. \tag{3.5}$$

Table 3.5 lists the surface energies of cuboctahedral clusters (E_{sur}) with closed layers, which are compared with E_o of the optimal configuration of atoms in the centered clusters of a given number of atoms (Table 3.1). As is seen, the cuboctahedral structure is not optimal for centered fcc clusters. Indeed, the assembling of a cuboctahedral cluster is finished with the occupation of its vertices with positions 0 mm (notation in Table 3.1), and vertex atoms have only five nearest neighbors. Table 3.1 shows that only the cuboctahedral structures with $m = 1$ and $m = 2$ can be optimal among the fcc structures with a centered atom.

Problem 3.7

Calculate the number of atoms and the atom binding energy for the sequence of cubic clusters with a short-range atomic interaction potential.

The cubic cluster can be either a centered or a noncentered fcc cluster. Its surface consists of six planes of direction {100}, and surface atoms are included in two square nets inserted in one another. The cube's vertices have coordinates mmm in the corresponding frame of reference, and the number m characterizes the figure number in the cube family. The atoms of the cube's edges have positions α, m, m, where $\alpha < m$ and its plane has the parity of m. Surface atoms inside squares correspond to shells α, β, m, where $\alpha, \beta < m$. Note that atomic layers of this figure that share a bound with surface layers do not have a square net. Projections of their atoms onto a neighboring surface layer are located at centers of squares that are formed by surface atoms. These two types of layers alternate in a cubic cluster, that is, the cluster has a complicated structure.

The cluster under consideration has $p = 8$ vertices, $q = 12$ edges, and $r = 6$ surface squares. Thus, a surface layer contains p vertex atoms, $q(m-1)$ edge atoms, and $r(2m^2 - 2m + 1)$ atoms inside squares. Each vertex atom has three nearest neighbors, each edge atom has five nearest neighbors, and each square atom has eight nearest neighbors. This gives

$$n = 4m^3 + 6m^2 + 3m + 1\,; \qquad E_{\mathrm{sur}} = 24m^2 + 18m + 6\,. \tag{3.6}$$

for the number of atoms n of a cubic cluster and its surface energy E_{sur}. In particular, this gives

$$A = E_{\mathrm{sur}}/n^{2/3} = 6 \cdot 2^{2/3} = 9.52$$

for the specific surface energy of large cube clusters. As is seen, the cube structure of clusters is far from the optimal structure of fcc clusters with a pairwise interaction of atoms.

Problem 3.8

Calculate the number of atoms and the atom binding energy for the sequence of octahedral clusters in the case of a short-range atomic interaction potential.

An octahedron as a geometric figure is represented in Figure 2.9c, and Figure 3.4 exhibits the octahedral cluster consisting of 44 atoms. The surface of the octahedral cluster consists of eight regular triangles, and an octahedral cluster can be either centered or noncentered. Atoms of this cluster occupy shells $\alpha\beta\gamma$, where $\alpha + \beta +$

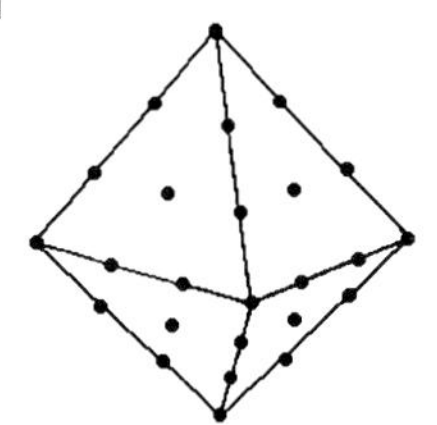

Fig. 3.4 Positions of atom centers in an octahedral cluster consisting of 44 atoms.

$\gamma \leq m$ and m is a whole number that characterizes the number of the figure in the octahedral series. The octahedral surface has $p = 6$ vertices and $q = 12$ edges and includes $s = 8$ regular triangles. There are $q(m-1)$ atoms in octahedral edges and $s(m^2/2 - 3m/2 + 1)$ atoms inside triangles. Vertex atoms relate to the shell $0, 0, m$; atoms inside edges occupy shells $0, p, m-p$, where p is a whole number; atoms inside triangles form shells $\alpha\beta\gamma$, where $\alpha + \beta + \gamma = m, \alpha, \beta, \gamma \geq 0$. Each vertex atom has four nearest neighbors, each edge atom has seven nearest neighbors, and each atom inside triangles has nine nearest neighbors. From this we obtain the following expressions for the total number of atoms n of an octahedral cluster and its surface energy E_{sur} [65]:

$$n = 2m^3/3 + 2m^2 + 7m/3 + 1\,, \quad E_{\text{sur}} = 6m^2 + 12m + 6\,. \tag{3.7}$$

The total binding energy of the cluster atoms is given by

$$E = 4m^3 + 6m^2 + 2m\,. \tag{3.8}$$

Table 3.6 lists the parameters of octahedral clusters that follow from these equations. Their energies are compared with E_o for the optimal configuration of atoms in the centered clusters (Table 3.1) for a given number of atoms.

Problem 3.9

Evaluate the number of atoms and the atom binding energy for a sequence of clusters with the structure of regular truncated octahedra for a short-range atomic interaction potential.

A regular truncated octahedron as a geometric figure is shown in Figure 2.10, and Figure 3.5 gives an example of such a cluster consisting of 201 atoms. The surface of this cluster consists of eight regular hexagons and six squares. Note that a regular truncated octahedron can be obtained only from octahedra for which m is a number divisible by three. This cluster can be either centered or noncentered or can be an octahedral cluster.

Introducing the number m of the family of regular truncated octahedra, we have that this cluster contains $p = 24$ vertices that occupy the shell $0, m, 2m$, and $q' = 12$ edges, which are common sides of hexagons and belong to shells $0\beta\gamma$, where $\beta + \gamma = 3m; \beta, \gamma > m$. Furthermore, this cluster has $q'' = 24$ edges on the boundary of hexagons and squares that relate to shells $\alpha, \beta, 2m$, where $\alpha + \beta = m$; $\alpha, \beta > 0$. Atoms inside squares belong to shells $\alpha, \beta, 2m$, where $\alpha + \beta < m$, and atoms inside hexagons form shells α, β, γ, where $\alpha + \beta + \gamma = 3m$ and $\gamma < 2m$.

Table 3.6 Parameters of octahedral clusters with a short-range atomic interaction; m is the number of a cluster in this family, n is the number of cluster atoms, E_{sur} is the cluster surface energy, and E_o is the minimum surface energy of an fcc cluster with a central atom (Table 3.1) and the same number of atoms.

m	n	E_{sur}	E_o
1	6	24	25
2	19	54	54
3	44	96	97
4	85	150	144
5	146	216	218
6	231	294	291
7	344	384	386
8	489	486	472
9	670	600	594
10	891	726	711
11	1156	864	858
12	1469	1014	987
13	1834	1176	1156

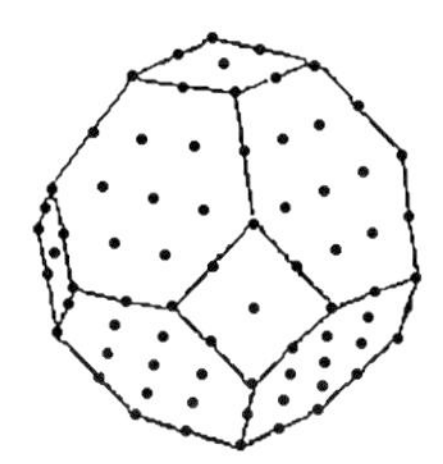

Fig. 3.5 Positions of atom centers in a cluster that has the form of a regular truncated octahedron and consists of 201 atoms.

Each atom of a vertex of a regular truncated octahedron has six nearest neighbors, each of $36(m-1)$ atoms of edges has seven nearest neighbors, each of $6(m-1)^2$ atoms inside squares has eight nearest neighbors, and each of $24m^2 - 24m + 8$ atoms inside hexagons has nine nearest neighbors. This gives

$$n = 16m^3 + 15m^2 + 6m + 1\,; \qquad E_{sur} = 48m^2 + 30m + 6 \tag{3.9}$$

for the number of atoms n and the surface energy E_{sur} of a cluster having the complete structure of a regular truncated octahedron. Table 3.7 contains these parameters for the first members of the series of clusters with the structure of a regular truncated octahedron. Note that this is the optimal structure of clusters with close packing (compare the data in Tables 3.1, 3.2, and 3.7).

Table 3.7 Parameters of clusters with the structure of a regular truncated octahedron with a short-range atomic interaction; m is the number of a cluster in this family, n is the number of cluster atoms, and E_{sur} is the cluster surface energy.

m	n	E_{sur}
1	38	84
2	201	258
3	586	528
4	1289	894
5	2406	1356
6	4033	1914

Problem 3.10

Determine the sphericity coefficients (2.45) and formula (2.46) for large fcc clusters of an octahedral structure.

The surface area S and the volume V of the octahedron are $S = 2\sqrt{3}l^2$, $V = \sqrt{2}l^3/3$, and formula (2.45) gives

$$\psi = 1 - \frac{\pi^{1/3}}{\sqrt{3}} = 0.154 .$$

Next, in the octahedron case we have for parameters of formula (2.47) $h = 2a = l/\sqrt{3}, \varphi_o = \pi/3, \tan\varphi_o = \sqrt{3}$. We have in this case

$$S = \frac{a^2\sqrt{3}}{2} , \quad \frac{\overline{r}}{a} = \frac{2}{3\sqrt{3}} \int_0^{\sqrt{3}} \frac{dx}{1+x^2}[(5+x^2)^{3/2} - 8] = 2.23 , \quad \frac{\overline{r^2}}{a^2} = 5 .$$

This gives

$$(\Delta r)^2 = \overline{r^2} - \overline{r}^2 = 0.0271a , \quad \frac{\delta r}{a} = 0.165 , \quad \frac{\delta r}{a} = 0.165 , \quad \eta = 0.074 .$$

As is seen, the sphericity parameter (2.45) for the octahedron is approximately double the value of that defined by formula (2.46).

Problem 3.11

Evaluate the number of atoms and the atom binding energy for the sequence of truncated octahedrons in the case of a short-range interaction potential of atoms.

To obtain a cluster of the truncated octahedral structure, we take the octahedral cluster and cut off six regular pyramids near its vertices; Figure 3.5 is an example of the resulting cluster. A formed cluster is characterized by the index m, the number of the octahedron in its family, and the index k, the number of atoms on the pyramid's edge. Using the above results, we get for the parameters of this cluster in the case of a short-range interaction of cluster atoms that

$$n = \frac{2}{3}m^3 + 2m^2 + \frac{7}{3}m + 1 - k(k+1)(2k+1); \quad E_{sur} = 6(m+1)^2 - 6k(k+1). \tag{3.10}$$

In the case $k = p, m = 3p$ formula (3.10) gives the parameters of the regular truncated octahedron in accordance with formula (3.9).

Problem 3.12

Determine the optimal character of the growth of large fcc clusters with a short-range interaction potential of atoms.

We first analyze the data in Tables 3.1 and 3.2. Using these data, Table 3.8 contains magic numbers of complete structures. It follows from this table that most of these magic numbers correspond to the structure of truncated octahedra. Let us represent positions of local minima in Table 3.8 in the form

$$n_{min} = \frac{16}{27}(l + \alpha)^3 , \tag{3.11}$$

Table 3.8 Parameters of complete structures of fcc clusters with a short-range interaction of atoms within the framework of the structure of a truncated octahedron (Figure 2.6). Asterisks mark the minimum of $A(n)$ as a function of magic numbers.

n	*A*	*m, k*	*n*	*A*	*m, k*
201*)	7.519	6, 2	1000	7.680	–
260	7.659	7, 3	1072*)	7.561	11, 3
314*)	7.533	7, 2	1126	7.650	11, 2
338	7.666	7, 1	1139	7.647	12, 5
369	7.814	–	1157	7.785	–
405	7.563	8, 3	1289*)	7.548	12, 4
459*)	7.562	8, 2	1385	7.581	12, 3
538	7.801	–	1504	7.587	13, 5
586*)	7.540	9, 3	1654*)	7.550	13, 4
640	7.594	9, 2	1750	7.602	13, 3
664	7.699	9, 1	1804	7.693	13, 2
675	7.719	–	1865	7.643	–
711	7.607	10, 4	1925	7.561	14, 5
807*)	7.545	10, 3	2075*)	7.561	14, 4
861	7.624	10, 2	2171	7.622	14, 3
885	7.746	10, 1	2190	7.614	15, 6
952	7.812	–	2225	7.710	14, 2
976	7.561	11, 4	2406*)	7.552	15, 5

where l is an integer. If we take $l = 6$ for $n = 201$, from formula (3.11) in the limit of large n it follows that $\alpha = 15/16 = 0.938$. From the treatment of the data in Table 3.6 we get $\alpha = 1.03 \pm 0.08$. If only parameters of regular truncated octahedra are taken from Table 3.8, we obtain $\alpha = 0.96 \pm 0.01$. Thus, the structures of large assembled clusters are similar to those of a regular or almost regular truncated octahedron.

Problem 3.13

Calculate the parameters of clusters of the hexahedron series for a short-range interaction of cluster atoms.

The simplest hexagonal cluster consisting of 13 atoms is similar to the cuboctahedral cluster of 13 atoms given in Figure 3.3. Hexagonal clusters have in the central layer a system of regular hexagons with a common center. Atom-balls of subsequent layers are located in the hollows of preceding ones, and, in accordance with the hexagonal symmetry, the projections of atoms from the opposite sides onto the central one coincide. The number of atoms of a subsequent layer is chosen such that it provides the maximum specific binding energy of cluster atoms. For comparison, Figure 3.6 shows clusters of 55 atoms with fcc structure (Figure 3.6a) and hexagonal structure (Figure 3.6b). These clusters have identical bases and side lengths of the large hexagon $2a$. The surface of the first of these figures, the cuboctahedron, consists of six squares and eight equilateral triangles. The centers of surface atom-balls are located on identical triangles of planes {111} or squares of planes {100}, and the centers of atoms on the boundary of a triangle and square lie strictly on one line. The surface atoms of the second figure, which is a hexahedron, do not lie on the same planes or lines.

Let us construct a series of hexahedra by taking as a parameter of the series m – the side length of the large hexagon that is the base of the figure and is expressed in units of distances between nearest neighbors. Figure 3.6b contains a hexahedron with $m = 2$, and Figure 3.7 represents the positions of surface atoms and the numbers of their bonds for a hexahedron with $m = 4$. For any m the number of hexahedral atoms n and the surface energy E_{sur} are given by the expressions [78]

$$n = 4m^3 + 6m^2 + 4m - 7\,, \quad E_{sur} = 21m^2 + 21m - 12\,. \tag{3.12}$$

Table 3.9 contains numerical parameters of the first terms of this series and the specific binding energies of the centered fcc clusters with the optimal configuration of atoms for a given number of atoms (Table 3.1).

Problem 3.14

Calculate the parameters of clusters that have the structure of a truncated hexahedron in the case of a short-range interaction potential of atoms [78, 79].

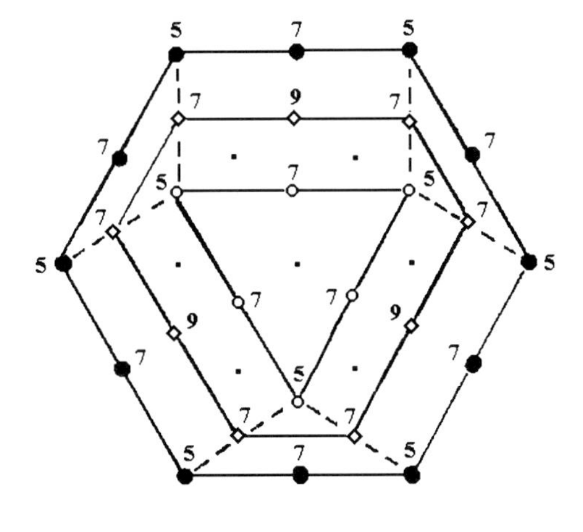

(a)

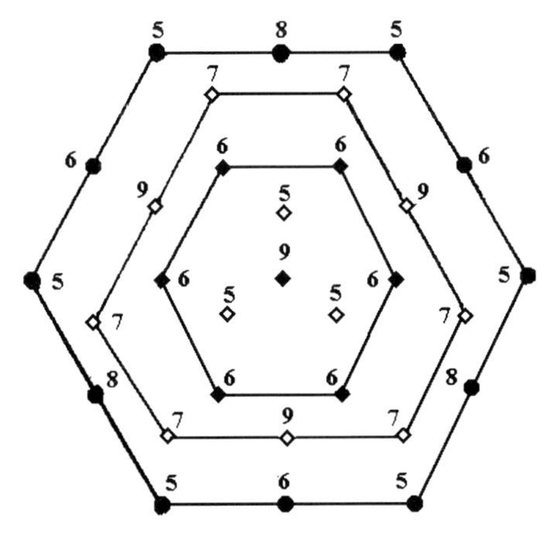

(b)

Fig. 3.6 Clusters with filled layers of an fcc structure, cuboctahedron (a) and hexahedral structure (b). The side length of the large central hexagon is $m = 2$ (a magic number of atoms is 55 for the cuboctahedron and 57 for the hexahedron). The upper part of the clusters is shown. Solid lines join the nearest surface atoms of one layer and dotted lines separate squares and triangles on the cuboctahedral surface. The values are the numbers of nearest neighbors of surface atoms. Small dark squares correspond to positions of nonvisible atoms of the cuboctahedron.

The hexahedron is not an optimal figure among those of hexagonal structure. Removing some layers of this figure can lead to an increase in the specific binding energy of the cluster. As an example, let us consider the truncated hexahedron such that the mth truncated hexahedron can be obtained from the $2m$th hexahedron by removal of some upper and lower layers, which gives a length m for the large hexagon of the upper and lower layers. The parameters for the series of the truncated hexahedra are the following:

$$n = 28m^3 + 21m^2 + 6m + 1 \, , \; E_{\text{sur}} = 72m^2 + 36m + 6 \, . \tag{3.13}$$

Table 3.10 contains the numerical parameters of the first terms of this series.

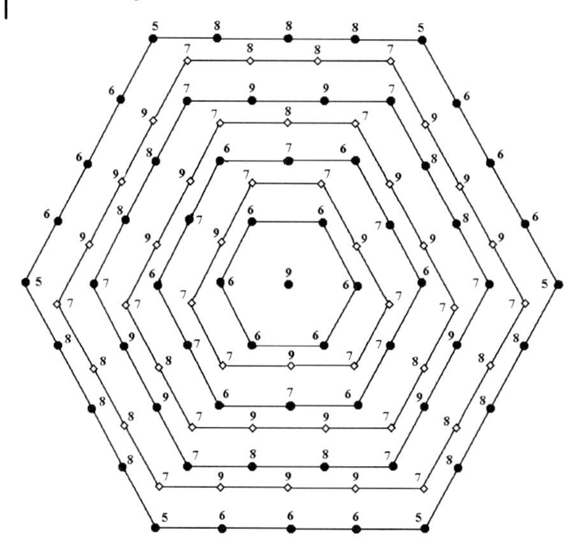

Fig. 3.7 The upper part of the cluster hexahedron with $m = 4$, which contains 361 atoms. Solid lines join nearest surface atoms of one layer and the values are the numbers of nearest neighbors of surface atoms.

Table 3.9 Parameters of clusters of the hexahedron family for a short-range interaction of atoms; m is the number of a cluster in this family, n is the number of cluster atoms, E_{sur} is the cluster surface energy, and E_o is the optimal surface energy of an fcc cluster with a central atom (Table 3.1) and the same number of atoms.

m	n	E_{sur}	E_o
1	7	30	29
2	57	114	117
3	167	240	234
4	361	408	398
5	663	618	591
6	1097	870	831
7	1687	1164	1082
8	2457	1500	1387 ± 11

One can see from the analysis of the hexahedral figure that removal of some edges from large clusters with the structure of a centered hexagon can lead to a decrease in the specific surface energy. The parameters of the series of figures obtained from the truncated hexahedron by removing three edges have the form

$$n = 28m^3 + 21m^2 + m - 2\,, \quad E_{sur} = 72m^2 + 30m\,. \tag{3.14}$$

Table 3.10 Parameters of truncated hexahedral clusters with a short-range interaction of atoms; m is the number of a cluster in this family, n is the number of cluster atoms, E_{sur} is the cluster surface energy, and E_o is the optimal surface energy of an fcc cluster with a central atom (Table 3.1) and the same number of atoms.

m	n	E_{sur}	E_o
1	57	114	117
2	323	366	373
3	967	762	753
4	2157	1302	1273

Table 3.11 Parameters of the family of clusters having the form of truncated hexahedra with three removed edges for a short-range interaction of atoms; m is the number of a cluster in this family, n is the number of cluster atoms, E_{sur} is the cluster surface energy, and E_o is the optimal surface energy of an fcc cluster with a central atom (Table 3.1) and the same number of atoms.

m	n	E_{sur}	E_o
1	48	102	103
2	308	348	362
3	946	738	745
4	2130	1272	1260

Table 3.11 contains numerical parameters for clusters that are the first terms of a series. The values of the specific binding energy are compared with those of centered fcc clusters with the optimal configuration of atoms at a given number of atoms (Table 3.1). At large m the parameters of these clusters and clusters that are truncated hexahedra are similar because the number of atoms removed $6m + 3$ is small compared with the number of surface atoms, and the change in the surface cluster energy is small compared with the surface cluster energy.

The total binding energy of atoms of a large cluster can be written in the following form according to formula (2.9)

$$E_n = 6n - An^{2/3}\,, \quad \varepsilon_{sub} = E_n/n = 6 - A/n^{1/3}\,. \tag{3.15}$$

The smaller the specific surface energy A, the more stable is a cluster of a given series. For hexahedral clusters in the limit of large size $A = 8.33$, for large truncated hexahedral clusters we have $A = 7.81$. Averaging over clusters with hexagonal structure, accounting for 80 magic numbers in the range $n = 100$–1000 (Figure 3.8), gives $A = 7.8 \pm 0.1$ for the specific surface energy. As is seen, the atom

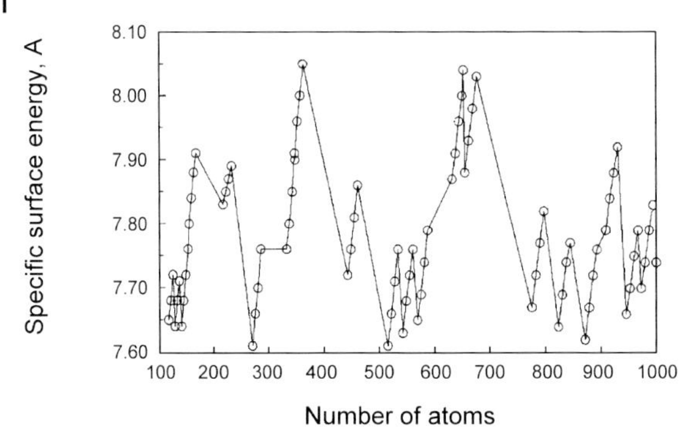

Fig. 3.8 Specific surface energy for hexagonal clusters with magic numbers of atoms. The average value of the specific surface energy A is 7.8 ± 0.1.

binding energies for large fcc clusters of optimal configurations of atoms are larger than for clusters of hexagonal structure. In particular, averaging over the data in Tables 3.1 and 3.2 using 60 magic numbers leads to $A = 7.60 \pm 0.05$ as the value of the specific surface energy for clusters of the completed fcc-structure.

3.2
Icosahedral Cluster Structures

Problem 3.15

Determine the parameters of an icosahedron.

The number of cluster structures is greater than the number of types of crystal lattices. Clusters with a pairwise interaction of atoms, including clusters of rare gases, demonstrate this fact. Indeed, clusters with a short-range interaction of atoms, like to bulk systems of bound atoms, can have a close packed structure that corresponds to an fcc or hexagonal lattice. Clusters with a short-range interaction of atoms also admit an icosahedral structure that is not realized for bulk crystals. Thus, clusters with pairwise interactions provide a convenient example for understanding the structural and energetic parameters of clusters.

The example represented in Figure 3.9 shows that the icosahedral structure [80] is similar to closed-packed structures when each internal atom has 12 nearest neighbors. But the close-packed structures are formed by atoms that are modeled by hard balls. This means that the distances between all the nearest neighbors of this structure are identical. In the case of the icosahedral structure, atoms are modeled by soft balls, so the distances between nearest atoms of different layers do not coincide with the distances of nearest atoms of the same layer. Hence, this structure is not optimal for a bulk system of atoms with a pair interaction and cannot be realized in the bulk crystal lattice. But because of its denser surface, this structure is of importance for large clusters, and it is analyzed below.

Fig. 3.9 Icosahedral cluster consisting of 561 atoms, that is, from five layers. Vertex and edge atoms are darkened.

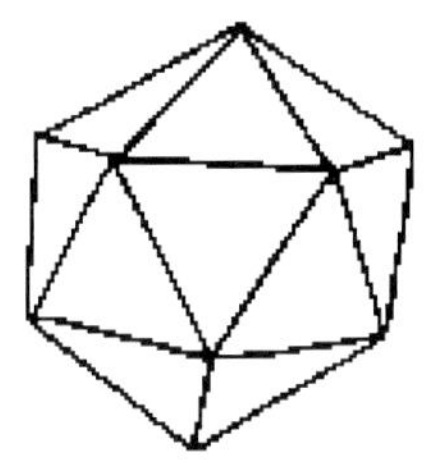

Fig. 3.10 Icosahedral geometric figure whose surface consists of 20 equilateral triangles.

All vertices of an icosahedron are situated on a sphere whose center coincides with the center of the figure (Figures 3.9, 3.10). By joining the nearest vertices, one can obtain a surface consisting of 20 equilateral triangles. The icosahedron has a high degree of symmetry $\mathbf{Y}_h$. It is characterized by six fivefold axes that pass through the center and opposite vertices of the sphere. Rotation through angle $2\pi/5$ around any of these axes preserves the figure. Along with this, the icosahedron is preserved as a result of turning through the angle $\pi/5$ around one of these axes and as a result of reflection with respect to the plane that is perpendicular to the axis and passes through the center. One more symmetry of the icosahedron corresponds to the inversion operation $x \longleftrightarrow -x; y \longleftrightarrow -y; z \longleftrightarrow -z$. Next, the icosahedron has reflectional symmetry with respect to the plane that passes through a given symmetry axis and two vertices of pentagons. This also holds for any axis of the icosahedron.

Let us construct the simplest icosahedral cluster consisting of 13 atoms (Figure 3.11). Let us take one of the cluster atoms as the center; the other 12 atoms are located on a sphere of radius R in the following way. Two of these are placed at the sphere's poles, that is, they are located on the line that connects these atoms and passes through the center. The other 10 atoms form two pentagons whose planes are perpendicular to this line – the icosahedral axis. The pentagons are inscribed in circles that are sections of planes and the sphere, and the pentagon's vertices are

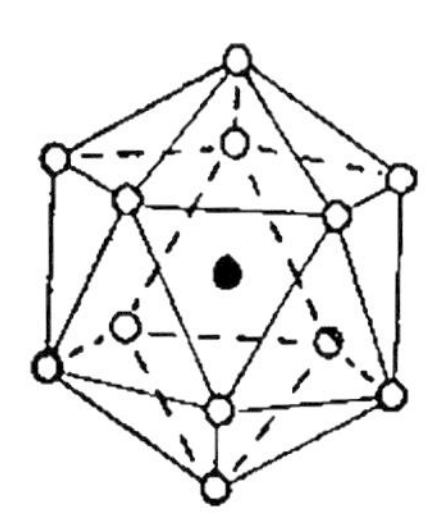

Fig. 3.11 Icosahedral cluster consisting of 13 atoms. Bonds of length R_o between nearest vertices of the icosahedron are drawn by lines. There are 12 bonds of length R between the center and each vertex.

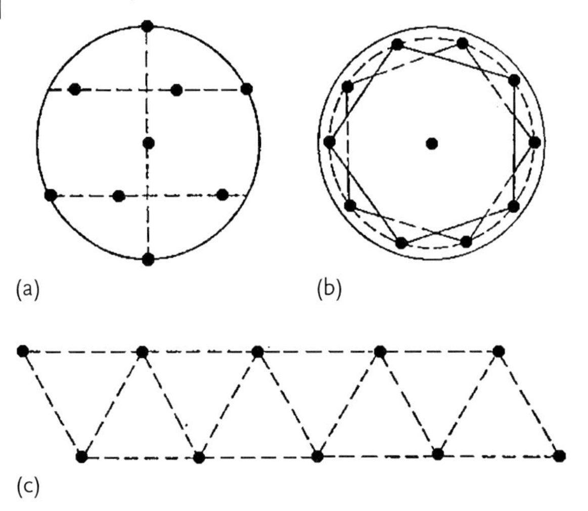

Fig. 3.12 Positions of atoms in an icosahedral cluster consisting of 13 atoms (filled circles mark the centers of the corresponding atoms). (a) Side view; (b) top view; (c) developed view of a cylinder in which pentagons are inscribed.

rotated by an angle of $\pi/5$ with respect to each other. These circles form a cylinder whose axis is the icosahedral axis.

By joining the nearest vertices of the icosahedron, we obtain 20 equilateral triangles. This means that the distances between nearest neighbors on the sphere are the same, and each atom has five nearest neighbors on the sphere. The nearest neighbors of polar atoms on the sphere are atoms of the nearest pentagon, and each atom of a pentagon has as nearest neighbors on the sphere one atom of the nearest pole, two nearest atoms of its own pentagon, and two atoms of the neighboring pentagon.

Let us find the relation between the sphere radius and the distance between nearest neighbors on the sphere for the icosahedron (Figure 3.12). Denote by R_o the distance between nearest neighbors on the sphere whose radius is R and by r the radius of the cylinder in which the pentagons are inscribed. One can obtain the following relations between the icosahedron's parameters. The pentagon has side length

$$R_o = 2r\sin(\pi/5)\,. \tag{3.16}$$

The distance between nearest neighbors that are vertices of different pentagons is

$$R_o = \sqrt{l^2 + [2r\sin(\pi/10)]^2}\,, \tag{3.17}$$

where l is the distance between pentagons. The distance between a pole and an atom of the nearest pentagon is

$$R_o = \sqrt{r^2 + (R - l/2)^2}\,. \tag{3.18}$$

In addition, the following relation holds between the radius of the sphere and that of the circle in which the pentagons are inscribed:

$$R^2 = r^2 + (l/2)^2 . \tag{3.19}$$

One can see that the first three equations give the relation between the icosahedron's parameters, and the fourth equation allows us to check the validity of the icosahedron's definition. The first and second equations give

$$r = l = R_o \cdot \sqrt{1/2 + 1/(2\sqrt{5})} = 0.851\,R_o . \tag{3.20}$$

From the third equation (3.18) we have

$$R = r\sqrt{5}/2 = 0.951\,R_o . \tag{3.21}$$

The last equation corroborates the above relations. Thus, the distance between the center and vertices of the icosahedron is approximately 5% less than the distance between its nearest vertices.

Problem 3.16

Determine the sphericity coefficients (2.45) and (2.46) for large icosahedral clusters with filled atomic shells.

The surface of the icosahedral figure consists of 20 equilateral triangles with side R_o, and the surface area of this figure is $S = 5\sqrt{3}R_o^2$. The perpendicular length from the icosahedron center to a facet is

$$h = \sqrt{R^2 - \frac{R_o^2}{3}} = 0.756\,R_o .$$

The icosahedron volume is the sum of volumes of pyramids whose bases are surface triangles, and the icosahedron center is their vertices, that is,

$$V = \frac{1}{3}hS = \frac{5}{\sqrt{3}}R_o^2\sqrt{R^2 - \frac{R_o^2}{3}} = 2.18\,R_o^3 .$$

From this we have, on the basis of formula (2.45), $\psi = 0.061$.

Applying formula (2.47) to the icosahedron case, we have for parameters of the equation $h = 0.756\,R_o, a = R_o/(2\sqrt{3}), \varphi_o = \pi/3, \tan\varphi_o = \sqrt{3}$, where R_o is the side length of surface triangles. Using as an elementary rectangular triangle in formula (2.47) that with the leg length $a = R_o/(2\sqrt{3})$, we obtain from this equation

$$S = \frac{R_o^2}{8\sqrt{3}}, \quad \overline{r^2} = h^2 + \frac{R_o^2}{12} = 0.655\,R_o^2 ,$$

and

$$\overline{r} = \frac{8R_o}{\sqrt{3}}\int_0^{\sqrt{3}}\frac{dx}{1+x^2}\left[\left(0.655 + \frac{x^2}{12}\right)^{3/2} - 0.432\right] = 0.809\,R_o ,$$

$$\Delta r = \sqrt{\overline{r^2} - (\overline{r})^2} = 0.023\,R_o , \quad \eta = \frac{\Delta r}{R_o} = 0.028 .$$

One can see that the icosahedral figure is closer to a spherical shape than other figures under consideration, and the sphericity coefficient η is significantly less than ψ.

Problem 3.17

Determine the total binding energy of an icosahedral cluster consisting of 13 atoms if the interaction potential between atoms has the form

$$U(R) = D\left[\frac{1}{p}\left(\frac{R_e}{R}\right)^p - \frac{1}{k}\left(\frac{R_e}{R}\right)^k\right]\left(\frac{1}{k} - \frac{1}{p}\right)^{-1} . \tag{3.22}$$

Compare this with the corresponding energy of a cuboctahedral cluster.

An icosahedral cluster consisting of 13 atoms has the following bond lengths, as follows from Problem 3.15. There are 12 bonds of length R, 30 bonds of length $R_o = 1.051R$, 30 bonds of length $1.701R$, and six bonds of length $2R$. This allows us to calculate the total binding energy of the cluster atoms. In particular, in the case of the Lennard-Jones interaction potential $k = 6$, $p = 12$, we have for the cluster binding energy

$$\frac{E}{D} = C_6\left(\frac{R_e}{R}\right)^6 - \frac{C_{12}}{2}\left(\frac{R_e}{R}\right)^{12} ,$$

where the parameters of this equation are

$$C_6 = 12 + \frac{30}{1.051^6} + \frac{30}{1.701^6} + \frac{6}{2^6} = 35.591\ ; \quad C_{12} = 28.568 .$$

Optimizing the cluster energy with respect to the distance R_{opt} between the central and other atoms, we get $R_{opt} = 0.964R_e$, $E = C_6^2/(2C_{12}) = 44.34D$. Table 3.10 lists the cluster binding energies of an icosahedral cluster consisting of 13 atoms for some values of the parameters k, p of the interaction potential (3.22).

One can perform the same operation for the cuboctahedral cluster of 13 atoms. This cluster has 36 bonds of length a, 12 bonds of length $a\sqrt{2}$, 24 bonds of length $a\sqrt{3}$, and six bonds of length $2a$, where a is the distance between nearest neighbors. Then, repeating the above operations for the cuboctahedral cluster of 13 atoms with the Lennard-Jones interaction potential, we get $C_6 = 38.48$, $C_{12} = 36.22$, so the optimal distance between nearest neighbors is $a = 0.990R_e$ and the total binding energy is $E = 40.885$. Table 3.12 contains the binding energies of the cuboctahedral cluster considered for some parameters of the interaction potential (3.22). As is seen, the icosahedral structure is preferable for this cluster compared with the cuboctahedral one.

Problem 3.18

Construct the series of complete icosahedral clusters and find the number of atoms for the mth cluster of the icosahedral family.

Table 3.12 The total binding energy in clusters of 13 atoms with interaction potential of atoms (3.22) and of icosahedral (ico) or cuboctahedral (co) structure [17, 75].

k, *p* of (3.22)	4–8	6–12	8–12	8–16	12–16
E_{ico}/D	50.23	44.34	42.45	43.60	39.56
E_{co}/D	47.27	40.88	38.79	38.10	36.62

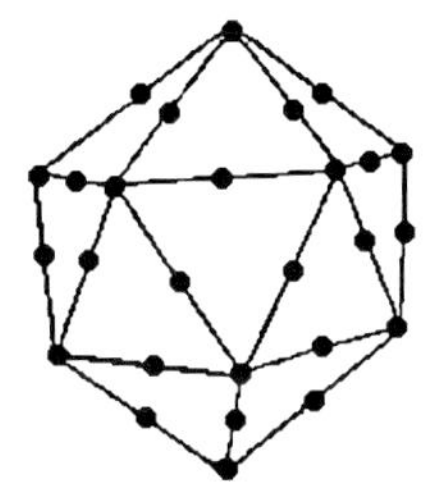

Fig. 3.13 Icosahedral cluster consisting of 55 atoms (two layers). Centers of surface atoms are indicated.

Figure 3.13 gives an example of an icosahedral cluster with filled layers. Let us take into account that the surface of a complete icosahedral cluster consists of $s = 20$ equilateral triangles, $p = 12$ vertices, and $q = 30$ common edges. We have that the mth icosahedral cluster has an edge length m times as great as the length of the first one. Therefore, one can construct the mth cluster of the icosahedral family in the following way. Dividing each icosahedron edge into m parts, draw lines parallel to the sides of the triangles through the points formed. Let us locate atoms at these points and the points of intersection of the lines. We obtain that the number of atoms at vertices is $p = 12$, on the sides of triangles is $q(m-1)$, and inside triangles is $s(m-1)(m-2)/2$. As an example, Figure 3.13 gives an icosahedron corresponding to a cluster of 55 atoms. Thus, the mth layer of the icosahedral cluster contains $\Delta n = 10m^2 + 2$ atoms. The total number of atoms for the mth icosahedral cluster is [80]

$$n = 10m^3/3 + 5m^2 + 11m/3 + 1\,. \tag{3.23}$$

As is seen, the numbers of atoms for the mth cluster of the icosahedral and cuboctahedral families are the same. This fact simplifies the comparison of the energetic parameters of these structures.

Problem 3.19

Determine the atom binding energy of the complete icosahedral cluster for a short-range interaction potential between atoms.

For this evaluation we use the fact that each surface atom placed at a vertex of the icosahedron has one bond of length R and five bonds of length R_o, each edge atom has two bonds of length R and six bonds of length R_o, and surface atoms inside triangles have three bonds of length R and six bonds of length R_o. Summarizing, we have found that the addition of the mth icosahedral layer increases the binding energy of the cluster atoms by the value

$$\begin{aligned}\Delta E &= p(\varepsilon_1 + 5\varepsilon_2/2) + q(m-1)(2\varepsilon_1 + 3\varepsilon_2) \\ &\quad + s(m-1)(m-2)(3\varepsilon_1 + 3\varepsilon_2)/2 \\ &= \varepsilon_1(30m^2 - 30m + 12) + \varepsilon_2 \cdot 30m^2 \,. \end{aligned} \tag{3.24}$$

Here $p = 12, q = 30, s = 20$ are the numbers of vertices, edges, and triangles of the icosahedron, respectively, and $\varepsilon_1 = -U(R), \varepsilon_2 = -U(R_o)$ are the binding energies of two atoms for distances R and R_o between them, respectively. We take into account that bonds of length R_o connect surface atoms, and such bonds are repeated twice as a result of the addition of a new layer. Thus, the total binding energy of the atoms of an icosahedron is given by

$$E = 10m^3(\varepsilon_1 + \varepsilon_2) + 15m^2\varepsilon_2 + m(2\varepsilon_1 + 5\varepsilon_2) \,. \tag{3.25}$$

Let us transform (3.25) into the form

$$E = X\varepsilon_1 + Y\varepsilon_2 = -X\,U(R) - Y\,U(R_o) \,,$$

where

$$X = 10m^3 + 2m \,, \quad Y = 10m^3 + 15m^2 + 5m \,, \tag{3.26}$$

and below we express the binding energies in units D. Let us choose the optimal relation between the parameters R, R_o, and R_e, which is the equilibrium distance between atoms of the diatomic molecule. Optimization of this relation allows one to choose the mean distance between nearest neighbors that leads to the maximum binding energy of the cluster atoms. This operation is simplified because parameters R and R_o are close to R_e, and formula (3.26) can be written in the form [81]

$$E = X + Y - \frac{1}{2}U''(R_e)\left[X(R - R_e)^2 + Y(R_o - R_e)^2\right] . \tag{3.27}$$

Optimization of this expression gives

$$X(R - R_e)\frac{\partial R}{\partial R_o} + Y(R_o - R_e) = 0 \,.$$

Since according to formula (3.21) we have $R = 0.951\,R_o$, this gives the following expressions for optimal distances [77, 81]:

$$R_o = R_e\left(1 + \frac{0.047X}{0.904X + Y}\right) , \quad R = R_e\left(1 - \frac{0.049Y}{0.904X + Y}\right) , \tag{3.28}$$

and the optimal binding energy of atoms in the cluster is given by [81]

$$E = X + Y - 1.2 * 10^{-3}\frac{XYR_e^2U''(R_e)}{0.904X + Y} \,. \tag{3.29}$$

Table 3.13 Parameters of icosahedral clusters of complete layers with a short-range interaction potential of atoms that has the Lennard-Jones core. Here X, Y are numbers of bonds of length R and R_o, respectively, E is the binding energy of the atoms of this cluster, and E_o is the binding energy of a cluster possessing an fcc structure with a central atom and optimal configuration of atoms for a given number of atoms (Table 3.1).

m	n	X	Y	E	E_o
1	13	12	30	41.2	36
2	55	84	150	229	216
3	147	276	420	681	662
4	309	648	900	1514	1492
5	561	1260	1650	2846	2835
6	923	2172	2730	4793	4809
7	1415	3444	4200	7474	7527
8	2057	5136	6120	11 005	11 116

Note that the second term is zero if $R = R_o = R_e$. Because these values are similar, the second term of the total binding energy in formula (3.29) is small compared with the first term. For example, in the case of the truncated Lennard-Jones interaction potential ($U''(R_e) = 72 D/R_e^2$) the second term is 2.3% of the first one for large clusters. Table 3.13 contains the results of calculations of the total binding energy of cluster atoms for clusters of an icosahedral series on the basis of formula (3.29) for the truncated Lennard-Jones interaction potential (i.e., a short-range interaction potential that has the Lennard-Jones potential form near the equilibrium distance). The energies of icosahedral and centered fcc clusters are compared in Table 3.13.

Problem 3.20

Obtain an asymptotic expression for the specific surface energy of an icosahedral cluster with filled layers for a short-range interaction of atoms.

The asymptotic form for the binding energy of cluster atoms is given by formula (2.9), which is an expansion over a small parameter $n^{-1/3}$, where n is the number of cluster atoms. Let us use formula (3.23) for the number of atoms for an icosahedral cluster with m filled layers $n = 10m^3/3 + 5m^2$, formula (3.29) for the binding energy of the atoms, and asymptotic expressions $X = 10m^3$, $Y = 10m^3 + 15m^2$, where we are restricted by terms with accuracy of order $1/m$ compared with the basic terms. In this approximation formula (3.29) gives for the cluster energy

$$E = 20m^3 + 15m^2 - (6.3m^3 + 7.5m^2)\,U'' .$$

Reducing this equation to the form (2.9), we get the parameters of this equation over a small parameter

$$\varepsilon_o = 6 - 1.89 \cdot 10^{-3}\,U'' , \quad A = 6.72 - 2.25 \cdot 10^{-3}\,U'' . \tag{3.30}$$

In particular, for the truncated Lennard-Jones interaction potential ($U'' = 72$) we obtain $\varepsilon_o = 5.864$, $A = 6.56$.

Let us consider this problem using another method. The cluster binding energy per atom is given by

$$\varepsilon = -3\,U(R) - 3\,U(R_o)\,,$$

because each internal atom has six nearest neighbors of the same layer at distance R_o, three nearest neighbors of the previous layer, and three nearest neighbors of the following layer at distance $R = 0.951\,R_o$. We take into account that each bond is shared between two atoms. Optimizing this specific binding energy, we obtain

$$\varepsilon = 6 - \frac{1}{2}U''(R_e)\cdot\left[(R_e - R)^2 + (R_e - R_o)^2\right]\,,$$

$$R_e - R + \frac{\partial R}{\partial R_o}(R_e - R_o) = 0\,.$$

From the last relation it follows that $R = 0.975\,R_e$, $R_o = 1.025\,R_e$, so the asymptotic expression for the specific binding energy has the form

$$\varepsilon_o = 6 - 0.00189\,U''\,.$$

The surface energy per atom for a short-range interaction is $\varepsilon_{\rm sur} = -\frac{3}{2}\,U(R)$. The number of surface atoms of an icosahedral cluster is $10m^2$, where m is the number of filled layers of the complete icosahedral cluster, and the total number of cluster atoms in this approximation is $n = 10m^3/3$. From this we obtain for the cluster surface energy

$$E_{\rm sur} = -15m^2\,U(R) = -15\cdot(0.3n)^{2/3}\,U(R)\,.$$

This gives for the cluster surface energy

$$A = 15\cdot(0.3)^{2/3}\left[1 - \frac{1}{2}(R - R_e)^2\,U''\right] = 6.72 - 0.0022\,U''\,,$$

which coincides with the above results.

Problem 3.21

Compare the density of atoms of a complete icosahedral cluster with that of a complete fcc cluster for a short-range interaction of atoms.

The number density of the fcc structure is equal to $\sqrt{2}/a^3$, where a is the distance between nearest neighbors, which is equal to R_e, the equilibrium distance of the diatomic molecule for a short-range interaction of atoms, and the above formula is valid for the complete figure. Let us find the number density of atoms in the complete icosahedral cluster. The volume of the icosahedral cluster with m complete layers is given by

$$V = 20\frac{m^3}{6}\cdot\frac{R_o^2\sqrt{3}}{4}\sqrt{R^2 - \frac{R^2}{3}} = 2.536\,m^3\,R^3\,.$$

Here we construct the icosahedron from 20 regular pyramids and consider the case $m \gg 1$. In this limiting case we have $X = Y$, that is, $R = 0.975 R_e$. This gives a cluster density $1.417/R_e^3$, which exceeds the density of the close-packed structure by 0.2%. Note that though the mean densities of icosahedral and close-packed structures are practically the same, the close-packed structure is characterized by the isotropic density, while the densities of the icosahedral structure in the radial and transverse directions differ by approximately 5%.

Problem 3.22

Find the parameters of icosahedral clusters with several filled surface triangles in the case of a short-range interaction of atoms.

On the basis of the above method, we assume that atoms of the icosahedral cluster with unfilled external layers possess the same positions as in this cluster with filled layers. Further, we estimate the error of this assumption. The problem is to find the configuration of the surface atoms for a given number of cluster atoms that provides the maximum number of bonds. Let us assume that filling some triangles of the icosahedral surface and the boundaries of these triangles corresponds to magic numbers of the cluster. We represent the parameters of the bonds for the cluster consisting of $(m-1)$ complete layers and k bound triangles above them in the form

$$\begin{aligned} \Delta X &= 3km^2/2 - X_1 m + X_2 \,; \\ \Delta Y &= 3km^2/2 - Y_1 m + Y_2 \,; \\ \Delta n &= km^2/2 - n_1 m + 1 \,, \end{aligned} \tag{3.31}$$

where parameters $\Delta X, \Delta Y, \Delta n$ denote an increase in the number of bonds of length R, R_o and the number of cluster atoms as a result of the addition of k surface triangles. Table 3.14 lists the values of the parameters of formula (3.31).

Problem 3.23

Determine the specific surface energy for the family of icosahedral clusters with k filled triangles in the case of large clusters and a short-range interaction of atoms.

Table 3.14 Parameters of formula (3.31) for filled surface triangles of icosahedral clusters [81].

k	X_1	X_2	Y_1	Y_2	n_1
5	25/2	6	25/2	5	5/2
8	18	8	15	4	3
10	21	9	15	3	3
12	24	10	15	2	3
15	55/2	11	25/2	0	5/2
18	31	12	10	2	2

Table 3.15 The specific surface energy for icosahedral clusters with a short-range interaction of atoms having the Lennard-Jones core. Here *m* is the filling layer number and *k* is the number of filled surface triangles [17, 81].

n	*m*, *k*	*A*	*n*	*m*, *k*	*A*
147	3, 0	6.499	637	5, 5	6.726
178	3, 5	6.801	688	5, 8	6.755
200	3, 8	6.845	724	5, 10	6.741
216	3, 10	6.823	760	5, 12	6.744
232	3, 12	6.818	817	5, 15	6.704
258	3, 15	6.759	874	5, 18	6.674
284	3, 18	6.721	923	6, 0	6.534
309	4, 0	6.519	1029	6, 5	6.657
360	4, 5	6.754	1099	6, 8	6.728
395	4, 8	6.785	1148	6, 10	6.721
420	4, 10	6.773	1197	6, 12	6.716
445	4, 12	6.768	1274	6, 15	6.686
485	4, 15	6.723	1351	6, 18	6.663
525	4, 18	6.693	1415	7, 0	6.542
561	5, 0	6, 523	1556	7, 5	6.679

Assuming the positions of atoms of an unfilled layer to be identical to those in the complete structure, we use formula (3.29) for the total binding energy of cluster atoms and formula (3.32) for additional cluster parameters with respect to complete layers. Table 3.15 contains the specific binding energies of clusters for the truncated Lennard-Jones interaction potential, that is, for a short-range interaction potential that takes into account only the interaction between nearest neighbors, and the core of this interaction potential coincides with the Lennard-Jones core.

To estimate the error of this assumption, let us determine the difference in the binding energies of the icosahedral cluster with one filled triangle above the complete layers if in the first case we use formula (3.29) and in the second case we take into account that the distance between the atoms of a triangle and the atoms of the previous layer is equal to R_e, the equilibrium distance between atoms of the diatomic molecule. The difference between these energies is given by

$$\begin{aligned} \Delta E &= \frac{3}{2} \cdot (m-1)(m-2) \cdot \frac{1}{2} U''(R_e)(R-R_e)^2 \\ &= 0.08 \Delta X \cdot \frac{Y^2}{(0904X+Y)^2} , \end{aligned} \tag{3.32}$$

where m is the number of the filling surface and ΔX is the number of bonds between atoms of the filling layer and atoms of the complete previous layer. Note that for clusters with numbers of atoms 150 ($m = 4$), 315 ($m = 5$), 571 ($m = 6$), 936 ($m = 7$), 1436 ($m = 8$), 2085 ($m = 9$), the contribution of the surface

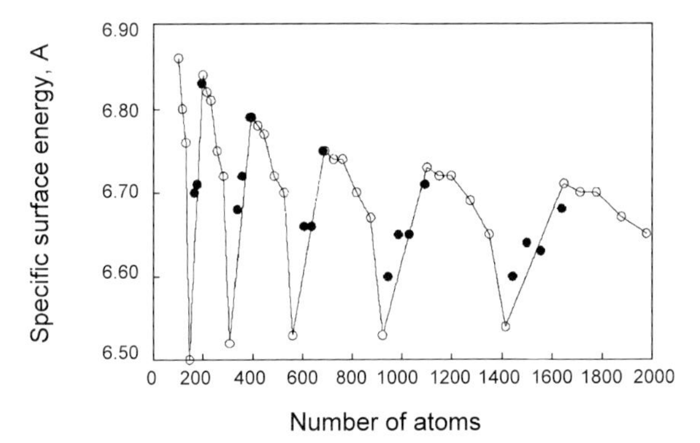

Fig. 3.14 The specific surface energy of icosahedral clusters with a short-range interaction of atoms. Open circles denote icosahedral clusters and dark triangles correspond to icosahedral clusters with an fcc structure of the surface.

atoms to the total binding energy of the cluster atoms lies in the range 1.9–2.3%. The ratio of the above difference of the binding energies corresponds to the error of the assumption that the binding energy of the surface atoms drops monotonically from 1.8 to 1% in this size range with an increase in size. Thus, the error from this assumption is relatively small.

Problem 3.24

Find the asymptotic behavior of the specific surface energy in the limit of large sizes for the family of icosahedral clusters with k filled triangles and a short-range interaction of atoms.

Figure 3.14 gives the size dependence of the specific surface energy of icosahedral clusters that is reduced to formula (2.9). As is seen, the optimal distribution of cluster atoms corresponds to complete cluster shells. Using the data in Table 3.15, one can determine the asymptotic form (2.9) of the cluster energy at large cluster sizes. As is seen, the specific surface energies of icosahedral clusters with complete and incomplete layers differ more strongly than in the case of clusters of close packed structures. But this difference decreases with cluster growth. Indeed, let us represent the total binding energy of the cluster atoms in the form $E + \Delta E$, where E corresponds to the complete cluster layers and ΔE is due to the atoms of the filling layer. Then we have on the basis of formula (3.29)

$$\Delta E = \Delta X + \Delta Y - 0.0012\, U'' \cdot \frac{XY}{0.904X + Y} \cdot \left(\frac{\Delta X}{X} + \frac{\Delta Y}{Y} + \frac{0.904\Delta X + \Delta Y}{0.904X + Y} \right), \tag{3.33}$$

where the additions $\Delta X, \Delta Y$ are due to the atoms of the filling layer.

Note that the number of atoms of the filling layer, as well as the parameters $\Delta X, \Delta Y$, are of order m^2. Then, restricting ourselves to terms proportional to m^2 on the basis of formula (3.31), we obtain $\Delta X = \Delta Y = 3\Delta n$. From this it follows that the addition of some surface atoms in the form of surface triangles to a complete icosahedral cluster does not change its specific surface energy A in the limit of large n.

Problem 3.25

Determine the specific surface energy of the icosahedral surface of a large cluster with the Lennard-Jones interaction potential of atoms.

The icosahedral surface is similar to the hexagonal one, for which the positions of atoms from a test atom are given in Table 2.9. Let us take into account that the distance between neighboring icosahedral layers is $\sqrt{R^2 - R_o^2/3}$ and the area per atom is $R_o^2\sqrt{3}/2$, where $R_o = 1.051R$. Then we take the distance between the atoms of an icosahedral structure in the same way as in the case of a hexagonal structure and replace the square of the distance between two atoms ma^2 by the value $R^2 + (m-1)R_o^2$ for atoms of the first layer, by the value $4R^2 + (m-4)R_o^2$ for atoms of the second layer, by the value $9R^2 + (m-9)R_o^2$ for atoms of the third layer, by $16R^2 + (m-16)R_o^2$ for atoms of the fourth layer, and so on. To determine the cluster surface energy, let us calculate the specific surface energy resulting from the creation of a cluster surface. If we draw a plane along a certain icosahedral layer and take out part of the cluster, the specific surface energy of the surface formed is half of the interaction energy per unit surface area for the interaction between atoms of the remainder and the removed cluster parts. According to formula (2.51), the specific surface energy in units D (the diatomic dissociation energy) is

$$\varepsilon_{\text{sur}} = \frac{1}{s}\left(\sum_{i,k} k n_{ik}\frac{R^6}{r_{ik}^6} - \sum_{i,k} k n_{ik}\frac{R^{12}}{2r_{ik}^{12}}\right),$$

where k is the number of layers between interacting atoms, so that for atoms of neighboring layers $k = 1$, r_{ik} is the distance between interacting atoms, n_{ik} is the number of such atoms for a given test atom, and $s = R_o^2\sqrt{3}/2$ is the surface area per atom. Our analysis is based on formula (2.51) using R instead of a.

On the basis of the above relations and the data in Table 2.9, we obtain for the surface parameters for the Lennard-Jones potential of atomic interaction by analogy with the derivation of formula (2.52)

$$B_6 = 4.133 + \frac{\pi R^2}{\sqrt{3}R_o^2 r_o^4} + \frac{4.126}{k_o^2} = 4.305\,, \quad B_{12} = 3.05\,.$$

Thus, the specific surface energy of a complete large icosahedral cluster is given by

$$\varepsilon_{\text{sur}} = \frac{2}{\sqrt{3}R_o^2}\left(\frac{4.305}{R^6} - \frac{3.05}{2R^{12}}\right). \tag{3.34}$$

Problem 3.26

Determine the specific energy and the specific surface energy for a large icosahedral cluster with the Lennard-Jones interaction potential between atoms.

By analogy with formulas (2.31) and (2.32), we have for the energy of interaction ε of an internal atom with other atoms of an icosahedral cluster in units of the binding energy D per bond

$$\varepsilon = \sum_{i,k} k n_{ik} \frac{R^6}{r_{ik}^6} - \sum_{i,k} k n_{ik} \frac{R^{12}}{2 r_{ik}^{12}} , \tag{3.35}$$

where k is the number of the layer from a test atom, r_{ik} is the distance between interacting atoms, n_{ik} is the number of such atoms for a given test atom, and $s = R_o^2 \sqrt{3}/2$ is the surface area per atom. We have

$$C_6 = \sum_{i,k} \frac{n_{ik}}{r_{ik}^6} = \frac{1}{R^6} \left(12.49 + \frac{4\pi \sqrt{2}}{3 r_o^{3/2}} \right) = \frac{12.59}{R^6} ;$$

$$C_{12} = \sum_{i,k} \frac{n_{ik}}{r_{ik}^{12}} = \frac{9.38}{R^{12}} .$$

This gives for the binding energy per atom

$$\varepsilon = 12.59 \frac{R_e^6}{R^6} - 9.38 \frac{R_e^{12}}{2 R^{12}} .$$

Optimizing this expression with respect to the atom binding energy, we get

$$R_{opt} = 0.952 R_e , \quad \varepsilon = 8.45 .$$

One can see that the large cluster of fcc structure with the Lennard-Jones interaction potential between atoms is characterized by a larger specific binding energy $\varepsilon = 8.61$ according to formula (2.33) than that of the icosahedral structure $\varepsilon = 8.45$. According to the above result, the binding energy due to interaction between nearest neighbors in a large icosahedral cluster is approximately 60% of the total binding energy, as well as for the Lennard-Jones crystal according to formula (2.32).

Let us calculate the surface energy of a large icosahedral cluster. Using the specific surface energy that was determined in the previous problem, we get $\varepsilon_{sur} = 3.51$ for the optimal distance between nearest atoms of neighboring layers $R = R_{opt} = 0.952 R_e$. The total area of a complete icosahedral cluster with m filled layers is $S = 5m^2 R_o^2 \sqrt{3}$ because the cluster surface consists of 20 equilateral triangles of area $m^2 R_o^2 \sqrt{3}/4$. Here the number of icosahedral layers m is connected to the total number of cluster atoms n by $m = (3n/10)^{2/3}$. From this we have for the cluster surface energy

$$E_{sur} = \varepsilon_{sur} S = 13.6 n^{2/3} .$$

Thus, the asymptotic expression for the binding energy of atoms in a large icosahedral cluster with the Lennard-Jones interaction potential of atoms has the form

$$E = 8.45n - 13.6n^{2/3} \,. \tag{3.36}$$

Problem 3.27

Determine the specific energy and the specific surface energy for an icosahedral bulk system with a Morse interaction potential between atoms given by formula (2.37).

The binding energy per atom ε for the Morse interaction potential between atoms in units of the binding energy D per bond is, according to formula (2.38),

$$\varepsilon = \exp(\alpha R_e) F(\alpha R) - \frac{1}{2} \exp(2\alpha R_e) F(2\alpha R) \,,$$

and, according to formula (2.39),

$$F(\alpha R) = \sum_k n_k \exp(-\alpha r_k) \,.$$

Here R_e is the equilibrium distance that corresponds to the minimum of the Morse interaction potential between two atoms, R is the distance between nearest neighbors of neighboring layers of the icosahedral structure, k is the number of an interacting atom, so that r_k is its distance from the test atom, and n_k is the number of such atoms. Note that the test atom is located inside the bulk icosahedron, but not at its center.

In the course of the calculations, we consider the icosahedral crystal lattice as a distorted hexagonal one. Let us represent $F(\alpha R)$ defined by formula (2.39) as

$$F(\alpha R) = F_{nn}(\alpha R) + F_{nnn}(\alpha R) \,,$$

where the first term corresponds to the interaction of nearest neighbors of the icosahedral structure and the second one corresponds to the interaction of nonnearest neighbors, so that we have

$$F_{nn}(\alpha R) = 6\exp(-\alpha R) + 6\exp(-\alpha R_o) \,,$$

and the first terms of the sum for $F_{nnn}(\alpha R)$ are

$$F_{nnn}(\alpha R) = 6\exp\left(-\alpha\sqrt{R^2 + R_o^2}\right) + 2\exp\left(-\alpha\sqrt{4R^2 - 4R_o^2/3}\right) + 12\exp\left(-\alpha\sqrt{R^2 + 2R_o^2}\right) + 6\exp\left(-\alpha R_o\sqrt{3}\right) + \ldots \,.$$

Here we replace the square of the distance of the kth atom from the test atom ma^2 for the hexagonal lattice by $l^2R^2 + (m - l^2)R_o^2$ for the icosahedral structure if this atom belongs m-th layer, and an interaction between nearest neighbors dominates

$F_{nn}(\alpha R) \gg F_{nnn}(\alpha R)$. Let us find the optimal distance between nearest neighbors on the basis of formula (2.41)

$$\exp(\alpha R_e) = \frac{F'(\alpha R)}{F'(2\alpha R)} \,.$$

Assuming in this equation $\alpha(R_o - R) \ll 1$, accounting for $R = 0.951 R_o$, and restricting ourselves in the above equation to the first term of the expansion over a small parameter, we get

$$R_e = 1.025 R = 0.975 R_o \,, \tag{3.37}$$

in accordance with formula (3.30). Next, restricting ourselves to two terms of the expansion over the small parameter $\alpha(R_o - R)$ in the expression for $F(\alpha R)$, we obtain

$$F(\alpha R) = 12 \exp(-\alpha R_e) \cdot \left(1 - \frac{1}{2}\alpha^2 R_e^2\right) \,. \tag{3.38}$$

In the case of the Morse interaction potential with a short-range interaction in a bulk icosahedral system this gives

$$\varepsilon = 6 - 0.00378 \alpha^2 R_e^2 \,, \tag{3.39}$$

in accordance with formula (3.30). The relation between the equilibrium distance of the Morse interaction potential R_e and the optimal distance between nearest neighbors R, as well as the specific energy of an icosahedral bulk in the general case of the Morse interaction, is given in Table 3.16.

Formula (2.55) for the specific surface energy per surface atom in the case of the Morse interaction potential between atoms has the form

$$g = \exp(\alpha R_e) G(\alpha R) - \frac{1}{2} \exp(2\alpha R_e) G(2\alpha R) \,,$$

Table 3.16 Parameters of an icosahedral bulk system of atoms with the Morse potential of atomic interaction.

αR	αR_e	$F_{nnn}(\alpha R)/F_{nn}(\alpha R)$	ε	$G(\alpha R)$	A	n_o
2	2.42	0.605	12.77	1.176	28.1	2680
3	3.32	0.332	9.82	0.303	18.3	1260
4	4.20	0.187	8.12	0.0889	13.0	1260
5	5.11	0.108	7.17	0.0288	10.2	1700
6	6.04	0.0636	6.61	$9.93 \cdot 10^{-3}$	8.56	1930
7	7.00	0.0383	6.27	$3.56 \cdot 10^{-3}$	7.64	1080
8	7.96	0.0235	6.05	$1.31 \cdot 10^{-3}$	7.05	260(116)
10	9.91	0.00920	5.77	$1.82 \cdot 10^{-4}$	6.34	62(41)
12	11.86	0.00374	5.58	$2.57 \cdot 10^{-5}$	5.82	26(23)

and

$$G(\alpha R) = \sum_{i,k} k n_{ik} \exp(-\alpha r_{ik}) ,$$

where k is the layer number for an interacting atom, r_{ik} is its distance from the test atom located in the zeroth layer, and n_{ik} is the number of such atoms. The connection between the specific surface energy for an icosahedral bulk A and the specific surface energy per atom g follows from the expression for the cluster surface energy (2.8),

$$E_{\text{sur}} = An^{2/3} = 10m^2 g ,$$

where n is the number of atoms of the icosahedral bulk system, m is the number of its layers, and, on the basis of formula (3.23),

$$A = 90^{1/3} g .$$

We have the following expression for the first terms of the series for $G(\alpha R)$, taking into account that the icosahedral lattice is a distorted hexagonal one:

$$G(\alpha R) = 3 e^{-\alpha R} + 3 \exp\left(-\alpha\sqrt{R^2 + R_{\text{o}}^2}\right) + 2 \exp\left(-\alpha\sqrt{4R^2 - \frac{4R_{\text{o}}^2}{3}}\right)$$
$$+ 6 \exp\left(-\alpha\sqrt{R^2 + 2R_{\text{o}}^2}\right) + 12 \exp\left(-\alpha\sqrt{4R^2 - \frac{R_{\text{o}}^2}{3}}\right) .$$

In particular, in the case of a short-range interaction one can restrict oneself to the first term of this series. Accounting for $\alpha(R_{\text{e}} - R) \ll 1$, we get

$$g = \frac{3}{2}\left[1 - \alpha^2(R_{\text{e}} - R)^2\right] , \quad \alpha R \gg 1 .$$

Using for the optimal distance between nearest neighbors $R = 0.975 R_{\text{e}}$, we obtain

$$A = 6.72 - 0.0044\alpha^2 R^2 , \tag{3.40}$$

in accordance with formula (3.30).

3.3
Competition between Cluster Structures

Problem 3.28

Consider a cluster with a filled icosahedral core and an fcc surface structure. Compare the energetic parameters of this and an icosahedral cluster for a short-range interaction of atoms.

As follows from the above analysis, clusters with a pairwise interaction of atoms whose basis is a short-range interaction between atoms may have three structures – fcc, hexagonal, and icosahedral. The competition between cluster structures chooses the optimal one for given parameters of the pairwise interaction potential, and this optimal structure depends on the number of cluster atoms. Therefore, at low temperatures, when a cluster is solid, one can find the optimal configuration of cluster atoms that leads to the maximum binding energy of cluster atoms. In a range of strong competition between structures, a change in the number of cluster atoms by one may lead to a change in the optimal structure.

Competition between cluster structures is not reduced to an alternation of optimal cluster structures with a variation in the number of cluster atoms. The optimal configuration of cluster atoms in a range of structure competition can contain elements of different structures. This phenomenon for a bulk crystal of a close-packed structure of atoms is known as twinning. Then the crystal lattice simultaneously contains elements of the fcc and hexagonal structures, which are altered in the lattice. In clusters with a pair interaction of atoms, the icosahedral structure can be realized together with closed-packed structures. Hence, new possibilities of structure mixing arise. In particular, in the case where the icosahedral structure is optimal, this mixing corresponds to a change in the structure of the surface layer from the face-centered one to the icosahedral one in the course of filling a layer, and the structure of internal filled layers remains icosahedral. Below we consider some cases of competition between structures for clusters with a pair interaction of atoms.

There are two possibilities for the location of surface atoms if their complete core has an icosahedral structure [82, 83] when these atoms are situated in hollows between atoms of the filled layer. Figure 3.15 shows these two cases of positions of surface atoms on an icosahedral substratum [81]. The first case corresponds to the growth of the icosahedral cluster. The second case corresponds to other positions

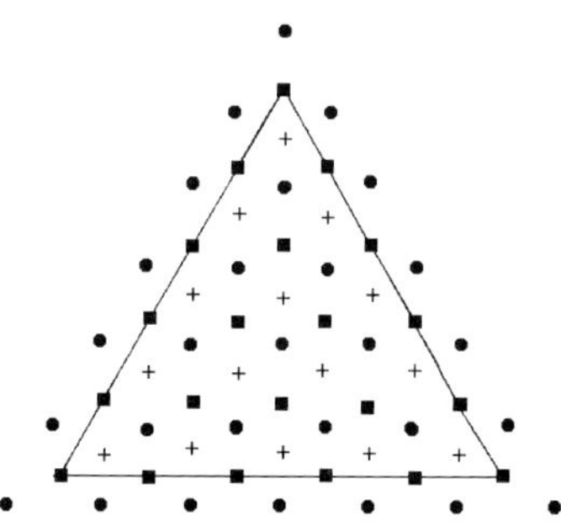

Fig. 3.15 Projections of surface atoms of an icosahedral cluster on a plane of surface triangles. Dark squares indicate the positions of atoms of the filled layer and the solid lines join boundary atoms. Dark circles mark positions of atoms for the icosahedral structure of the filling layer, and crosses indicate their fcc structure.

Table 3.17 The specific surface energy for clusters with an icosahedral core and fcc structure of a surface layer for a short-range interaction of atoms. Here m is the number of the filling layer and k is the number of filled surface triangles.

n	m, k	A	n	m, k	A
153	3, 1	6.634	636	5, 5	6.659
165	3, 3	6.697	681	5, 8	6.749
177	3, 5	6.707	944	6, 1	6.618
195	3, 8	6.833	986	6, 3	6.651
319	4, 1	6.630	1028	6, 5	6.648
339	4, 3	6.682	1091	6, 8	6.715
359	4, 5	6.675	1443	7, 1	6.612
389	4, 8	6.795	1499	7, 3	6.636
576	5, 1	6.624	1555	7, 5	6.627
606	5, 3	6.669	1639	7, 8	6.683

of the surface atoms, and this structure of surface atoms is called the fcc structure of the surface.

Table 3.17 gives the specific binding energies of clusters with an icosahedral core and an fcc structure of the surface layer for a short-range interaction of atoms. The second derivative of the atomic interaction potential corresponds to the Lennard-Jones interaction potential of two atoms near their equilibrium distance ($U''(R_e) = 72D/R_e^2$). Figure 3.14 contains the specific surface energies of clusters with an icosahedral core and a short-range interaction of atoms when the structure of the surface layer may be both icosahedral and fcc. The optimal structure of the surface layer is chosen such that it gives the larger binding energy of cluster atoms.

Thus, we consider filled icosahedral layers as a substratum, and the surface atoms are located in its hollows. If we take the icosahedral structure to be analogous to the hexagonal one, a filling layer can have both fcc and hexagonal (icosahedral) structures. As follows from a comparison of the data in Tables 3.15 and 3.17, the fcc structure of a filling layer is energetically profitable until the unfilled surface layer includes small numbers of filled surface triangles. The reason is that an fcc triangle contains a greater number of atoms than an icosahedral one [$m(m-1)/2$ and $(m-1)(m-2)/2$ for the fcc and icosahedral triangles, respectively, for the mth filling layer]. A comparison of these structures shows that at the first stage of filling of a new layer, the fcc structure of the layer is preferable, and after filling eight triangles, the icosahedral structure of the filling layer provides the maximum binding energy of the cluster atoms.

An advantage of the icosahedral structure for a partially filled surface layer is due to edge atoms that are situated in an optimal way for the icosahedral structure. In the case of an fcc surface structure, the distance of the edge atoms from neighboring atoms of triangles is not optimal and is given by

$$R_c = 0.714 \cdot \sqrt{L^2 - R^2} + 0.567R\,, \tag{3.41}$$

where R is the distance between nearest neighbors of neighboring layers of the icosahedral cluster and L is the distance between triangles and substratum atoms. This length differs from the equilibrium distance R_e more than the lengths of the X- and Y-bonds (the lengths R and R_o). For example, for large clusters when $R = 0.975R_e$ and $R_o = 1.025R_e$, we have at $L = R_e$ that $R_c = 1.128R_e$. Then $U(R_c) = -0.734D$, that is, the difference with the optimal parameters is remarkable for these bonds, which will be labeled below as C-bonds.

Because the number of C-bonds is relatively small, optimization can be made by neglecting the C-bonds. Then the distance between the nearest atoms of the triangle is equal to R_o, the same as the distance between neighboring atoms of the previous layer, and the distance between atoms of the triangle and the substratum is equal to the equilibrium distance R_e. To understand the contribution of the surface atoms to the cluster's energy, let us consider the number of cluster atoms when the surface atoms form five close-packed triangles that have five boundary edges. Then the number of surface atoms is equal to $5m(m-1)$, and the number of C-bonds is $5(m-1)$, where m is the number of the figure in the series. The contribution to the total binding energy of the surface atoms decreases with increasing m, and the contribution of the C-bonds falls more sharply. If m increases from 4 to 8, then the numbers of atoms in the corresponding figures are 177, 359, 636, 1028, and 1555. At these numbers of atoms the contributions of the surface atoms to the total binding energy are 16, 14, 12, 10, and 9%, respectively, and the contributions of the C-bonds are 1.3, 0.8, 0.6, 0.4, and 0.3%, respectively. These values justify the method used for calculating the energy of the C-bonds.

Problem 3.29

Compare the binding energies of large cuboctahedral and icosahedral clusters with complete layers and a short-range interaction of atoms.

The numbers of atoms in complete clusters of cuboctahedral and icosahedral structures are identical according to formulas (3.3) and (3.23). This simplifies the comparison of these structures. The surface energy of the cuboctahedral cluster with m filled layers is given by formula (3.4):

$$E_{sur}^{cubo} = 18m^2 + 18m + 6\,.$$

Let us separate the binding energy of an icosahedral cluster by the usual method:

$$E^{ico} = \varepsilon_o^{ico} n - E_{sur}^{ico}\,,$$

where n is the number of cluster atoms and the asymptotic expression for ε_o^{ico} is given by formula (3.30). Using formula (3.29), we obtain

$$E_{sur}^{ico} = \varepsilon_o^{ico} n - E = 15m^2 + 16m + 6 - 0.0028\frac{n}{m}U''\,.$$

The difference between the surface energies of the cuboctahedral and icosahedral clusters of the same size is

$$\Delta E = E_{sur}^{cubo} - E_{sur}^{ico} = 3m^2 + 0.0042\,U''n^{2/3} = (1.344 + 0.0094\,U'')n^{2/3}\,.$$

From this we obtain a cluster size of n_o at which the binding energies of the icosahedral and cuboctahedral clusters are equalized:

$$n_o = \left(\frac{710}{U''} + 5\right)^3\,. \tag{3.42}$$

As is seen, this value depends on the form of the atomic interaction potential. In particular, if the core of this interaction potential is identical to the Lennard-Jones potential ($U'' = 72$), we get $n_o = 1800$. This result contains an error because we used asymptotic expressions for the cluster binding energies. In this case, for clusters of $m = 7$ complete layers ($n = 1415$) we have $E_{cubo} = 7476$ for the total binding energy of the cuboctahedral cluster (Table 3.5), and for the icosahedral cluster of this size we have $E_{ico} = 7474$ (Table 3.13). For the next complete layer $m = 8$ we have $E_{cubo} = 11\,040$, $E_{ico} = 11\,005$.

Problem 3.30

Compare the binding energies of large fcc and icosahedral clusters with a short-range atomic interaction.

Let us use an asymptotic expression for the specific binding of a large fcc cluster [formulas (2.9) and (3.15)] in the form $\varepsilon = 6 - 7.60/n^{1/3}$, and for a large icosahedral cluster (Problem 3.20) $\varepsilon = \varepsilon_o - An^{-1/3}$, where $\varepsilon_o = 6 - 0.0019\,U''$, $A = 6.72 - 0.0022\,U''$ according to formula (3.30). From this it follows for the cluster size n_o at which the binding energies of the structures are equalized that

$$n_o = \left(\frac{470}{U''} + 1.2\right)^3\,. \tag{3.43}$$

In particular, for the truncated Lennard-Jones interaction potential of atoms ($U'' = 72$) this equation gives $n_o = 450$. Because asymptotic expressions are used for the binding energies, this result contains some errors. In particular, for $n = 395$ the binding energy of the icosahedral cluster is 1951 (Table 3.15), whereas that of the fcc cluster is 1959 (Table 3.1). Note that formula (3.43) is valid for values U'' that are not large. Then the cluster size in the competition range is large enough.

Problem 3.31

Compare the binding energies of large fcc and icosahedral clusters for the Morse interaction potential between atoms.

We use the asymptotic expression (2.9) for the binding energy of the cluster atoms. The parameters of this equation for large fcc clusters are determined by formulas (3.15) and (3.30) for large icosahedral clusters. From this one can find the cluster size n_o at which the binding energies of the fcc and icosahedral clusters become the same. It is given by the formula

$$n_o = \left(\frac{A_{\rm fcc} - A_{\rm ico}}{\varepsilon_{\rm fcc} - \varepsilon_{\rm ico}} \right)^3 . \quad (3.44)$$

Here $\varepsilon_{\rm fcc}$ and $\varepsilon_{\rm ico}$ are values of the parameter ε_o in formula (2.9) for the fcc and icosahedral structures, respectively, and $A_{\rm fcc}$, $A_{\rm ico}$ are the specific surface energies of these structures.

Values of the parameter n_o as a function of αR are given in Table 3.16. For large αR this size is determined by formula (3.43):

$$n_o = \left(\frac{235}{\alpha^2} + 1.2 \right)^3 . \quad (3.45)$$

Values of n_o obtained on the basis of this formula are represented in Table 3.16 in parentheses. As follows from the data in Table 3.16, the competition cluster size n_o depends strongly on the Morse potential parameter α. The function $n_o(\alpha)$ is nonmonotonic for small α.

Problem 3.32

Assuming a nonregularity of $\Delta E = 10$ in the binding energy of icosahedral and fcc clusters with a short-range interaction of atoms, determine the region of cluster sizes where the competition between these structures takes place. The atomic interaction potential has a Lennard-Jones core.

The peculiarity of clusters compared with bulk particles consists in a nonmonotonic dependence of the binding energy of atoms on the number of cluster atoms. Nevertheless we approximate the cluster binding energy by the average monotonic dependence (2.9). Then the difference between the accurate and approximated cluster energies, which is given in Figure 3.16 for fcc-clusters, characterizes the degree of nonregularity in the cluster binding energy. As follows from Figure 3.16, it is correct to take $\Delta E = 10$ in average.

Under these conditions, the condition of equality of the binding energies for the icosahedral and fcc clusters has the form

$$0.0019\, U'' n_o \pm 10 = (0.84 + 0.0022\, U'') n_o^{2/3} . \quad (3.46)$$

Here we use the same asymptotic expressions for cluster binding energies as in Problem 3.29. Using the Lennard-Jones core of the short-range interaction potential of atoms, we transform this relation into

$$0.136\, n_o \pm 10 = n_o^{2/3} .$$

If we take the plus sign in this expression, the solution of the equation yields $n_o = 150$; in the case of the minus sign we get $n_o = 600$. Thus, in this case the

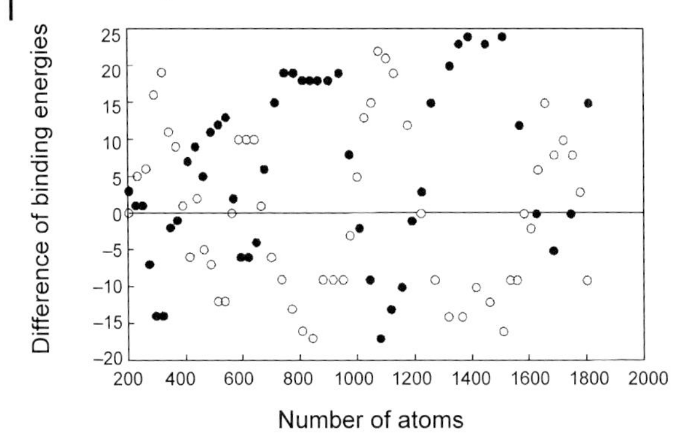

Fig. 3.16 The difference between the binding energy of fcc clusters and the monotonic approximation of this energy as a function of the number of cluster atoms. Open circles correspond to clusters with a central atom and dark circles indicate clusters without a central atom. From this it follows for the irregular part of the cluster energy that $\Delta E \approx 10$.

competition between the icosahedral and fcc structures takes place in the region $n_o = 150$–600. This means that in this region of sizes any of these structures can be energetically profitable, depending on the cluster size. That is why the expressions for the binding energies of the icosahedral and fcc structures are similar. Hence, the competition between these structures occurs in a wide region of cluster sizes.

Problem 3.33

Analyze the competition between the fcc and icosahedral structures of large clusters for the Lennard-Jones interaction of atoms.

Use the asymptotic expressions for the specific binding energies of cluster atoms given by formula (3.15) for fcc clusters and by formula (3.30) for icosahedral clusters (Problems 3.14 and 3.20). By analogy with formula (3.46), equality of these energies for the Lennard-Jones interaction potential of atoms with a nonregularity $\Delta E = 10$ for each structure leads to the equation

$$0.16 n_o \pm 20 = 1.5 n_o^{2/3} .$$

The solution of this equation yields $n_o = 820$, with the region of competition between these structures $n_o = 300$–1200. As is seen, the region of competition between these structures is rather wide, since the expressions for the energies of these structures as a function of size are similar. For this reason a small error in each energy can change the result considerably. In addition, competition between these structures depends strongly on the form of the interaction potential between atoms.

Problem 3.34

Analyze the character of the growth of large fcc clusters for a short-range interaction of atoms.

A large fcc cluster with a short-range interaction of atoms has a structure that is similar to the regular or almost regular truncated octahedron. Such a cluster has as facets six squares in the {100} direction and eight regular or almost regular hexagons in the {111} direction. Starting from the figure of the regular truncated octahedron and filling its facets with atoms, we analyze the character of the cluster growth at large cluster sizes. For this analysis it is more convenient to remove surface atoms of this figure instead of adding them. If all the atoms of the same square surface are removed from the cluster that is the p-regular truncated octahedron of its family, then we obtain a decrease Δn in the number of atoms and ΔE_{sur} in the surface cluster energy:

$$\Delta n = (p+1)^2 , \quad \Delta E_{\text{sur}} = 2p + 2 . \tag{3.47}$$

If all the atoms of one surface hexagon are removed, these parameters are equal:

$$\Delta n = 3p^2 + 3p + 1 , \Delta E_{\text{sur}} = 6p + 3 . \tag{3.48}$$

If the atoms of an adjacent square and hexagon are removed, then the change in the above parameters is given by

$$\Delta n = (2p+1)^2 , \quad \Delta E_{\text{sur}} = 4(2p+1) . \tag{3.49}$$

If the atoms of a square and two adjacent hexagons are removed and these hexagons do not share a border, then we have for these parameters

$$\Delta n = 7p^2 + 6p + 1 , \quad \Delta E_{\text{sur}} = 14p + 6 . \tag{3.50}$$

From this we get for the change in the specific surface energy of the cluster as a result of removing cluster elements that

$$\Delta A = \frac{E_{\text{sur}} - \Delta E_{\text{sur}}}{(n - \Delta n)^{2/3}} - \frac{E_{\text{sur}}}{n^{2/3}} = A\left(\frac{2\Delta n}{3n} - \frac{\Delta E_{\text{sur}}}{E_{\text{sur}}}\right) . \tag{3.51}$$

Formulas (3.47)–(3.51) give in the limit of $p \gg 1$ that $\Delta A \sim p^{-2}$, i.e., the irregular part of the surface energy that is the difference between the cluster energy and its smooth dependence on the numbers of cluster atoms n does not depend on n in the limit $n \to \infty$.

This conclusion is true only for filled facets. If atoms of a facet are partially removed, the irregular part of the cluster surface energy depends on its size. We demonstrate this in the case of removal of a part of the surface square. Let us remove a part of a facet {100} that has a square form and whose edge contains i atoms. Then for the above parameters with accuracy proportional to $1/i$ we have

$$\Delta n = i^2 , \Delta E_{\text{sur}} = 2i , \tag{3.52}$$

and from formula (3.51) within the accuracy proportional to $1/p$ we have

$$\Delta A = A\frac{i^2 - ip}{24p^3} . \tag{3.53}$$

The extremum of this value corresponds to $i = p/2$ and is given by

$$\Delta A = -\frac{1}{4 \cdot 2^{1/3} \cdot n^{1/3}} = -\frac{0.2}{n^{1/3}} .$$

Thus, the irregular part of the total energy of large fcc clusters, which is the difference between the correct binding energy for a given number of atoms and the approximation of this value by a smooth function of the number of cluster atoms, does not increase with cluster growth. Though this conclusion is drawn for clusters with a short-range interaction of atoms, it is true for any pairwise interaction potential of atoms because a long-range interaction gives a smooth dependence of the cluster energy on the configuration of the cluster atoms.

Problem 3.35

Determine the irregular part in the total binding energy of the cluster atoms for large fcc clusters with a short-range interaction of atoms.

Let $E(n)$ be the total binding energy of the cluster atoms, which is approximated for large fcc clusters by the dependence (2.9)

$$E = 6n - An^{2/3} ,$$

where A is a constant in the range of cluster sizes considered. We define the irregular part ΔE of the cluster binding energy as

$$\Delta E^2 = \int_{n_1}^{n_2} \frac{\left[E(n) - 6n + An^{2/3}\right]^2}{n_2 - n_1} dn , \tag{3.54}$$

where the range $n_2 - n_1$ includes many oscillations of the cluster energy.

The irregular part of the cluster energy is the sum of two parts. The first one accounts for oscillations in the cluster energy at magic numbers of atoms when some facets and other elements of the cluster's surface are filled. The other part is due to the addition of the atoms to cluster structures with magic numbers. Below we evaluate the first part of the irregular cluster energy and take into account that the second part increases it by one or two.

To calculate the irregular part of the cluster energy, we take 20 magic cluster numbers in the range of atom numbers $n = 1289$–1979 and change the integral (3.54) by the sum over these magic numbers. As a result, we obtain for this size range $A = 7.60$, $\Delta E = 6$. Adding to this the second part of the irregular cluster energy due to nonmagic numbers of atoms, we get the following estimation:

$$\Delta E_{sur} \approx 10 . \tag{3.55}$$

Above we used this estimation for the analysis of large fcc clusters.

Problem 3.36

Generalize formula (2.9) for the binding energy of atoms in a large solid cluster taking into account its nonmonotonic dependence on the number of atoms.

Let us introduce an irregular part in the total binding energy $E(n)$ of cluster atoms for a large cluster. Indeed, taking the average values of the parameters ε_o and A in formula (2.9) for a given character of atomic interaction and a range of atom numbers, we have that the real dependence $E(n)$ oscillates around the average dependence. Then, if we randomly take a number of cluster atoms in this range, we can determine the binding energy of atoms as

$$E(n) = -\varepsilon_o + An^{2/3} \pm \Delta E \,, \tag{3.56}$$

where ΔE is an irregular part of the total binding energy of cluster atoms.

For example, if we use the data of Tables 3.1 and 3.2 for large fcc clusters with a short-range interaction of atoms and a range of cluster sizes $n = 1200$–1800, we obtain the following parameters of formula (3.56):

$$\varepsilon_o = 6 \,, \quad A = 7.60 \,, \quad \Delta E \approx 10 \,.$$

Thus, formula (3.56) is a generalized form of formula (2.9). In it we use the above result that the irregular part ΔE of the total binding energy of cluster atoms is independent of the number of cluster atoms in the limit of large n.

4
Elementary Processes and Processes in Gases Involving Clusters

An elementary process of collision of atomic particles (atoms, electrons, ions) with clusters may be divided into two stages, where the first stage is a contact cluster–atomic particle that leads to a strong interaction between colliding particles. A given process results from this strong interaction in the second stage. For large clusters, when the cluster size is large compared with the atomic size, the distance of strong interaction between the colliding cluster and atomic particle is small compared with the cluster size. Therefore, the cross section of contact between colliding particles is determined by the cluster size and is equal to the cluster cross section.

For some processes of cluster collisions a colliding atomic particle may form a bound state with a cluster. This may be important for cluster growth processes, when an atom attaches to a cluster with a certain probability (the sticking probability). In other cluster processes, the bound state of an atomic particle with a cluster surface may be important for subsequent processes involving this atomic particle. This takes place in chemical processes with such an atomic particle, where the cluster surface plays the role of a catalyst. Indeed, let us assume that the chemical process with this atomic particle is characterized by some activation energy that determines the rate of the chemical process. This activation energy is different for free and bound atomic particles, and in the case of a higher efficiency of a chemical process involving a bound atomic particle, clusters may be used as catalysts in this chemical process.

In some cases when clusters partake in chemical processes directly, as takes place in combustion processes, one can divide the chemical process into two parts, so that in the first stage an incident atom or molecule forms a bound state with the cluster surface, and reconstruction of this surface takes place in the second stage. This leads to the formation of a molecule as a reaction product, and this molecule leaves the cluster surface. In this manner, oxidation proceeds for organic clusters or organic particles. As a result, gaseous CO or CO_2 is formed in the course of combustion of organic particles in a gas that includes oxygen.

Cluster Processes in Gases and Plasmas. Boris M. Smirnov

ISBN: 978-3-527-40943-3

4.1 Cluster Collision Processes

Problem 4.1

Within the framework of the hard sphere model for atom–cluster collisions and the liquid drop model for the cluster structure, determine the cross section of collision of a charged cluster with a charged atomic particle.

For the hard sphere cluster model (Figure 2.1), the cross section of cluster collision with a neutral atom is given by formula (2.3) and the cross section of their contact is determined by formula (2.4), $\sigma = \pi r_o^2$. In the case of collision of a charged cluster with a charged atomic particle, the relation between the impact parameter of collision ρ and the distance of closest approach r_{min} follows from the law of conservation of the angular momentum of particles and has the form [84]

$$1 - \frac{\rho^2}{r_{min}^2} = \frac{U(r_{min})}{\varepsilon} ,$$

where $U(R)$ is the interaction potential of colliding particles at a distance R between them, ρ is the impact parameter of collision, r_{min} is the distance of closest approach, and ε is the energy of colliding particles in the reference frame of the center of mass. Taking the charge of an atomic particle to be e and the cluster charge to be $-Ze$, we have for the cross section of their contact $r_o = r_{min}$

$$\sigma = \pi \rho^2(r_{min}) = \pi r_o^2 + \pi r_o \frac{Ze^2}{\varepsilon} , \tag{4.1}$$

where $\rho(r_{min})$ is the impact parameter of collision for which the distance of closest approach is equal to r_{min}.

Problem 4.2

Within the framework of the hard sphere model for electron–cluster collisions determine the diffusion cross section for the scattering of a zero-energy electron on a large spherical cluster.

The interaction potential between the electron and the cluster is given by formula (2.1). Indeed, an electron cannot penetrate inside a cluster, and the interaction between them takes place at atomic distances from the cluster surface. The potential is significant for distances from the cluster of the order of atomic sizes that are small compared with the cluster size. The differential and diffusion cross sections for classical scattering of an atomic particle on the cluster for the hard sphere model are described by formulas (2.2) and (2.3). We consider now the quantum case of electron–cluster scattering for the hard sphere model.

Using the apparatus of quantum mechanics (e.g., [85]), we represent the electron wave function in the form of incident and scattered waves, and at large distances r from a cluster the electron wave function has the form

$$\Psi = \exp(i\mathbf{kr}) + \frac{f(\vartheta)}{r} \exp(ikr) ,$$

where k is the electron wave vector, ϑ is the scattering angle, and $f(\vartheta)$ is the scattering amplitude, so the differential scattering cross section is

$$d\sigma = 2\pi |f(\vartheta)|^2 d\cos\vartheta .$$

Transferring to the case under consideration, we note that the electron wave function coincides with the above asymptotic expression at any distance r that exceeds the cluster radius r_o, and on the cluster surface the electron wave function is zero. Next, owing to the low energy of electrons, the electron wave function does not depend on angles and has the form

$$\Psi = \frac{\exp(ikr) - \exp(-ikr)}{2ikr} + \frac{f(\vartheta)}{r}\exp(ikr) ,$$

where the first term corresponds to a plane wave. The requirement that this wave function be zero at the cluster surface gives for the scattering amplitude

$$f = \frac{\exp(-2ikr_o) - 1}{2ik} = -r_o ,$$

where r_o is the scattering length for a slow electron colliding with a cluster. From this we obtain for the differential and diffusion cross sections of slow electron scattering on a cluster within the hard sphere model instead of formulas (2.2)–(2.3) in the classical case

$$d\sigma = 2\pi r_o^2 d\cos\vartheta , \quad \sigma^* = \int (1 - \cos\vartheta) d\sigma = 2\pi r_o^2 . \tag{4.2}$$

As is seen, the scattering cross section in the quantum case exceeds that in the classical case by a factor of two.

Problem 4.3

Using the hard sphere model for atom–cluster collisions and the liquid drop model for the cluster structure, determine the rate constant of collision resulting from the contact of two neutral clusters.

The cross section of collision is $\sigma = \pi(r_1 + r_2)^2$, where r_1, r_2 are the radii of the corresponding clusters. The rate constant of this collision is [86]

$$k = \langle v \cdot \pi(r_1 + r_2)^2 \rangle = \sqrt{\frac{8T}{\pi\mu}} \cdot \pi(r_1 + r_2)^2 = k_o \sqrt{\frac{n_1 + n_2}{n_1 n_2}} \cdot \left(n_1^{1/3} + n_2^{1/3}\right)^2 , \tag{4.3}$$

where k_o is given by formula (2.7):

$$k_o = \pi \cdot \sqrt{\frac{8T}{\pi m}} \cdot \left(\frac{3m}{4\pi\rho}\right)^{2/3} = 1.93 T^{1/2} m^{1/6} \rho^{-2/3} .$$

Here v is the relative atom and cluster velocity, μ is the reduced cluster and atom mass, m is the atom mass, the angle brackets indicate averaging over cluster velocities, and n_1 and n_2 are the numbers of cluster atoms; above we used formula (2.5) for the number of cluster atoms.

Problem 4.4

Determine the rate constant of mutual neutralization of charged clusters resulting from the collision of positively and negatively charged clusters.

Taking the charges of the clusters to be $\pm e$, we obtain on the basis of formulas (4.1) and (4.3) that [17]

$$k = \left\langle v \cdot \pi (r_1 + r_2)^2 \right\rangle = k_o \sqrt{\frac{n_1 + n_2}{n_1 n_2}} \cdot \left(n_1^{1/3} + n_2^{1/3} \right)^2 (1 + \zeta) , \tag{4.4}$$

where

$$\zeta = \frac{e^2}{(r_1 + r_2) T} . \tag{4.5}$$

Problem 4.5

Determine the rate of ionization of a positively charged small particle by electron impact in a plasma with the Maxwell velocity distribution function of electrons. Assume that each contact of the colliding electron and particle leads to electron attachment.

First we evaluate the rate constant of electron attachment to a small particle within the framework of the liquid drop model. Then each collision of closest approach less than the particle radius leads to electron attachment. The cross section of the process is given by formula (4.1), and the rate constant of electron attachment, in the limit $Ze^2/r \ll T_e$, is

$$k_{at} = \langle v \sigma \rangle = Z e^2 r \sqrt{\frac{8\pi m}{T_e}} .$$

Here Z is the positive charge of the particle expressed in electron charges, r is the particle radius, v is the electron velocity, and T_e is the electron temperature. Brackets mean averaging over velocities on the basis of the Maxwell distribution function of electrons.

Let us denote by ν_{ion} the ionization rate of a particle by electron impact in a plasma, that is, the number of ionization acts per particle per unit time. In the case of equilibrium in a plasma containing electrons and charged particles, we have from the equality of rates of formation and decay of charged particles of a given charge that

$$N_e N_{Z-1} k_{at}(Z - 1) = N_Z k_{ion} ,$$

where N_e is the number density of electrons, N_Z and N_{Z-1} are the number densities of particles of charges Z and $Z-1$, respectively, and $k_{at}(Z-1)$ is the rate constant of electron attachment to a particle of charge $Z-1$. Next, because of the thermodynamic equilibrium in a plasma, the number densities of charged particles are connected by the Saha equation

$$\frac{N_e N_{Z-1}}{N_Z} = 2\left(\frac{m_e T_e}{2\pi\hbar^2}\right)^{3/2} \exp\left(-\frac{I_Z}{T_e}\right),$$

where m_e is the electron mass, I_Z is the particle ionization potential if its charge equals Z. Thus, we have for the ionization rate [86]

$$\nu_{ion} = \frac{2}{\pi} Z e^2 r \frac{m_e^2 T_e}{\hbar^3} \exp\left(-\frac{I_Z}{T_e}\right). \quad (4.6)$$

In fact, we used above the principle of detailed balance to evaluate the ionization rate. Hence, thermodynamic equilibrium in a plasma is used only as a method, so that the ionization rate (4.6) is valid in the absence of thermodynamic equilibrium between charged particles and electrons. Next, we used the Maxwell energy distribution of plasma electrons, that is, it is the assumed equilibrium inside the electron subsystem that allows one to introduce the electron temperature.

Problem 4.6

Determine the ionization cross section of a metal cluster by electron impact within the framework of the statistical model assuming a strong interaction of an incident electron with one cluster electron. Then an incident electron penetrates inside a cluster and exchanges energy with a cluster electron many times, so the energy distribution of these electrons in the end is given by statistical law.

Denote the cluster charge by Z, its energy by ε, and its radius by r_o. Hence, near the cluster surface an incident electron has energy $\varepsilon + Ze^2/r_o$. Considering a system of cluster electrons as a degenerate electron gas, we find that an incident electron may exchange energy only with electrons near the Fermi surface because only they can transfer into unoccupied states after energy exchange. Therefore, our analysis is based on a model according to which an incident electron that has penetrated into a cluster interacts strongly with an electron located on the Fermi surface. Exchange of energy between electrons proceeds randomly, according to statistical laws. If the energy of a cluster electron exceeds the cluster ionization potential J, this electron is released. But this electron may be removed from the charged cluster if the energy transferred exceeds $J + Ze^2/r_o$.

Hence, the ionization cross section in this case is

$$\sigma_{ion} = \sigma_{cap} P,$$

where the cross section of the contact between an incident electron and a cluster is

$$\sigma_{cap} = \pi r_o^2 \left(1 + \frac{Ze^2}{r_o \varepsilon}\right),$$

and P is the probability of release of both electrons as a result of their interaction. Within the framework of the statistical model, under given conditions this probability is

$$P(J/\varepsilon, Ze^2/r_o\varepsilon) = \frac{\int\limits_{Ze^2/r_o}^{\varepsilon-J} \left[\varepsilon'(\varepsilon + Ze^2/r_o - \varepsilon')\right]^{1/2} d\varepsilon'}{\int\limits_{0}^{\varepsilon+Ze^2/r_o} \left[\varepsilon'(\varepsilon + Ze^2/r_o - \varepsilon')\right]^{1/2} d\varepsilon'} . \tag{4.7}$$

We take into account that the statistical weight of free electron states in an energy range from ε to $\varepsilon + d\varepsilon$ is $\varepsilon^{1/2} d\varepsilon$. In addition, we assume, similarly to the Thomson model, that a bound electron is motionless before collision with another electron, so the total energy of both electrons is ε. The energy of an incident electron after electron interaction is ε' near the cluster surface, and, correspondingly, the energy of an initially bound electron becomes $\varepsilon + Ze^2/r_o - \varepsilon'$ after interaction. Ionization takes place if the energy of an incident electron is between Ze^2/r_o and $\varepsilon - J - Ze^2/r_o$ after its removal. In this case, after ionization the cluster charge is $Z + 1$, and we take into account that a fast released electron is screened by a slow one, but the ionization potential J relates to an electron released from a neutral cluster which becomes charged after electron removal. We assume that the ionization potential as the energy of electron liberation near the cluster surface is independent of the cluster charge until the electron is located near the surface. Of course, this holds true for a relatively low Z.

Thus, formula (4.7) requires additional assumptions along with the model ones. The expression (4.7) may be reduced to the form

$$\sigma_{\text{ion}} = \sigma_{\text{cap}} \cdot P(x, y), \quad P(x, y) = \frac{8}{\pi} \int\limits_{x}^{1-y} \sqrt{t(1-t)}dt ,$$
$$x = \frac{Ze^2/r_o}{\varepsilon + Ze^2/r_o}, \quad y = \frac{J + Ze^2/r_o}{\varepsilon + Ze^2/r_o} . \tag{4.8}$$

One can see that $y + x \leq 1$, $y > x$. The function $P(x, y)$ is represented in Figure 4.1 for some relations between y and x.

Problem 4.7

Determine the recombination rate constant of two oppositely charged clusters in a dense buffer gas.

Under these conditions the recombination process of two clusters of charges $Z_+ e$ and $Z_- e$ in a dense gas is restricted by the cluster approach because of a large frictional force. The velocities of positively v_+ and negatively v_- charged particles are determined by an electric field that is created by another charged particle. The electric field strength of charge e at the point where a particle is located is $E = e/R^2$,

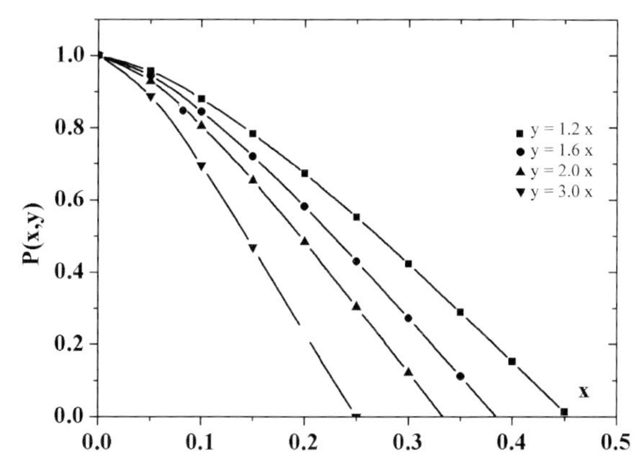

Fig. 4.1 Defined by (4.8), the probability $P(x, y)$ of cluster ionization by a captured electron; $y = 1.2x$ – dark squares, $y = 1.6x$ – dark circles, $y = 2x$ – upright triangles, $y = 3x$ – inverted triangles.

and the velocities of the particles are $v_+ = EK_+, v_- = EK_-$, where K_+ and K_- are the mobilities of clusters. Thus the number of negatively charged clusters that fall on a test positive cluster per unit time is $J_- = 4\pi R^2(v_+ + v_-)N_- = 4\pi e(K_+ + K_-)N_-$, where N_- is the number density of negative clusters. According to the definition, the recombination coefficient k_{rec} of ions is

$$\frac{dN_+}{dt} = -k_{\text{rec}} N_+ N_- = -J_- N_+ ,$$

where N_+ is the number density of positively charged clusters. From this we get the Langevin formula [87] for the recombination coefficient of positively and negatively charged clusters in a dense gas:

$$k_{\text{rec}} = 4\pi e(K_+ + K_-) ,$$

where K_+ and K_- are, respectively, the mobilities of positively and negatively charged clusters in a gas, and the cluster charges are equal to the electron charge e. From this, in the limit of relatively small clusters, one can represent the recombination coefficient for oppositely charged clusters in the form

$$k_{\text{rec}} = \frac{3\sqrt{\pi} Z_+ Z_- e^2}{2\sqrt{2mTN}} \left(\frac{1}{r_+^2} + \frac{1}{r_-^2} \right) , \quad \lambda \gg r_+, r_- , \tag{4.9}$$

where r_+ and r_- are radii of the positive and negative clusters, respectively, and λ is the mean free path of atoms in a gas. As is seen, the recombination coefficient $k_{\text{rec}} \sim 1/r_o^2$, where $r_o \sim r_+, r_-$. In particular, taking $r_+ = r_- = r_o$ for air at room temperature and atmospheric pressure, we have $k_{\text{rec}} r_0^2 = 2.6 \cdot 10^{-20}\,\text{cm}^5/s$.

4.2
Attachment of Atoms to Clusters and Cluster Evaporation

Problem 4.8

Express the rate of atom evaporation from a plane surface through the rate of atom attachment to the surface on the basis of the principle of detailed balance.

According to the definition, the saturated vapor pressure p_{sat} at a given temperature T provides the equilibrium between processes of atom attachment to a plane bulk surface and atom evaporation from this surface. The temperature dependence of the saturated vapor pressure according to formula (2.59) has the form

$$p_{sat} \sim \exp\left(-\frac{\varepsilon_o}{T}\right) , \tag{4.10}$$

where ε_o is the binding energy of a surface atom. Introducing the probability ξ that the contact of an atom with the bulk surface leads to atom attachment, we have for the flux of atoms attached to this surface

$$j_{at} = \sqrt{\frac{T}{2\pi m_a}} \cdot N\xi . \tag{4.11}$$

This is the product of three factors, and the first factor is the average velocity of atoms in the direction perpendicular to the surface, m_a is the atomic mass, N is the atom number density, ξ is the probability of an atom joining with the surface after contact. The flux of evaporating atoms may be determined by the principle of detailed balance and is given by the equation

$$j_{ev} = C\exp\left(-\frac{\varepsilon_o}{T}\right) , \tag{4.12}$$

where ε_o is the binding energy for atoms of a bulk surface, and C depends weakly on the temperature and is determined by properties of the surface. If the atom number density is equal to the number density of saturated vapor N_{sat} at this temperature, the attachment flux becomes equal to the evaporation flux:

$$j_{ev} = j_{at} = \sqrt{\frac{T}{2\pi m_a}} N_{sat}(T)\xi ,$$

where $N_{sat}(T) = N_o \exp(-\varepsilon_o/T)$. Comparing these expressions for the evaporation flux, we find the factor C in formula (4.12) for the evaporation flux

$$C = \sqrt{\frac{T}{2\pi m_a}} N_o\xi ,$$

which gives the following formula for the flux of the evaporated atoms:

$$j_{ev} = \sqrt{\frac{T}{2\pi m_a}} N_o\xi \exp\left(-\frac{\varepsilon_o}{T}\right) . \tag{4.13}$$

Problem 4.9

Within the framework of the liquid drop model for a cluster, obtain the expression for the rate of evaporation of a large cluster.

Assume the cluster surface to be identical to the bulk surface. Then the character of atom evaporation from the cluster surface is similar to that for a macroscopic plane surface, and the difference between them is connected to the atom binding energy. Then, replacing the atom binding energy ε_o for a bulk surface with the atom binding energy ε_n for a cluster consisting of n atoms, we have for the flux of evaporating atoms from the cluster surface [35, 88]

$$j_{ev} = \sqrt{\frac{T}{2\pi m}} N_{sat}(T)\xi \exp\left(-\frac{\varepsilon_n - \varepsilon_o}{T}\right) , \tag{4.14}$$

where $N_{sat}(T)$ is the atom number density at the saturated vapor pressure p_{sat} at a given temperature. Using the dependence

$$N_{sat}(T) = N_o \exp\left(-\frac{\varepsilon_o}{T}\right) , \tag{4.15}$$

we obtain for the rate of atom evaporation ν_{ev} from the cluster surface within the framework of the liquid drop model

$$\nu_{ev} = \pi r_o^2 \sqrt{\frac{T}{2\pi m_a}} N_o \xi \exp\left(-\frac{\varepsilon_n}{T}\right) = k_o N_o n^{2/3} \xi \exp\left(-\frac{\varepsilon_n}{T}\right) , \tag{4.16}$$

where r_o is the cluster radius, and the rate constant k_o is given by formula (2.7).

This expression may be derived from the principle of detailed balance for the processes of atom attachment and atom evaporation of a cluster:

$$A_n + A \leftrightarrow A_{n+1} . \tag{4.17}$$

The rate of atom attachment is given by formula (2.7):

$$\nu_n = N k_o n^{2/3} \xi ,$$

where we add to formula (2.7) the probability ξ of atom attachment at its contact with the cluster. Assuming the main factor of atom evaporation to be the atom binding energy ε_n for binding with the cluster and introducing this energy into formula (4.15) for the saturated vapor pressure with respect to this cluster, we obtain from this equilibrium for the evaporation rate

$$\nu_{n+1}^{ev} = k_o N_o n^{2/3} \xi \exp\left(-\frac{\varepsilon_{n+1}}{T}\right) , \tag{4.18}$$

which corresponds to formula (4.16).

Problem 4.10

Analyze the equilibrium between large clusters and a parent vapor.

An equilibrium between clusters and a parent vapor is realized according to formula (4.17) and takes place when the rates of the processes of atom attachment to the cluster surface and cluster evaporation are equalized. For large clusters consisting of n atoms, on the basis of formulas (4.11), (4.15), and (4.16) we have

$$\varepsilon_n - \varepsilon_o = T \ln \left[\frac{N_{sat}(T)}{N} \right] = T \ln S \,, \tag{4.19}$$

where the supersaturation degree for the gas of attaching atoms is given by formula (2.58):

$$S = \frac{N_{sat}}{N} \,.$$

Note that this equilibrium for a macroscopic cluster is given by $S = 1$, while for a cluster of finite size the equilibrium equation has the form

$$S = S_n \,,$$

where S_n is the supersaturation degree when n is the critical number of cluster atoms for a given number density of free atoms, that is, this number of cluster atoms and the cluster radius are given by formulas (2.63) and (2.64). As is seen, this equilibrium is possible only for a supersaturated vapor.

Problem 4.11

Considering metal clusters consisting of 1000 atoms within the framework of a liquid drop model, find the equilibrium temperatures for which the equilibrium is supported at the number densities of free metal atoms of $10^{12}\,\text{cm}^{-3}$, $10^{14}\,\text{cm}^{-3}$, and $10^{16}\,\text{cm}^{-3}$.

This equilibrium means that the rates of cluster evaporation and attachment of free atoms to this cluster are equal and is described by relation (4.19). On the basis of the liquid drop model for this cluster, we use formula (2.9) for the binding energy of cluster atoms, so the binding energy of an evaporating atom is

$$\varepsilon_n = \varepsilon_o - \frac{2A}{3n^{1/3}} \,.$$

Using the values of the energetic parameters ε_o and A given in Table 2.3 and also the parameters of formula (2.59) for the saturated vapor pressure given in Table 2.3, we calculate the values of the equilibrium temperatures for the indicated number densities of free metal atoms. These values are represented in Table 4.1.

Table 4.1 The equilibrium temperatures in 10^3 K for metal clusters of 10^3 atoms when the equilibrium number densities of free metal atoms are equal to 10^{12} cm^{-3} (a), 10^{14} cm^{-3} (b), and 10^{16} cm^{-3} (c).

Metal	*a*	*b*	*c*	Metal	*a*	*b*	*c*
Ti	1.62	1.88	2.24	Ag	1.06	1.25	1.53
V	1.74	2.03	2.44	Ta	2.75	3.20	3.82
Cr	1.37	1.61	1.96	W	2.93	3.41	4.08
Fe	1.42	1.68	2.05	Os	2.70	3.14	3.76
Co	1.46	1.72	2.08	Ir	2.23	2.61	3.13
Cu	1.26	1.49	1.81	Pt	1.93	2.27	2.75
Zr	2.18	2.55	3.09	Au	1.35	1.59	1.95
Mo	2.22	2.60	3.14	U	1.88	2.24	2.76

Problem 4.12

Show that the flux of ions from the surface of a metal cluster is small compared with the flux of evaporating atoms.

The flux of evaporating atoms from a cluster surface is determined by formula (4.14) and depends on the cluster temperature T as $j_{ev} \sim \exp(-\varepsilon_n/T)$, where ε_n is the binding energy of a surface atom of a cluster consisting of n atoms. At high temperatures, clusters are charged, and, accordingly, a typical cluster charge Z is estimated as $Z \sim r_o T/e^2$, where r_o is the cluster radius and e is the electron charge. Let us consider a large cluster whose charge is high, $Z \gg 1$. If an ion is released from the surface of a large cluster, it obtains additional energy of order Ze^2/r_o because of the Coulomb interaction between the cluster and the ion removed. This gives that the ion binding energy with the cluster surface is equal to $\varepsilon_n - Ze^2/r_o$, and the ratio of the ion flux j_{ion} from the cluster surface to that of the atoms j_{at} is estimated as

$$\frac{j_{ion}}{j_{at}} \sim \frac{Z}{n^{2/3}} \exp\left(\frac{Ze^2}{r_o T}\right) .$$

Above we accounted for the cluster charge Z on the cluster surface and the number of surface atoms of order $n^{2/3}$, and $Z \ll n^{2/3}$. For the mean cluster charge $Z \sim r_o T/e^2$ this ratio is

$$\frac{j_{ion}}{j_{at}} \sim \frac{Z}{n^{2/3}} ,$$

and the flux of evaporating atoms is high compared with the ion flux.

Problem 4.13

Cluster growth results from attachment of atoms to a cluster when attaching atoms are located in a buffer gas and their concentration in the buffer gas is small. Determine the rate of cluster growth if the cluster radius is large compared with the mean free path of attaching atoms in a buffer gas.

We are guided by the case when an admixture of metal atoms is located in a buffer gas and metal clusters may be formed from metal atoms. Attachment of metal atoms to a cluster creates a gradient of concentration c in the vicinity of the cluster surface for metal atoms in a buffer gas, and motion of atoms attaching to a cluster results from diffusion of these atoms in a buffer gas when the concentration of metal atoms is small. If we draw a sphere of a radius R centered on a cluster, the rate J of attaching atoms passing through this sphere is given by

$$J = -4\pi R^2 D N \frac{dc}{dR} , \tag{4.20}$$

where D is the diffusion coefficient for attaching atoms in a buffer gas and Nc is the number density of attaching atoms, so that N is the number density of buffer gas atoms. Because attaching atoms are not formed and absorbed in a space outside the cluster, this relation may be considered as the equation for the concentration $c(R)$ of attaching atoms. Solving this equation under the boundary condition $c(r_o) = 0$, where r_o is the cluster radius, and introducing c_∞ as the concentration of attaching atoms far from the cluster, we find for the concentration of attaching atoms at a distance R from the cluster

$$c(R) = c_\infty \left(1 - \frac{r_o}{R}\right) \tag{4.21}$$

and, correspondingly, from this we obtain for the rate of atom attachment to the surface of a neutral cluster [89]

$$J = 4\pi r_o N c_\infty = 4\pi D r_o N_m . \tag{4.22}$$

This is the Smoluchowski formula, and $N_m = N c_\infty$ is the number density of free atoms attached to a cluster.

Problem 4.14

Determine the time when a growing cluster reaches a given size as a result of growth on the basis of an ion as a nucleus of condensation in a supersaturated parent vapor. Assume the cluster radius to be small compared with the mean free path of attaching atoms in a buffer gas.

Under given conditions one can neglect the evaporation processes, so the balance equation for the number n of drop atoms takes the form

$$\frac{dn}{dt} = \frac{4\pi r_o^2 \rho}{m_a} \cdot \frac{dr_o}{dt} = 4\pi r_o^2 j_{at} = 4\pi r_o^2 \sqrt{\frac{T}{2\pi m_a}} N \xi ,$$

where within the framework of the liquid drop model the number n of cluster atoms is

$$n = \frac{4\pi r_o^3 \rho}{3m_a},$$

and we used formula (4.11) for the flux j_{at} of attaching atoms under conditions when attaching atoms reach the cluster surface freely. Solving the balance equation, we have for the drop radius at time t

$$r_o = \frac{1}{\rho}\sqrt{\frac{Tm_a}{2\pi}} N\xi \cdot t$$

when it exceeds significantly the initial drop radius. We assumed above that growth of the liquid drop does not change the number density N of free atoms.

Problem 4.15

Determine a typical time for cluster growth on the nuclei of condensation in a buffer gas for the diffusion regime of cluster growth when the cluster radius is small compared with the mean free path of attaching atoms in a buffer gas.

In this case one can ignore cluster evaporation, and we base our analysis on the Smoluchowski formula (4.22) for the rate of atom attachment to the cluster surface for a large cluster size when cluster growth is restricted by atom diffusion in a buffer gas. In this case the balance equation for a number of cluster atoms n has the form

$$\frac{dn}{dt} = J_{at} = 4\pi r_o D N c_\infty .$$

From the solution of this equation we obtain for the evolution of the cluster radius $r_o(t)$

$$r_o = \sqrt{\frac{m_a D N c_\infty t}{2\rho}} .$$

Problem 4.16

Analyze the character of atom attachment to a large neutral cluster when attaching atoms are located in a buffer gas and the liquid drop model is used for a cluster.

This character depends on the relation between the mean free path of attaching atoms in a buffer gas and a cluster radius r_o. Let us consider the problem in a general form, introducing c_o, the concentration of attaching atoms near the cluster surface, and c_∞, their concentration far from the cluster. Solving equation (4.20) by using $J =$ const and the boundary condition far from the cluster, we obtain

$$c(R) = c_\infty + \frac{J}{4\pi D N R} .$$

On the other hand, considering the attachment process near the cluster and introducing ξ, the probability of atom attachment as a result of its contact with the cluster surface, we obtain for the rate of attachment events

$$J = 4\pi r_o^2 \cdot \sqrt{\frac{T}{2\pi m}} N c_o \cdot \xi \,, \tag{4.23}$$

where r_o is the cluster radius, the first factor in this product is the cluster surface area, and the second factor is the flux of atoms attaching to a cluster surface. Comparing this and the Smoluchowski formula (4.22), we have for the atom concentration near the cluster surface

$$c_o = \frac{c_\infty}{1+\alpha} \,,$$

where the parameter α is of fundamental importance for this process and has the form

$$\alpha = \sqrt{\frac{T}{2\pi m}} \cdot \frac{r_o \xi}{D} \,. \tag{4.24}$$

The parameter α characterizes the competition between the kinetic and diffusion regimes of the attachment process. Indeed, in the case

$$\alpha \ll 1$$

the diffusion regime is realized for the attachment process and the Smoluchowski formula (4.22) is valid for this process. In other words, the attachment process is restricted in this case by the diffusional approach of atoms to the cluster in a dense buffer gas. In another limiting case

$$\alpha \gg 1$$

the kinetic regime is realized, and atoms attach to the cluster as a result of their pairwise collisions. Note that the kinetic regime exists under the criterion

$$\lambda \gg \frac{r_o}{\xi} \,, \tag{4.25}$$

that is, along the mean free path λ of atoms in a buffer gas and a cluster radius r_o, the probability ξ of atom attachment as a result of atom–cluster contact is taken into account by this criterion.

Correspondingly, the rate of atom attachment to the cluster in a general case is given by

$$J = 4\pi r_o^2 \cdot \sqrt{\frac{T}{2\pi m_a}} \cdot \frac{N c_\infty \xi}{1+\alpha} = 4\pi r_o D \frac{N_m \alpha}{1+\alpha} \,, \tag{4.26}$$

where $N_m = N c_\infty$ is the number density of attached atoms far from the cluster.

Problem 4.17

Excited atoms are located in a buffer gas, and quenching of these excited atoms proceeds with probability ξ as a result of contact of an excited atom with the cluster surface. Determine the rate of atom quenching on the cluster surface.

Denote by N_* the number density of excited atoms far from the cluster. We have, then, an analogy between the concentration of excited atoms $c(R) = N_*(R)/N$ in a buffer gas and the rate of atom quenching, as it takes place for atom attachment and is given by formula (4.26), that takes the form

$$J = 4\pi r_o^2 \sqrt{\frac{T}{2\pi m}} \cdot \frac{N_* \xi}{1+\alpha} = 4\pi r_o D \left(1 + \frac{1}{\alpha}\right) . \tag{4.27}$$

Then this formula gives for the rate of atom quenching on the surface of one cluster in the kinetic regime of atom quenching $\alpha \gg 1$

$$J = \pi r_o^2 \sqrt{\frac{8T}{\pi m}} N_* \xi .$$

In the diffusion regime $\alpha \ll 1$ we obtain the Smoluchowski formula (4.22) for the rate of atom quenching

$$J = 4\pi D r_o N_* \xi ,$$

where D is the diffusion coefficient of an excited atom in a buffer gas.

Problem 4.18

Derive the balance equation for the cluster size taking into account the processes of cluster evaporation and atom attachment in the kinetic regime of these processes.

We use the equilibrium (4.17) between clusters of different sizes, and we represent the balance equation for a large cluster size in the form

$$\frac{dn}{dt} = 4\pi r_o^2 (j_{\rm at} - j_{\rm ev}) ,$$

where r_o is the cluster radius, and the fluxes of atom attachment to cluster $j_{\rm at}$ and cluster evaporation $j_{\rm ev}$ are given by formulas (4.11) and (4.13), respectively. On the basis of these formulas, we reduce the above balance equation to the form

$$\frac{dn}{dt} = 4\pi r_o^2 \sqrt{\frac{T}{2\pi m_a}} \xi \cdot \left[N - N_{\rm sat}(T) \exp\left(-\frac{\varepsilon_n - \varepsilon_o}{T}\right)\right]$$
$$= 4\pi r_o^2 \sqrt{\frac{T}{2\pi m_a}} N \xi \cdot \left(1 - \frac{1}{S_n}\right) , \tag{4.28}$$

where S_n is the supersaturation degree, and the critical cluster size is defined by formula (2.63) at a given number n of cluster atoms and number density N of attaching atoms.

From this equation it follows that clusters are divided into two groups, such that small clusters evaporate, whereas large clusters grow as a result of attachment of atoms. The boundary cluster size is the critical cluster size that is determined by formula (2.63). Note that this regime of cluster evolution exists only for a supersaturated vapor $N > N_{sat}(T)$ or $S > 1$.

In the derivation the balance equation we are based on formula (4.11) for the flux of atoms attaching to a cluster surface. This equation relates to the kinetic regime of cluster evolution when the cluster size is small compared with the mean free path of attaching atoms in a buffer gas. One can extend this equation to a general case on the basis of formula (4.26) for the rate of atom attachment. Then the balance equation (4.28) is reduced to the form

$$\frac{dn}{dt} = 4\pi r_o^2 \sqrt{\frac{T}{2\pi m_a}} N\xi \cdot \left(1 - \frac{1}{S_n}\right) \cdot \frac{1}{1+\alpha} . \qquad (4.29)$$

Problem 4.19

In the kinetic regime of cluster evaporation determine a typical time of evaporation of a neutral cluster of small size compared with the cluster critical radius for a given pressure of the parent vapor.

The rate of atom evaporation from the cluster surface follows from formula (4.18):

$$J_{ev} = 4\pi r_o^2 j_{ev} = 4\pi r_o^2 \sqrt{\frac{T}{2\pi m_a}} N_{sat}(T)\xi \exp\left(\frac{\Delta\varepsilon}{T n^{1/3}}\right)$$
$$= 4\pi r_o^2 \sqrt{\frac{T}{2\pi m_a}} N_{sat}(T)\xi \exp\left(\frac{2\gamma m_a}{\rho T r_o}\right) ,$$

where the binding energy of a cluster atom in accordance with formulas (2.8) and (2.10) is taken in the form

$$\varepsilon_o - \varepsilon_n = -\frac{dE_n}{dn} = \frac{\Delta\varepsilon}{n^{1/3}} = \frac{2A}{3n^{1/3}} = \frac{8\pi r_W^3 \gamma}{3r_o} = \frac{2m_a\gamma}{r_o\rho} . \qquad (4.30)$$

This gives the balance equation for the number of atoms in an evaporating cluster

$$\frac{dn}{dt} = \frac{4\pi\rho r_o^2}{m_a}\frac{dr_o}{dt} = 4\pi r_o^2 \sqrt{\frac{T}{2\pi m_a}} N\xi \left[\frac{1}{S}\exp\left(\frac{2\gamma m_a}{\rho T r_o}\right) - 1\right] ,$$

or

$$\frac{dr_o}{dt} = \frac{1}{\rho}\sqrt{\frac{T m_a}{2\pi}} N_{sat}(T)\xi \left[\exp\left(\frac{2\gamma m_a}{\rho T r_o}\right) - \exp\left(\frac{2\gamma m_a}{\rho T r_{cr}}\right)\right] , \qquad (4.31)$$

where the critical cluster radius r_{cr} is given by formula (2.64).

Let us consider the limiting cases. When $r_{cr} - r_o \ll r_o$, one can expand the expression in the parentheses, and the balance equation is

$$\frac{dr_o}{dt} = \frac{1}{\rho}\sqrt{\frac{Tm_a}{2\pi}}N\xi \cdot \frac{2\gamma m_a}{\rho T}\left(\frac{1}{r_o} - \frac{1}{r_{cr}}\right) = \frac{r_{cr} - r_o}{\tau_o},$$

where

$$\tau_o^{-1} = \frac{1}{\rho}\sqrt{\frac{Tm_a}{2\pi}}N\xi \cdot \frac{2\gamma m_a}{\rho T r_{cr}^2}.$$

From this we find the total time of cluster evaporation

$$\tau_{ev} = \tau_o \ln \frac{r_{cr}}{r_{cr} - r_o}.$$

In the other limiting case, $2\gamma m_a/(\rho T) \ll r_o \ll r_{cr}$, we have for the evaporation time

$$\frac{1}{\tau_{ev}} = \frac{1}{r_o^2} \cdot \frac{1}{\rho}\sqrt{\frac{Tm_a}{2\pi}}N\xi \cdot \frac{4\gamma m_a}{\rho T}.$$

In the limiting case $r_o \ll 2m_a\gamma/(T\rho)$, one can neglect the second term in the balance equation, and we have for the total evaporation time

$$\tau_{ev} = \tau_o \cdot \frac{r_o^2}{r_{cr}^2} \cdot \exp\left(-\frac{2m_a\gamma}{T\rho r_o}\right).$$

Problem 4.20

Analyze the character of cluster evolution in the case where the binding energy of the cluster atoms is a stepwise function of the number of cluster atoms n with maxima at magic numbers.

The critical cluster size is introduced through the relation (4.19):

$$S_n = \exp\left(-\frac{\varepsilon_n - \varepsilon_o}{T}\right).$$

If the binding energy of cluster atoms ε_n is a monotonic function of n, the critical size divides the cluster sizes n into two ranges. Clusters with $n > n_{cr}$ grow, and clusters with $n < n_{cr}$ evaporate. Correspondingly, the cluster flux in the space n is directed to large sizes for $n > n_{cr}$ and to zero for $n < n_{cr}$, and the flux is zero at the critical cluster size.

In the case where ε_n is a stepwise function of n, the range where the cluster growth is not continuous. For such n according to the above relation we have

$$S > S_n \equiv \exp\left(-\frac{\varepsilon_n - \varepsilon_o}{T}\right). \quad (4.32)$$

The same conclusion corresponds to the size range where clusters evaporate. Assume that in the size range, where clusters evaporate, relation (4.32) is satisfied only at several magic numbers. Then the evolution of clusters in the course of their evaporation has the following character at a constant number density of free atoms. When the cluster reaches a magic number of atoms, an atom becomes attached, and the cluster being formed evaporates. This process is repeated several times, and the probability P_n of emitting one atom for a cluster consisting of n atoms is given by

$$\frac{1}{P_n} = 1 + S \exp\left(\frac{\varepsilon_n - \varepsilon_0}{T}\right) = 1 + \frac{S}{S_n} .$$

Analogously, the total time of cluster evaporation is in this case given by

$$\tau_{\text{ev}} = \int \frac{dn}{\nu_n^{\text{ev}}} \left(1 + \frac{S}{S_n}\right) ,$$

where ν_n^{ev} is the rate of evaporation of a cluster of n atoms.

In this case one can see the increase in the evaporation time compared to the case where ν_n is a continuous function of n. If the numbers of cluster atoms for which criterion (4.32) is fulfilled or is violated are distributed randomly, the character of the cluster evolution has a more complex character. In this case the diffusion character of the motion in n-space takes place, and each step of this diffusion is not random but is determined by the competition between processes of cluster evaporation, and cluster growth for a given cluster size. Thus, each step is determined by the probability P_n, which is not a continuous function of n.

Problem 4.21

Clusters of an identical size n are located in a supersaturated vapor and criterion (4.32) is satisfied for them. Analyze the character of the size evolution of clusters if q free atoms correspond to each cluster and the variation in the number of cluster atoms as a result of this evolution is relatively small. Assume that the binding energy of cluster atoms is determined by formula (2.9).

Cluster growth finishes when the number density of attaching atoms is equal to the saturated vapor number density (4.19) and the cluster radius is equal to the critical radius (2.64) for a given pressure of attaching atoms. Neglecting the size fluctuations during cluster growth, we assume clusters to be of identical sizes in the course of their growth. Cluster growth is accompanied by a decrease in the number density of free atoms, and this process stops when the cluster size corresponds to the critical size at a given number density of free atoms. If at the end of the process q' free atoms correspond to one cluster, the condition of the cluster critical size has the form

$$\frac{q'}{q} S = \exp\left(\frac{\Delta\varepsilon}{T(n + q - q')^{1/3}}\right) ,$$

where q is the initial number of free atoms per cluster. Because the relative variation of cluster atoms is small, that is, $n \gg q - q'$, from this relation it follows that

$$q' = \frac{q}{S} \exp\left(\frac{\Delta\varepsilon}{T n^{1/3}}\right) .$$

Correspondingly, the number of cluster atoms is $n' = n + q - q'$ at the end of the process.

Problem 4.22

Determine a typical time of cluster evaporation in a dense buffer gas for the diffusion regime of cluster evaporation in a buffer gas.

The concentration c of atoms of the parent vapor in a buffer gas is small. Because of the processes of evaporation and attachment of atoms to the cluster, the concentration of metal atoms varies near the cluster. Let us denote their concentration far from the cluster by c_∞. The flux of attachment atoms to the cluster surface is given by

$$j = -D N \nabla c .$$

Since atoms are not absorbed in the volume, we have $J = 4\pi R^2 j = \text{const}$, where R is the distance from the cluster. This is the equation for the concentration of metal atoms of a parent vapor in a space

$$J = -4\pi R^2 D \frac{dc}{dR} ,$$

and the solution of this equation is

$$c(R) = c_\infty + \frac{J}{4\pi D N R} . \tag{4.33}$$

The total flow rate of atoms toward the cluster is

$$J = J_{\text{at}} - J_{\text{ev}} = J_{\text{ev}} \frac{S - S_n}{S_n} ,$$

where J_{ev} is expressed by formula (4.16), $S = N/N_{\text{sat}}$ is the supersaturation degree (2.58), and S_n is the supersaturation degree when n corresponds to the critical number of cluster atoms according to formula (2.63). In the case of a dense buffer gas we have $J_{\text{at}} \approx J_{\text{ev}} \gg J$. Hence, near the cluster surface $S = S_n$, that is, the number density of atoms of the parent vapor reaches such a value when the number of cluster atoms is the critical one. Then the concentration of free atoms of the parent gas at a distance R from the cluster is $c(R) = c_\infty + (c_* - c_\infty) r_o/R$, and the rate of cluster evaporation is given by

$$J_{\text{ev}} = 4\pi D N r_o (c_* - c_\infty) ,$$

where c_* is the concentration of parent atoms in the buffer gas at which the cluster radius is equal to the critical radius. Assuming $c_* \gg c_\infty$, we obtain the balance equation for the number of cluster atoms:

$$\frac{dn}{dt} = \frac{4\pi r_o^2 \rho}{m_a} \cdot \frac{dr_o}{dt} = 4\pi r_o D N_{sat}(T) \exp\left(\frac{2\gamma m_a}{\rho T r_o}\right) .$$

Taking $r_o \gg 2\gamma m_a/(\rho T)$, we get from this for the evaporation time

$$\tau_{ev} = \frac{r_o^2 \rho}{2 m_a D N_{sat}(T)} . \tag{4.34}$$

From the above relation we obtain for the atom concentration near the cluster surface

$$c_o = c_\infty + \alpha(c_* - c_o) ,$$

and the parameter α, which characterizes the competition between the kinetic and diffusion regimes of the cluster growth process, is given by formula (4.24)

$$\alpha = \sqrt{\frac{T}{2\pi m_a}} \cdot \frac{\xi r_o}{D} .$$

Thus we have

$$c_o = \frac{c_\infty + \alpha c_*}{1 + \alpha} ,$$

and the criterion of the diffusion regime for cluster evaporation corresponds to $c_\infty \ll c_o$. Under the condition $c_\infty \ll c_*$, the criterion

$$\alpha = \sqrt{\frac{T}{2\pi m_a}} \cdot \frac{\xi r_o}{D} \gg 1$$

holds true and means the diffusion regime of cluster evaporation. This analysis is valid if the cluster radius is large compared with the mean free path λ of free parent atoms in the buffer gas, because we use the diffusion character of motion of atoms near the cluster surface. Using an estimation for the diffusion coefficient of metal atoms in a buffer gas $D \sim \lambda \sqrt{T/m_a}$, we obtain

$$\alpha \sim \frac{\xi r_o}{\lambda} .$$

Thus, the diffusion regime of cluster growth and evaporation ($\alpha \gg 1$) takes place for a large cluster radius

$$r_o \gg \frac{\lambda}{\xi} .$$

Problem 4.23

Determine the typical time of cluster evaporation in a dense buffer gas for the diffusion regime of the evaporation process if the number density of buffer gas atoms is lower than the equilibrium number density of free metal atoms given by formula (4.19).

In the diffusion regime of evaporation, when the departure of evaporating atoms is restricted by atom diffusion, the concentration of atoms of a parent gas near the cluster surface is equal to the equilibrium one c_*, which corresponds to the critical concentration for a given cluster radius determined by formula (4.19) and exceeds the concentration far from the cluster surface, that is, $c_* \gg c_\infty$. But in the case when the equilibrium number density of atoms of a parent gas (metal atoms) exceeds the number density of buffer gas atoms, the equilibrium for attachment and evaporation processes is not fulfilled at the cluster surface and the concentration of metal atoms at the cluster surface is $c_o = 1$. Then on the basis of formula (4.33) one can find the rate of cluster evaporation in this case:

$$J = 4\pi D N r_o (1 - c_\infty) \, .$$

Solving the balance equation $dn/dt = -J$, we find the evaporation time in this case:

$$\tau_{ev} = \frac{r_o^2 \rho}{6 m_a D N (1 - c_\infty)} \, .$$

Problem 4.24

Determine a typical time of drop evaporation in a dense buffer gas taking into account both the kinetic and the diffusion regimes of this process. The number density of buffer gas atoms exceeds the equilibrium number density of metal atoms according to formula (4.19).

We are based on formula (4.34) for the evaporation time in the diffusion regime of cluster evaporation. For generalization of this equation we use the balance equation (4.26), which includes both regimes of cluster evolution. Then roughly the generalization is reduced to multiplication the evaporation time (4.34) by the factor $(1 + \alpha)$, and the evaporation time is given by

$$\tau_{ev} = \frac{r_o^2 \rho}{2 m_a D N_{sat}(T)} \cdot (1 + \alpha) \, . \tag{4.35}$$

4.3
Cluster Heat Processes in Gases

Problem 4.25

A cluster whose size is small compared with the mean free path of atoms in a surrounding buffer gas is irradiated by a laser beam. The energy flux of laser radiation is J, and the cross section of absorption of laser radiation is σ_{abs}. Assuming the power of absorbed radiation goes into cluster heating, find the cluster temperature T_{cl} within the framework of a model of equal temperatures.

The model of equal temperatures assumes that after contact with a cluster an atom acquires the mean kinetic energy $3T_{cl}/2$. This takes place for a strong interaction between an incident atom and a cluster as a result of their contact, in particular, if in the course of contact with the cluster surface an atom is located for some time on the cluster surface, and this time is long enough for strong energy exchange. Then, on average, an atom in contact with the cluster surface takes the energy $3(T_{cl} - T)/2$, where T is the gas temperature. This leads to the following balance equation:

$$J\sigma = \frac{3}{2}(T_{cl} - T)\nu_n = \frac{3}{2}(T_{cl} - T)Nk_o n^{2/3} ,$$

where the rate of contact between an atom and the cluster surface is given by formula (2.7) in the limit $r_o \ll \lambda$; we used a model of equal temperatures for atoms in contact with the cluster surface. From this we obtain for the difference of the cluster and gas temperatures

$$T_{cl} - T = \frac{2J\sigma}{3Nk_o n^{2/3}} . \tag{4.36}$$

Problem 4.26

Metastable atoms located in a buffer gas are quenched on the cluster surface in the diffusion regime when the cluster size is large, and heat release results from the thermal conductivity of a gas. Determine the increase in the cluster temperature under these conditions.

As a result of quenching of a metastable atom on the cluster surface, energy $\Delta\varepsilon$ is liberated on the cluster surface or close to it and is expended on heating the cluster or surrounding buffer gas. The heat power resulting from the quenching process is $P = J\Delta\varepsilon$, where J is the number of quenching events per unit time. Assuming the energy released is removed at large distances from the cluster by gaseous thermal conductivity, we obtain the following equation for the heat balance at a distance R from the cluster:

$$P = 4\pi R^2 \kappa \frac{dT}{dR} ,$$

where κ is the thermal conductivity coefficient of the gas, T is the gas temperature, and $q = -\kappa dT/dR$ is the heat flux from the cluster. We use here the fact that heat is released from the cluster surface and propagates far from the cluster.

We have an identical equation for the flux of quenched atoms:

$$J = -4\pi R^2 D N_a \frac{dc}{dR} ,$$

where N_a is the number density of buffer gas atoms and c is the concentration of metastable atoms in a gas. Using the connection between these values, we get

$$\kappa \frac{dT}{dR} = -\Delta\varepsilon \cdot D N_a \frac{dc}{dR} .$$

It is convenient to rewrite this equation in the form

$$\frac{dT}{dc} = -\Delta\varepsilon \frac{DN}{\kappa} .$$

If the right-hand side of this equation does not depend on the temperature, then the solution of the equation yields

$$\Delta T \equiv T(r_o) - T(\infty) = \frac{DN_a}{\kappa}(c_\infty - c_o)\Delta\varepsilon ,$$

where c_o is the concentration of metastable atoms on the cluster surface, c_∞ is the concentration of metastable atoms far from the cluster, $T(\infty)$ is the gas temperature at large distances from the cluster, $c_o = 0$ in the diffusion regime, and we have

$$\Delta T \equiv T(r_o) - T(\infty) = \frac{DN_a}{\kappa}(c_\infty - c_o)\Delta\varepsilon . \tag{4.37}$$

Let us estimate the diffusion coefficient of metastable atoms in a buffer gas as $D \sim v_T\lambda$, where v_T is the thermal atom velocity, λ is the mean free path of metastable atoms in a buffer gas, and $\kappa \sim v_T N_a \lambda$ is the thermal conductivity coefficient of a buffer gas, so that N_a is the number density of buffer gas atoms, and we assume the mean free path for a metastable atom and that for a buffer gas atom to be of the same order of magnitude. This gives $DN_a/\kappa \sim 1$, and we obtain the following estimation for cluster heating due to quenching of metastable atoms on the cluster surface:

$$\Delta T \sim \Delta\varepsilon c_\infty \sim \Delta\varepsilon \frac{N_*}{N_a} .$$

Problem 4.27

Metastable atoms located in a buffer gas are quenched on the cluster surface in the kinetic regime when the cluster size is small compared with the mean free path of atoms in a buffer gas. Determine the increase in the cluster temperature with respect to the buffer gas temperature far from a cluster within the framework of the model of equal temperatures. Assume that the probability of quenching a metastable atom as a result of its contact with a cluster surface is one.

The cluster temperature increase follows from the heat balance that is determined by the process of quenching of metastable atoms on the cluster surface and the process of heat transport due to gas thermal conductivity. The heat power resulting from the quenching process is $P = J\Delta\varepsilon$, where $\Delta\varepsilon$ is the excitation energy for a metastable atom and J is the rate of quenching of metastable atoms on the cluster surface per unit time, which in this case is given by

$$J = N_* \cdot \sqrt{\frac{8T}{\pi m_a}} \cdot \pi r_o^2 .$$

Here N_* is the number density of metastable atoms and m_a is the mass of a metastable atom. For the model of equal temperatures, each atom of a buffer gas that collides with a cluster exchanges with it an energy $3(T_{cl} - T)/2$, where T_{cl} is the cluster temperature and T is the buffer gas temperature. This model is valid if an atom is located on the cluster surface long enough. The heat balance equation in this case has the form

$$\Delta\varepsilon N_* \cdot \sqrt{\frac{8T}{\pi m_a}} \cdot \pi r_o^2 = \frac{3}{2}(T_{cl} - T) N_a \cdot \sqrt{\frac{8T}{\pi m}} \cdot \pi r_o^2 , \tag{4.38}$$

where m is the mass of a buffer gas atom, and we assume that the energy of quenching of metastable atoms transfers to a buffer gas as a result of collision of buffer gas atoms with the cluster surface. From this we get for cluster heating

$$\Delta T \equiv T_{cl} - T = \frac{\Delta\varepsilon N_*}{N_a} \sqrt{\frac{m_a}{m}} . \tag{4.39}$$

4.4
Combustion and Catalytic Processes in Gases Involving Clusters

Problem 4.28

An organic cluster is heated owing to oxidation in a buffer gas that contains oxygen. The probability of oxidation χ as a result of the contact of an oxygen molecule with the cluster surface depends on the cluster temperature T_{cl} according to the Arrhenius law:

$$\chi(T_{cl}) = \chi_o \exp\left(-\frac{E_a}{T_{cl}}\right) , \tag{4.40}$$

where E_a is the activation energy of the combustion process. Find the criterion of thermal explosion of a cluster in the kinetic regime using the model of equal temperatures for the exchange by energy between a buffer gas atom and a cluster as a result of their contact.

Let us denote by $\Delta\varepsilon$ the energy released as a result of the combustion process involving an oxygen molecule and a cluster. In the kinetic regime we have the following heat balance equation by analogy with formula (4.38):

$$\Delta\varepsilon\chi(T_{\rm cl})N_{\rm o}\cdot\sqrt{\frac{8T}{\pi m_{\rm a}}}\cdot\pi r_{\rm o}^2=\frac{3}{2}(T_{\rm cl}-T)N_{\rm a}\cdot\sqrt{\frac{8T}{\pi m_{\rm a}}}\cdot\pi r_{\rm o}^2\,, \tag{4.41}$$

where $\chi(T_{\rm cl})$ is given by formula (4.40), $N_{\rm o}$ is the number density of oxygen molecules, $N_{\rm a}$ is the number density of buffer gas atoms, $m_{\rm o}$ is the mass of the oxygen molecule, and m is the mass of a buffer gas atom.

Let us represent this heat balance equation in the form

$$T_{\rm cl}-T=A\exp\left(-\frac{E_{\rm a}}{T_{\rm cl}}\right)\,,\quad A=\frac{2}{3}\Delta\varepsilon\chi_{\rm o}\sqrt{\frac{m}{m_{\rm o}}}\,. \tag{4.42}$$

Figure 4.2 gives the dependence of the left-hand side and right-hand side of this equation on the cluster temperature at different values of parameter A. At large values of this parameter this equation has no solution. This means that in this case the heat released cannot be removed as a result of collision with buffer gas atoms because of a strong temperature dependence for the probability (4.40) of the combustion process and corresponds to the thermal instability of this process. At the threshold of the thermal instability it is necessary to take into account the equality of both the rates and the derivatives for heat release and heat transport. Therefore, denoting the cluster temperature at the threshold of the thermal instability by $T_{\rm cr}$, we have the following relations for this temperature:

$$T_{\rm cr}-T_\infty=A\exp\left(-\frac{E_{\rm a}}{T_{\rm cr}}\right)\,,\quad \frac{T_{\rm cr}^2}{E_{\rm a}}=A\exp\left(-\frac{E_{\rm a}}{T_{\rm cr}}\right)\,. \tag{4.43}$$

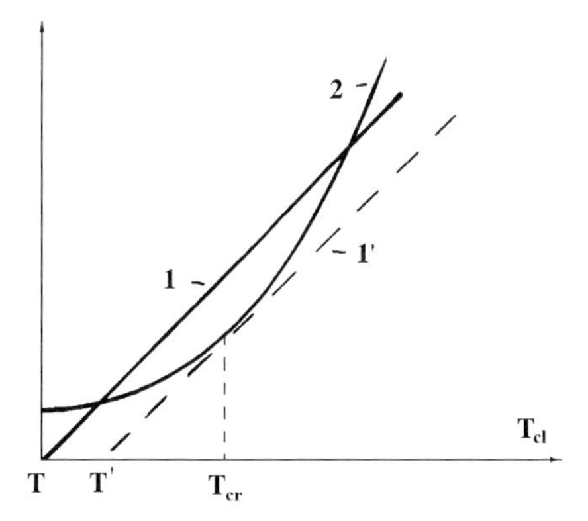

Fig. 4.2 Equation (4.42) connecting gas temperatures far from a cluster T and on its surface $T_{\rm cl}$. Curves 1, 1′ are the left-hand side of equation (4.42), where T, T' are gas temperatures far from the cluster, and curve 2 is the right-hand side of equation (4.42). $T_{\rm cr}$ is the cluster temperature that corresponds to the threshold of thermal explosion.

From this it follows that $T_{cr} - T = T_{cr}^2/E_a$, and since $T_{cr} \ll E_a$, the temperature difference $T_{cr} - T$ is relatively small, which justifies the above assumptions.

Problem 4.29

Find the criterion of the thermal explosion of a porous spherical coal particle that is burned in a gaseous mixture containing oxygen.

The heat flux from a burning particle is $\mathbf{q} = -\kappa \nabla T$. Assuming $\kappa(t) = \text{const}$ in the temperature range under consideration and taking into account the absence of heat release in a gaseous space, we obtain for the gas temperature

$$T(R) = T(\infty) + \frac{C}{R},$$

where R is the distance from the particle's center and T_∞ is the gas temperature far from the particle. The total power of heat released is given by

$$P = 4\pi R^2 q = 4\pi\kappa C = \frac{4\pi}{3} r_o^3 p_o \exp\left(-\frac{E_a}{T_{cl}}\right),$$

where we use for the heat flux $q = -\kappa dT/dR$, and the specific power of the combustion process per particle is $p_o \exp(-E_a/T_o)$. Taking from this $C = P/(4\pi\kappa)$, we obtain for the temperature on the particle's surface in the diffusion regime

$$T(R) = T(\infty) + \frac{p}{4\pi\kappa R}. \tag{4.44}$$

From this we have for the temperature of a spherical particle with radius r_o

$$T_{cl} = T(\infty) + \frac{r_o^2 p_o}{3\kappa} \exp\left(-\frac{E_a}{T_o}\right).$$

Note that this relation is valid for the diffusion regime $\lambda \ll r_o$, when heat transport from the particle to a gas results from the thermal conductivity of a gas.

Let us determine from this equation the threshold of the thermal instability whose nature is such that the heat release grows strongly with an increase in temperature, and for certain values of the particle's temperature the released power cannot be transported as a result of gaseous thermal conductivity. From the equality of the power derivatives we obtain the relation

$$\frac{r^2 p_o}{3\kappa} \exp\left(-\frac{E_a}{T_{cr}}\right) = \frac{T_{cr}^2}{E_a} = T_{cr} - T_\infty \tag{4.45}$$

for the particle temperature T_{cr} that corresponds to the instability threshold. Note that according to equation at the threshold of thermal instability, the difference between the particle's temperature and the gas temperature far from the particle is relatively small:

$$T_{cr} - T_\infty = \frac{T_{cr}^2}{E_a} \ll T_{cr},$$

because $T_{cr} \ll E_a$. This relation coincides with that (4.43) for the kinetic regime of heat transport.

Table 4.2 The threashold temperature for the thermal instability of an activated coal particle of a radius r_o located in atmospheric air.

T_{cr}, K	1000	1200	1400	1600	1800
r_o, µm	99	31	14	7.9	5.2

Problem 4.30

Apply relation (4.45) for combustion of a porous spherical particle of activated porous coal.

We use the following parameters for combustion of activated birchen coal in air [90–93]:

$$E_a = 34 \pm 1\,\text{kcal/mol}\,, \quad p_o = 4 \cdot 10^{10.0 \pm 0.3}\,\text{W/cm}^3\,,$$
$$a = 0.8\,, \quad \rho = 0.8 \pm 0.2\,\text{g/cm}^3\,, \qquad (4.46)$$

where p_o is the preexponential factor for the specific power of heat release resulting from this combustion process, a is the gray coefficient for radiation of a hot coal surface, and ρ is the coal density. Substituting these parameters into formula (4.45), we get the connection between the radius of an activated coal particle and the threshold temperature of the thermal instability as a result of its combustion. The results are given in Table 4.2, where we ignore the radiation of a burning particle. Taking this fact into account, we obtain higher temperature values for the instability threshold.

Problem 4.31

Clusters are injected into a buffer gas with a number density of N_{cl}, and their average radius r_o greatly exceeds the mean free path of atoms or molecules of the buffer gas. Chemically active molecules are located in the buffer gas and decay as a result of contact with the surface of clusters. Neglecting the heat release of this process, determine the character of decay of the active molecules in a buffer gas with clusters.

Clusters are catalyzers in this process, and their presence causes decay of chemically active molecules. The approach of chemically active molecules to the cluster surface corresponds to the diffusion regime of molecule motion, and the rate of decay of active molecules on the cluster surface of an individual particle is determined by the Smoluchowski formula (4.22) and is given by

$$\nu = 4\pi D N_{mol} r_o\,.$$

Here N_{mol} is the number density of active molecules, D is their diffusion coefficient in a buffer gas, and r_o is the cluster radius. From this we have the balance equation for the number density of active molecules:

$$\frac{d N_{mol}}{dt} = 4\pi D r_o N_{cl} N_{mol}\,,$$

where N_{cl} is the number density of clusters. This gives the time dependence for the number density of active molecules in the form

$$N(t) = N(0)\exp(-t/\tau)\,, \tag{4.47}$$

where

$$\frac{1}{\tau} = 4\pi D r_o N_{cl}\,. \tag{4.48}$$

As is seen, at a given mass of particles per unit volume, the time for this process increases proportionally to $1/r_o^2$ with a decrease in cluster size.

Problem 4.32

A spherical cluster is burned in a gaseous mixture containing oxygen and its surface emits radiation as a gray surface that is characterized by the gray coefficient a. Determine the part of the energy released in the form of radiation.

We have the heat balance equation for the burning particle

$$-4\pi R^2 \kappa \frac{dT}{dR} = P(T_{cl}) + 4\pi r_o^2 a\sigma T_{cl}^4\,,$$

where the left-hand side of the equation corresponds to the power of heat transport in a gas at a distance R from a cluster as a result of its thermal conductivity. The first term on the right-hand side of this equation corresponds to the power of the combustion process, and the second term accounts for the radiation of the burning particle, so that a is the gray coefficient of its surface and σ is the Stefan–Boltzmann constant. From this equation we find that the following part of the released energy is consumed on radiation

$$\eta = 4\pi r_o^2 a\sigma T_{cl}^4 / P(T_{cl})\,.$$

It is possible that $\eta \geq 1$ if the energy is taken from a gas. Because the power of the combustion process is $P(T_{cl}) \sim \exp(-E_a/T_{cl})$, the contribution of radiation to the released energy drops with an increase in the gas temperature in the diffusion regime considered.

Problem 4.33

The process of combustion of a porous organic particle in a gaseous mixture containing oxygen proceeds over all the particle's volume. Determine the gas temperature at which all the released energy is consumed in radiation of the particle.

In this case the gas temperature is equal to the particle's temperature, so the particle and the surrounding gas do not exchange energy. Let us write the balance equation under these conditions. The energy of the particle radiation is

$$P = 4\pi r_o^2 \cdot a\sigma T_{cl}^4\,,$$

Table 4.3 The particle's temperature at a given particle radius if the energy, resulted from combustion of activated coal, is transformed into particle's radiation. Parameters of combustion are given by formula (4.46).

r_o, μm	1	3	10	30	100
T, K	1620	1390	1220	1100	990

where a is the gray coefficient of the particle surface and σ is the Stefan–Boltzmann constant. The power released as a result of combustion of the particle is

$$P = \frac{4\pi}{3} r_o^3 p \,, \quad p = p_o \exp\left(-\frac{E_a}{T}\right) ,$$

where p is the specific power of the combustion process, that is, the power released per unit volume, and E_a is the activation energy of the combustion process. From the equality of these powers we obtain the connection between the particle's radius and temperature under the conditions of the combustion regime considered:

$$r_o = \frac{3a\sigma T_{cl}^4}{p_o} \exp\left(-\frac{E_a}{T_{cl}}\right) .$$

Table 4.3 contains the connection between the particle's radius and temperature for these parameters.

Problem 4.34

Derive the criterion of the kinetic regime for the combustion process of a porous spherical coal particle in a gaseous mixture containing oxygen.

In the case of the diffusion regime of combustion, the combustion power is restricted by the diffusion flux of oxygen molecules to the burning particle, and all these oxygen molecules are further used for combustion of the particle. The flux of oxygen molecules is equal to $j = -DN\nabla c$, where D is the diffusion coefficient for oxygen molecules and N is the total number density of gas molecules. Using the formula for the concentration of active molecules at a distance R from the particle in the diffusion regime $c(R) = c_\infty(1 - r_o/R)$, where c_∞ is the oxygen molecule concentration far from the particle, we obtain for the power of heat release in the diffusion regime of combustion

$$P_{\text{dif}} = 4\pi r^2 |j| \Delta\varepsilon = 4\pi r_o D N c_\infty \Delta\varepsilon \,,$$

where $\Delta\varepsilon \approx 4\,\text{eV}$ is the energy released per oxygen molecule in the case of the total combustion of the fuel. For the kinetic regime, the power of heat release for a porous particle is

$$P_{\text{kin}} = \frac{4\pi}{3} r_o^3 p_o \exp\left(-\frac{E_a}{T}\right) ,$$

where $p = p_o \exp(-E_a/T)$ is the power of heat release per unit volume. The kinetic regime is realized if the heat release process is restricted by the rate of chemical reactions, so $P_{\text{kin}} \ll P_{\text{dif}}$. This regime occurs at small particle sizes and in this case the gas temperature satisfies the criterion

$$r_o \ll R_o = \exp\left(\frac{E_a}{2T_{\text{cl}}}\right)\sqrt{\frac{3DNc_\infty \Delta\varepsilon}{p_o}},$$

where T_o is the temperature of the burning particle. In particular, for combustion of porous coal in air at atmospheric pressure ($c_\infty = 0.21$) we obtain from this formula the following values of the critical particle sizes on the basis of the combustion rates for porous birchen coal (4.46): $R_o = 10\ \mu\text{m}$ for $T = 1910\ \text{K}$, $R_o = 20\ \mu\text{m}$ for $T = 1510\ \text{K}$, and $R_o = 30\ \mu\text{m}$ for $T = 1230\ \text{K}$.

Problem 4.35

Organic clusters are burned in air, and this process proceeds in the kinetic regime. An increase in air temperature as a result of the combustion process leads to propagation of the thermal wave in neighboring air regions. Estimate the parameters of the thermal wave accounting for the activation character of the combustion process and assuming the complete combustion of a fuel in this process.

Let us denote by E_a the activation energy of the total combustion process; T_i is the initial air temperature, T_f is its final temperature after burning of all clusters, χ is the air thermal diffusivity coefficient, N_{cl} is the number density of clusters, and r_o is their radius. The cluster number density is relatively small, so that

$$N_{\text{cl}} r_o^3 \ll 1 .$$

Under these conditions, the main contribution to the air heat capacity is from regions located far from the clusters compared with their size r_o at the beginning. For simplicity, we take the cluster radius to be identical for all clusters, so this parameter varies in the same way for different clusters at a given space point in the course of their burning. Then the air temperature T is connected with a current radius r by the relation

$$T = T_f - (T_f - T_i) \cdot \left(\frac{r}{r_o}\right)^3 ,$$

where r_o is the initial cluster radius.

We consider the combustion regime when the combustion process proceeds in the form of a thermal wave and heat transport from a region where the combustion process is finished to a cold region results from the gas thermal conductivity. An increase in the air temperature due to heat transport leads to acceleration of the combustion process in this region. Hence, the combustion process develops as a result of heat transport, and this process propagates through space in the form of a thermal wave.

The parameters of the thermal wave of combustion are Δx, the width of a transition zone between cold and hot regions of a given space, and u, the wave speed. The heat balance equation that accounts for the combustion process and heat transport has the form

$$\frac{\partial T}{\partial t} + \chi \frac{\partial^2 T}{\partial x^2} = \frac{P}{C_p} .$$

Here x is the direction of wave propagation, $\chi = \kappa / C_p$ is the air thermal diffusivity coefficient, P is the power per unit volume resulting from the combustion of individual clusters, and C_p is the air heat capacity. For simplicity, we assume that clusters provide a small contribution to the heat capacity of air with clusters. Representing this process in the form of a wave $T = T(x - ut)$, we get

$$-u \frac{dT}{dx} + \chi \frac{d^2 T}{dx^2} = \frac{P}{C_p} . \tag{4.49}$$

The solution of this equation allows one to determine the profile of the air temperature $T(x - ut)$ and the wave speed u, which is the eigenvalue of this equation. We use the activation character of the combustion process according to formula (4.40) and consider on the kinetic regime of combustion of an individual cluster assuming the temperatures of air and clusters to be identical. In the course of the combustion process the air temperature $T < T_{cr}$ during the basic time of combustion, where T_{cr} corresponds to the threshold of the thermal instability in accordance with formula (4.43). A typical time τ of cluster burning of a particle depends on its typical temperature T_o through the Arrhenius dependence (4.40):

$$\tau = \tau_o \exp \left(\frac{E_a}{T_o} \right) .$$

On the basis of formula (4.23) for the rate of atom collision with the cluster surface in the kinetic regime we have

$$\frac{1}{\tau_o} = \pi r_o^2 \sqrt{\frac{8 T_o}{\pi m}} \cdot [O_2] \xi_o \frac{\Delta \varepsilon}{P} \backsim \frac{1}{r_o} .$$

Here m is the mass of the oxygen molecules, $[O_2]$ is the number density of oxygen molecules, $\Delta \varepsilon$ is the released energy per oxygen molecule, the probability of reaction of an oxygen molecule and a cluster ξ after their contact as determined in accordance with formula (4.40) is

$$\xi = \xi_o \exp \left(-\frac{E_a}{T_o} \right) ,$$

and $P \sim r_o^3$ is the total power of the energy released per cluster.

On the basis of these relations we rewrite equation (4.49) in the form

$$-u \frac{dT}{dx} + \chi \frac{d^2 T}{dx^2} = \frac{T_f - T_i}{\tau_o} \exp \left(-\frac{E_a}{T_o} \right) .$$

From this we get an estimation for the speed u of the thermal wave [17, 146]

$$u \sim \frac{\Delta x}{\tau_o} \exp\left(-\frac{E_a}{T_o}\right) \sim \frac{\chi}{\Delta x},$$

where Δx is the width of the transition range from initial to final system parameters. This gives the following estimations for the parameters of the thermal wave [17, 146]:

$$u \sim \sqrt{\frac{\chi}{\tau_o}} \exp\left(-\frac{E_a}{2T_o}\right), \quad \Delta x \sim \sqrt{\chi \tau_o} \exp\left(\frac{E_a}{2T_o}\right). \tag{4.50}$$

5
Clusters in External Fields

We consider two cases of cluster interaction with an external field, when a cluster is located in a constant electric field and when a cluster is located in the field of an electromagnetic wave that may be absorbed as a result of a transition between cluster states. These interactions are different for metal and dielectric clusters, and according to the definition, a metal cluster has a continuous spectrum. Similar to this definition of a bulk metal for which the Mott–Hubbard correlation energy $U = I - EA$ is zero (I is the ionization potential, EA is the electron affinity) [94–98], for metal clusters the energy of electron transition between lowest unoccupied molecular orbital (LUMO) and highest occupied molecular orbital (HOMO) is zero [99]. Some clusters are dielectric when they are small and metallic when large. In particular, the transition from dielectric to metallic for mercury clusters proceeds at $n \approx 400$ cluster atoms [103].

Macroscopic models for clusters are suitable for analyzing the behavior of clusters in external fields. The clusters are modeled as bulk particles with a macroscopic dielectric constant. This approach is better for metal clusters since an external field is shielded at practically atomic distances because of the high electron density inside clusters. For dielectric clusters the intermediate region outside and inside the cluster exceeds the atomic size, and the possibility of using macroscopic models for clusters requires additional analysis.

Applying macroscopic models to a cluster located in the field of an electromagnetic wave or light wave, we have that a cluster is small compared to the wavelength of the radiation. Therefore, the cross section of light scattering is proportional to r_0^6, where r_0 is the cluster radius and the cross section of light absorption is proportional to r_0^3, that is, to the cluster volume [48]. In addition, the basis of this process results from the interaction of an electromagnetic wave with individual atoms, with the correction on the partaking of surrounding atoms, or from the interaction of an electromagnetic wave with a cluster plasma. The choice between these alternative versions can follow from a comparison with an experiment. If clusters absorb radiation effectively, hot clusters become effective light sources.

Cluster Processes in Gases and Plasmas. Boris M. Smirnov

ISBN: 978-3-527-40943-3

5.1
Electric Properties of Large Clusters

A small particle consists of many atoms, which provide its strong interaction with external fields. In electric fields, a dipole moment is induced on a small particle, and its value is proportional to the particle's volume, or to the number of atoms that constitute the particle. Because a small particle consists of a large number of atoms, the induced electric dipole moment of the particle may not be small even at low electric fields. This determines the particle's behavior in a plasma and the interaction between small particles located in external fields. Below we evaluate the electric parameters of small particles under various conditions.

Problem 5.1

Estimate the polarizability of a small metal particle of size l.

Take the particle's size in the field direction to be of order l and the electric field strength to be E. Under the action of the electric field, a charge q is induced on opposite sides of the particle, and the interaction force of induced charges is of order q^2/l^2. This force is compensated by the force of interaction of charges with the electric field qE. From this we obtain the value of the induced charge $q \sim El^2$. Introducing the polarizability α of the particle on the basis of the relation

$$\mathbf{D} = \alpha \mathbf{E} , \tag{5.1}$$

where $D \sim ql$ is the particle's dipole moment, we obtain the following estimation for the particle's polarizability:

$$\alpha \sim l^3 . \tag{5.2}$$

We assume the size of the particle in three orthogonal directions to be of the same order of magnitude (of the order of l).

Problem 5.2

Determine the polarizability of a metal cluster within the framework of a liquid spherical drop.

Let us consider the case of a weak electric field of strength E and of a high conductivity of a cluster material. The charge of the cluster surface is induced by the external electric field and creates a cluster dipole moment. The electric field potential outside the cluster is

$$\varphi = \varphi' - \mathbf{ER} ,$$

where φ' is the electric potential created by the surface cluster charge, the second term corresponds to the electric potential of the external electric field, and $\mathbf{R}$ is a point coordinate if the origin of the frame of reference is located at the cluster center. Because of the absence of charge outside the cluster, the Poisson equation for the electric potential outside the cluster has the form

$$\Delta\varphi' = 0 .$$

Solving this equation and taking into account that the electric potential drops far from the cluster, we get

$$\varphi' = \sum_{l=1}^{\infty} \frac{A_l}{R^l} P_l(\cos\theta) ,$$

where R, θ are the spherical coordinates of a point where the direction of the electric field passes through the cluster center and is the polar axis of the frame of reference. Because the cluster charge is zero, the term with $l = 0$ is absent in this sum. The condition that the total electric potential is zero at the cluster surface gives the following coefficients for this expansion:

$$A_l = E r_o^3 \delta_{l1} ,$$

where δ_{ik} is the Kronecker symbol and r_o is the cluster radius. Comparing the electric potential of the cluster electric field with that of the dipole moment D far from it ($\varphi' = \mathbf{DR}/R^3$), we obtain for the cluster dipole moment ($\mathbf{D} = \mathbf{E} r_o^3$) and the cluster polarizability α

$$\alpha = r_o^3 . \tag{5.3}$$

Let us find the surface charge density σ induced by the electric field on the cluster surface. We have for the dipole moment

$$D = \int r\cos\theta \cdot \sigma d\mathbf{s} ,$$

where $d\mathbf{s} = 2\pi r^2 d\cos\theta$ is the area of a cluster surface element. Since the cluster surface charge is induced by the external field, its angle dependence has the form $\sigma = \sigma_o \cos\theta$. This gives

$$D = \frac{4\pi}{3} r_o^3 \sigma_o = E r_o^3 , \quad \sigma = \frac{3}{4\pi} E\cos\theta .$$

The above results are valid in the limit of a small electric field strength. Denoting by N_e the number density of conductive electrons of the metal cluster, we have that the induced surface cluster charge must be considerably less than $e N_e r$, where e is the electron charge, so

$$E \ll e N_e r_o .$$

Problem 5.3

Determine the polarizability of a small spherical dielectric cluster modeling it by a uniform spherical drop.

The Poisson equation for the electric potential outside and inside the cluster is $\Delta\varphi = 0$, and the boundary condition on the cluster surface has the form

$$\varepsilon \frac{\partial\varphi}{\partial R}(R \to r_o - 0) = \frac{\partial\varphi}{\partial R}(R \to r_o + 0) ,$$

where R is the distance from the cluster, r_o is the cluster radius, and ε is the dielectric constant of the cluster material. The electric potential outside the cluster is

$$\varphi = -\mathbf{ER} + \frac{\mathbf{DR}}{R^3} .$$

Here the first term is due to an external electric field, and the second term is determined by an induced surface cluster charge. Because the electric potential inside the cluster is restricted, the solution of the Poisson equation $\Delta\varphi = 0$ inside the cluster may be represented in the form

$$\varphi = C\mathbf{ER} ,$$

where C is a numerical coefficient. This coefficient and the cluster polarizability can be found from the continuity condition for the electric potential φ, and the electric induction $\varepsilon\mathbf{E}$ arises when we intersect the cluster surface. This yields

$$C = \frac{3\varepsilon}{\varepsilon + 2} , \quad \alpha = \frac{\varepsilon - 1}{\varepsilon + 2} r^3 . \tag{5.4}$$

In the case $\varepsilon = 1$, these equations give $C = 1, \alpha = 0$ because the cluster is uniform and surface charges are absent. The other limit relates to a metal cluster $\varepsilon \to \infty$ that corresponds to a high density of internal charges.

The above equations can be used for an alternative electric field if a time of variation of the electric field is large compared with a time of movement of internal electrons. In particular, if a cluster is located in a harmonic electric field $\mathbf{E}\cos\omega t$, the polarizability can be represented in the form

$$\alpha(\omega) = \frac{\varepsilon(\omega) - 1}{\varepsilon(\omega) + 2} r_o^3 , \tag{5.5}$$

and this equation requires the validity of the condition

$$\omega \ll \Sigma , \quad \alpha = r_o^3 , \tag{5.6}$$

where Σ is the conductivity of the cluster's material.

Problem 5.4

Determine the charge of a small spherical dielectric cluster that is located in a unipolar plasma in a constant electric field.

Ions of a unipolar plasma are captured by a cluster until the cluster electric field starts to be directed away from the cluster. As a result, the lines of force will not pass through the cluster, and ions moving along the lines of force will not be captured by the cluster. The smaller electric field strength corresponds to a point of the cluster surface for which the radius vector from the cluster center is directed opposite the electric field. The electric field strength at this point is zero.

We have for the cluster charge q the relation

$$\frac{q}{r_o^2} = E' ,$$

where E' is the electric field strength resulting from the electric field $\mathbf{E}$ of an external source. Using formula (5.4), we have

$$E' = \frac{3\varepsilon}{\varepsilon + 2} E , \quad q = E r_o^2 \frac{3\varepsilon}{\varepsilon + 2} . \tag{5.7}$$

Problem 5.5

A chain aggregate is a system of several bound solid particles (Figure 5.1 [100]). Taking a cylindrical particle as the model of a chain aggregate, determine the polarizability of a chain aggregate and the distribution of induced charges over the aggregate surface when the electric field is directed perpendicular to the cylindrical axis. Assume that the aggregate is similar to a metallic object and that its length greatly exceeds the cylinder radius.

Under these conditions, the electric potential of the cylinder is constant. Using the cylindrical coordinates z, ρ, Φ, we have for the electric potential outside the

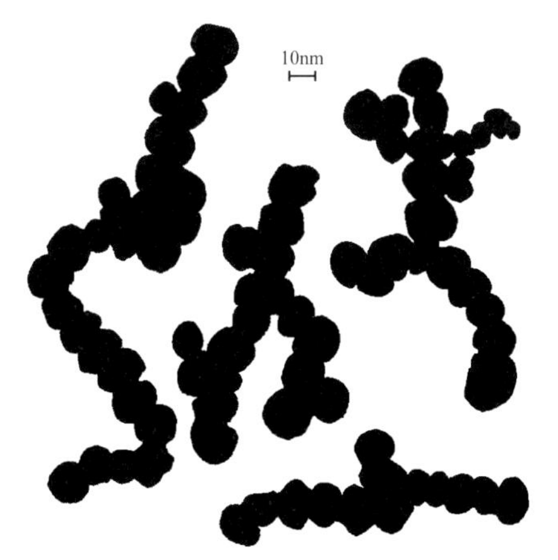

Fig. 5.1 Chain aggregates formed as a result of combustion of magnesium dust [100].

particle

$$\varphi = -E\rho \cos \Phi + \varphi' \,,$$

where φ' is the electric potential created by induced charges of the cluster under the action of an external electric field of strength E directed along the x axis (perpendicular to the cylindrical z axis). Because of the absence of charges outside the cluster, this potential satisfies the Laplace equation

$$\Delta\varphi' = 0 \,.$$

The solution of this equation outside the cluster for which the potential drops far from the particle is as follows:

$$\varphi' = \sum_{m=-\infty}^{\infty} A_m e^{im\varphi} \rho^{-|m|} \,.$$

Taking the cylinder's length to be large compared with its radius, we obtain for the potential of its surface at points that are not close to its edges

$$\varphi' = \frac{r^2}{\rho} E \cos \Phi \,.$$

We take the total electric potential on the cluster's surface to be zero and require the electric potential of induced charges to drop far from the particle. Thus, we have for the surface induced charge of the particle

$$\sigma = \frac{1}{4\pi} \frac{\partial \varphi'}{\partial r} = \frac{E}{2\pi} \cos \Phi \,.$$

This leads to the following dipole moment of the particle:

$$D = \int_0^{2\pi} d\Phi \int_{-l}^{l} dz \cdot r_o \frac{E \cos \Phi}{2\pi} \cdot r_o \cos \Phi = E r_o^2 l \,,$$

where $2l$ is the length of the cylinder. From this we have for the particle's polarizability

$$\alpha = \frac{D}{E} = r_o^2 l \,. \qquad (5.8)$$

Because we neglect the edge effects in deriving this equation, it is valid in the limit $l \gg r_o$.

Problem 5.6

Determine the polarizability of a chain aggregate modeling it by a cylindrical metallic particle if the electric field is directed along its axis and the cylinder length greatly exceeds the cylinder radius.

We take as before the cylinder's length and radius to be $2l$ and r_o, respectively. Using cylindrical coordinates with the origin at the cylinder's center of mass, we take for the electric potential on the cylinder's surface

$$\varphi(z) = \int_0^{2\pi} d\Phi' \int_{-l}^{l} dz' \cdot r\sigma(z') \cdot \left[(z - z')^2 + 4r^2 \sin^2 \frac{\Phi'}{2}\right]^{-1/2} - Ez ,$$

where r, z', Φ' are the cylindrical coordinates of the cylinder's surface, $r_o d\Phi' dz'$ is the surface element, and $\sigma(z)$ is an induced charge per unit surface area. We use the axial symmetry of the problem so that the electric potential does not depend on the azimuthal angle. Hence, we evaluate the electric potential at $\Phi = 0$.

Taking the surface electric potential to be zero, we get the equation for the surface density of induced charges $\sigma(z)$

$$Ez = \int_0^{2\pi} d\Phi' \int_{-l}^{l} dz' \cdot r_o\sigma(z') \cdot \left[(z - z')^2 + 4r^2 \sin^2 \frac{\Phi'}{2}\right]^{-1/2} . \tag{5.9}$$

We now assume the condition $l \gg r_o$ to be valid. We have that the integral in the above equation is determined by $|z - z'| \sim r_o$, whereas the surface charge density $\sigma(z)$ varies as a result of a shift in z by a typical value of approximately l. This allows us to replace $\sigma(z')$ in the above equation by $\sigma(z)$. Thus, we have

$$Ez = r_o\sigma(z) \cdot \int_0^{2\pi} d\Phi' \int_{-l}^{l} dz' \cdot \left[(z - z')^2 + 4r_o^2 \sin^2 \frac{\Phi'}{2}\right]^{-1/2}$$
$$\approx 2\pi r_o\sigma(z) \cdot 2\ln(l/r_o) ,$$

where we assume the value $\ln(l/r_o)$ to be large. This gives for the charge density $\sigma(z)$ per unit surface area and the charge density dq/dz per unit length

$$\sigma(z) = \frac{Ez}{4\pi r_o \ln\left(\frac{l}{r_o}\right)} , \quad \frac{dq}{dz} = 2\pi r_o\sigma(z) = \frac{Ez}{2\ln\left(\frac{l}{r_o}\right)} . \tag{5.10}$$

From this we get for the dipole moment of the cylindrical particle

$$D = \int_0^{2\pi} d\Phi \int_{-l}^{l} r_o dz \cdot z\sigma(z) = \frac{El^3}{3\ln(l/r_o)} ,$$

so we have for the particle's polarization in the direction of the electric field [48]

$$\alpha = \frac{D}{E} = \frac{l^3}{3\ln\left(\frac{l}{r_o}\right)} . \tag{5.11}$$

A more correct accounting for edge effects leads to a change the value $\ln(l/r_o)$ by $\ln(l/r_o) + \text{const}$. Let us make a more accurate calculation of the polarizability of a long cylindrical metallic particle if the electric field is directed along its axis. We repeat the above operations using the following order of expansion over a small parameter $1/\ln(l/r_o)$. On the basis of condition (5.9) that the potential on the surface of the particle is zero, we rewrite the equation for the surface potential in the form

$$Ez = \int_0^{2\pi} d\Phi' \int_{-l}^{l} \frac{dz' \cdot r_o\sigma(z)}{\sqrt{(z-z')^2 + 4r_o^2 \sin^2 \frac{\Phi'}{2}}} + \int_0^{2\pi} d\Phi' \int_{-l}^{l} \frac{dz' \cdot r_o\left[\sigma(z') - \sigma(z)\right]}{\sqrt{(z-z')^2 + 4r_o^2 \sin^2 \frac{\Phi'}{2}}} .$$

We take the first integral, accounting for $\int_0^{2\pi} d\Phi' \ln \sin \Phi' = -\pi \ln 2$, so that

$$\int_0^{2\pi} d\Phi' \int_{-l}^{l} \frac{dz' \cdot r_o\sigma(z)}{\sqrt{(z-z')^2 + 4r_o^2 \sin^2 \frac{\Phi'}{2}}} = 2\pi r_o \sigma(z) \ln\left[\frac{4(l^2 - z^2)}{r_o^2}\right] .$$

In the second integral we neglect terms proportional to r_o and take into account that in the main region of the surface $\sigma(z)/z = \text{const}$. Then the second integral becomes

$$\int_0^{2\pi} d\Phi' \int_{-l}^{l} \frac{dz' \cdot r_o\left[\sigma(z') - \sigma(z)\right]}{|z-z'|} = 2\pi r_o \frac{\sigma(z)}{z} \int_{-l}^{l} \frac{dz' \cdot (z'-z)}{|z-z'|} = -4\pi\sigma(z) .$$

Thus, we have

$$2\pi r_o \sigma(z) = Ez\left[\ln\frac{4(l^2 - z^2)}{r_o^2} - 2\right]^{-1} .$$

This gives

$$D = \int_{-l}^{l} z' dz' \cdot 2\pi r_o \sigma(z') = \frac{El^3}{3}\left[\ln\left(\frac{4l}{r_o}\right) - \frac{7}{3}\right]^{-1} ,$$

for the dipole moment of the particle, if we account for the first two terms of the expansion over a small parameter $\ln(l/r_o)$, that is, the polarizability of the cylindrical metallic particle is [48]

$$\alpha = \frac{l^2 r_o}{3}\left[\ln\left(\frac{4l}{r_o}\right) - \frac{7}{3}\right]^{-1} . \tag{5.12}$$

One can see that this expression differs that given by formula (5.11) by a factor in the argument of the logarithm. This corresponds to taking into account an additional term of expansion over a small parameter $1/\ln(l/r_o)$.

Problem 5.7

Determine the polarizability tensor of a chain aggregate modeling it by a cylinder dielectric particle with dielectric constant ε.

Under the action of an external electric field, an internal electric field is created inside the particle, and the electric induction inside the particle is

$$\varepsilon \mathbf{E}' = \mathbf{E}' + 4\pi \mathbf{P} ,$$

where ε is the dielectric constant of the particle's material and $\mathbf{P}$ is the dipole moment per unit volume. Introducing the particle dipole moment $\mathbf{D}$ and its volume V, we obtain from this relation

$$(\varepsilon - 1)\mathbf{E}' = \frac{4\pi \mathbf{D}}{V} .$$

Because an internal field and the dipole moment are created by the external electric field, these vectors have the same direction, so

$$a\mathbf{E}' + b\mathbf{D} = \mathbf{E} .$$

The coefficients a, b depend on the particle's geometry and do not depend on its dielectric constant because in the above relations the dependence on the dielectric constant is separated. On the basis of this fact, in this relation we take first $\varepsilon = 1$. In this case the polarization of the particle is absent, that is, $\mathbf{D} = 0$, $\mathbf{E} = \mathbf{E}'$. This gives $a = 1$. If we take a metal particle, the internal electric field is absent, and the induced dipole moment is $\mathbf{D} = \alpha_m \mathbf{E}$, where α_m is the polarizability of a metal particle of a given form. This gives $b = 1/\alpha_m$, so the above relation is

$$\mathbf{E}' + \frac{\mathbf{D}}{\alpha_m} = \mathbf{E} .$$

Taking the strength of the internal electric field from the relation between this value and the particle dipole moment, we get

$$\left[\frac{4\pi}{V(\varepsilon - 1)} + \frac{1}{\alpha_m} \right] \mathbf{D} = \mathbf{E} .$$

From this it follows for the particle polarizability $\alpha(\mathbf{D} = \alpha \mathbf{E})$ that

$$\alpha = \left[\frac{4\pi}{V(\varepsilon - 1)} + \frac{1}{\alpha_m} \right]^{-1} . \tag{5.13}$$

In particular, formula (5.13) yields for a spherical particle ($V = 4\pi r_o^3/3$, $\alpha_m = r_o^3$) the following expression for the polarizability:

$$\alpha = r_o^3 \frac{\varepsilon - 1}{\varepsilon + 2} .$$

This coincides with formula (5.4). In this case of a cylindrical dielectric particle, formula (5.13) gives for the component of the polarizability tensor directed perpendicular to the cylindrical axis ($\alpha_m = r_o^2 l$, $V = \pi r_o^2 L$)

$$\alpha_\perp = r_o^2 l \frac{\varepsilon - 1}{\varepsilon + 2} . \tag{5.14}$$

In the case where the electric field is directed along the cylindrical axis, for the relation between the cylindrical parameters $l \gg r_o$, from formula (5.11) we have $\alpha_m \gg V\varepsilon$. This allows us to neglect the second term in formula (5.13), and this polarizability component is

$$\alpha_\parallel = V \cdot \frac{\varepsilon - 1}{4\pi} = r_o^2 l \frac{\varepsilon - 1}{2} . \tag{5.15}$$

Problem 5.8

Modeling a chain aggregate by a metal cylinder, determine the average polarizability of a chain aggregate in a uniform electric field if the angle between the electric field and the longest aggregate axis is determined by Boltzmann's law.

According to Boltzmann's law, the probability of an angle θ between the electric field and the direction of the cylindrical axis is proportional to the factor $\exp(-U/T)$, where T is the system temperature, and the interaction potential of the cylindrical particle with an electric field is given by

$$U = -\alpha_\parallel E^2 \cos^2\theta/2 - \alpha_\perp E^2 \sin^2\theta/2 .$$

According to formulas (5.8) and (5.11) we have

$$\alpha_\perp/\alpha_\parallel = 3r_o^2 \ln(l/r_o)/(l^2) \ll 1 ,$$

so one can neglect the transverse component of the polarizability. Hence, the average particle polarizability is $\overline{\alpha} = \alpha_\parallel \langle\cos^2\theta\rangle$, where the average is made over angles θ between the direction of the electric field and the particle's axis. Then, for the average particle polarizability according to Boltzmann's law we get

$$\overline{\alpha} = \alpha_\parallel g\left(\frac{\alpha_\parallel E^2}{2T}\right) ,$$

where

$$g(x) = \frac{\int_0^1 d\cos\theta \cdot \cos^2\theta \cdot \exp(x\cos^2\theta)}{\int_0^1 d\cos\theta \cdot \exp(x\cos^2\theta)} .$$

In the limiting case of low temperatures $\alpha_\parallel E^2/(2T) \ll 1$, when the directions of the field and the cylindrical axis are almost coincident, this formula gives $\overline{\alpha} = \alpha_\parallel$. In the other limiting case of high temperatures, when the distribution on angles is isotropic, $\alpha_\parallel E^2/(2T) \ll 1$. From this formula it follows that $\overline{\alpha} = \alpha_\parallel/3$.

Problem 5.9

Analyze the character of interaction of a chain aggregate that is modeled by a cylindrical metal particle with a spherical metal particle of the same radius in a uniform electric field. The interaction potential results from interaction of induced dipole moments, and the distance between particles greatly exceeds their radii.

Under the above conditions, when the distance between particles is large compared with the radius of the spherical particle, one can consider the spherical particle to be located at a point and to have an induced dipole moment $\mathbf{D}_1 = r_o^3\mathbf{E}$, where r_o is the particle radius. Interaction of this dipole moment with an induced charge of the chain aggregate determines the interaction potential of particles. If a length of the chain aggregate is large, the axis is directed along the electric field. We assume this condition to be fulfilled.

According to formula (5.10) the induced surface charge on the cylindrical metal particle is proportional to the distance z from its center. Here we choose the axes such that the origin is the center of the cylindrical particle, the z axis is directed along the axis of the particle and the electric field, and ρ is the distance from the z axis. Let $z = \pm l$ be the poles of the cylindrical particle. Take the induced charge per unit length of the cylindrical particle to be $\sigma = Cz$. Because the dipole moment of the particle is

$$D_2 = \int_{-l}^{l} \sigma z dz\,,$$

we have for the coefficient C the expression

$$C = \frac{3D_2}{2l^3}\,,$$

where $D_2 = \alpha_2 E$ is the induced dipole moment of the cylindrical particle, and according to formula (5.11) the polarizability of the cylindrical particle along its axis is $\alpha_2 = l^3/[3\ln(l/r_o)]$, where r_o is the cylinder radius.

The interaction potential of a charge e and dipole moment $\mathbf{D}$ is $U(R) = e\mathbf{D}\mathbf{n}/R^2$, where R is the distance between these objects and $\mathbf{n}$ is the unit vector along $\mathbf{R}$. In this case we have $R = \sqrt{(z-z')^2 + \rho^2}$, where z' is the charge coordinate along the polar axis and z, ρ are the dipole coordinates. From this we obtain for the interaction potential of the induced charges of a cylindrical particle and the induced dipole moment of a spherical particle that

$$U(R) = \int_{-l}^{l} Czdz\frac{\mathbf{D}_1\mathbf{n}}{R^2} = \frac{3D_2}{2l^3}\int_{-l}^{l} zdz\frac{\mathbf{D}_1\mathbf{n}}{R^2}\,.$$

In particular, in the limiting case $R \gg l$, from this it follows that the interaction potential of the two dipole moments is

$$U(R) = \frac{\mathbf{D}_1\mathbf{D}_2 - 3(\mathbf{D}_1\mathbf{n})(\mathbf{D}_2\mathbf{n})}{R^3}\,.$$

Let us consider the other limiting case $l \gg R$ in the region of attraction near the ends of a cylindrical particle, that is, $z > l$. Then we get the following expression for the interaction potential:

$$U(R) = \frac{3D_2}{2l^3} \int_{-l}^{l} z' dz' \frac{\mathbf{D}_1 \mathbf{n}}{(R')^2} = \frac{3D_1 D_2}{2l^3} \int_{-l}^{l} z'(z' - z) \frac{dz'}{(R')^3} \, . \tag{5.16}$$

In this limiting case the interaction potential is determined by a region of the cylindrical particle near its end. This allows us to replace z' under the integral with l, so we get

$$U(R) = -\frac{3D_1 D_2}{2l^2 R} \, ,$$

where $R = \sqrt{z^2 + \rho^2}$ is the distance of the spherical particle from the end of the cylindrical one, and this equation is valid for $z > l$.

Thus, from formula (5.16) for the interaction potential of induced charges of the spherical and cylindrical metal particles we have at distances between them that are large compared with the particles' radii and small compared with the cylinder's length that the sign of interaction varies as we move to the ends of the cylindrical particle. A rough analysis gives that repulsion of the particles proceeds at $|z| < l$ and attraction takes place at $|z| > l$, where z is the coordinate of the spherical particle along the axis of the cylindrical particle. A general form of the interaction potential of the particles is given in Figure 5.2, where the boundaries between regions of attraction and repulsion are represented.

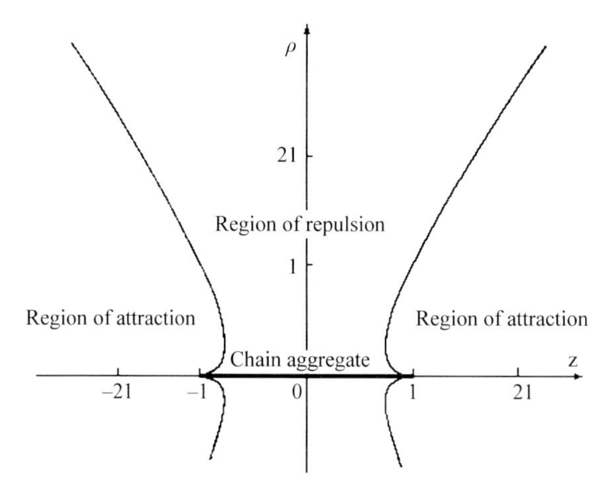

Fig. 5.2 The character of interaction between a cylindrical particle and a spherical one of the same radius. The length of a cylindrical particle is significantly larger than its radius; the cylindrical axis is directed along an external electric field, and the interaction potential results from the interaction of induced charges of particles [17].

5.2 Radiative Processes Involving Small Particles

Problem 5.10

Determine the cross section of scattering of radiation on a small spherical particle.

We base our consideration on a small parameter r_o/λ, where r_o is the particle radius and λ is the wavelength of the radiation. This allows us to restrict the dipole interaction potential between the electromagnetic wave field and the particle (dipole approximation), so the intensity of the radiation resulting from the induced dipole moment $\mathbf{D}$ is given by the vector product

$$dJ = \frac{1}{4\pi c^3}\left[\frac{d^2\mathbf{D}}{dt^2}\right]\cdot \mathbf{n}\,d\Omega\,, \tag{5.17}$$

where $\mathbf{n}$ is the unit vector along the direction of propagation of the radiation and $d\Omega$ is the element of the solid angle. Let us represent the electric field strength of the electromagnet wave in the form

$$\mathbf{E} = \mathbf{E}_o e^{i\omega t} + \mathbf{E}_o^* e^{-i\omega t}\,,$$

where ω is the wave frequency. This gives

$$c[\mathbf{EH}]/(4\pi) = c|E_o|^2/(2\pi)$$

for the specific flux of the incident radiation averaged over a long time compared with the period of oscillations. The dipole moment induced by the electromagnetic field is given by

$$\mathbf{D} = \alpha(\omega)\mathbf{E}_o e^{i\omega t} + \alpha^*(\omega)\mathbf{E}_o^* e^{-i\omega t}\,,$$

where $\alpha(\omega)$ is the particle's polarizability. Using the relation

$$\frac{d^2\mathbf{D}}{dt^2} = -\omega^2\mathbf{D}$$

in formula (5.17) and properties of the vector product, we have

$$[\mathbf{Dn}]^2 = ([[\mathbf{Dn}]\mathbf{D}^*]\mathbf{n}) = |\mathbf{D}|^2 - (\mathbf{Dn})(\mathbf{D}^*\mathbf{n}) = |D|^2\sin^2\theta\,,$$

where θ is the angle between vectors $\mathbf{D}$ and $\mathbf{n}$. Thus, we obtain the cross section of scattering of radiation on the particle as the ratio of the scattering probability dJ per unit time to the flux of incident radiation $c|E_o|^2/(2\pi)$:

$$d\sigma = dJ\cdot\frac{2\pi}{c|E_o|^2}\,.$$

Because

$$|D|^2 = \overline{\left[\alpha(\omega)\mathbf{E}_o e^{i\omega t} + \alpha^*(\omega)\mathbf{E}_o^* e^{-i\omega t}\right]^2} = 2|\alpha(\omega)|^2|E_o|^2\,,$$

where the bar indicates averaging over a long time, we get

$$d\sigma = \left(\frac{\omega}{c}\right)^4 \cdot |\alpha(\omega)|^2 \sin^2\theta \, d\Omega \, .$$

This type of radiation scattering is known as Rayleigh scattering. From this formula it follows that for the total cross section of scattering we have

$$\sigma_{\text{tot}} = \int d\sigma = \frac{8\pi}{3}\left(\frac{\omega}{c}\right)^4 \cdot |\alpha(\omega)|^2 \, . \tag{5.18}$$

In particular, in the case of scattering on a spherical metal particle of radius r_o this formula gives [48]

$$\sigma_{\text{tot}} = \frac{8\pi}{3}\left(\frac{\omega}{c}\right)^4 r_o^6 \cdot \left|\frac{\varepsilon(\omega) - 1}{\varepsilon(\omega) + 2}\right|^2 \, . \tag{5.19}$$

Because $\lambda \sim c/\omega \gg r_o$, the scattering cross section has order $\sigma_{\text{tot}} \sim r_o^2 (r_o/\lambda)^4$.

Problem 5.11

Determine the absorption cross section for a small spherical particle in the dipole approximation.

Because of the small parameter r_o/λ, where λ is the wavelength, the absorption cross section is small compared with the particle cross section. This small parameter allows us to account for the dipole interaction of an electromagnetic field and particle only. The interaction potential between the induced dipole moment **D** and the field of an electromagnetic wave is $-\mathbf{ED}$, where **E** is the wave electric field strength, and the power absorbed by the particle is given by

$$P = -\left\langle \mathbf{E}\frac{d\mathbf{D}}{dt} \right\rangle ,$$

where brackets indicate averaging over a long time compared with the period of wave oscillations.

Take the electric field strength of a monochromatic wave as we did earlier in the form

$$\mathbf{E} = \mathbf{E}_o e^{i\omega t} + \mathbf{E}_o^* e^{-i\omega t} ,$$

where ω is the wave frequency. This gives

$$\mathbf{D} = \alpha(\omega)\mathbf{E}_o e^{i\omega t} + \alpha^*(\omega)\mathbf{E}_o^* e^{-i\omega t} ,$$

for the induced dipole moment by the electromagnetic field, where $\alpha(\omega)$ is the particle's polarizability. From this it follows that for the absorbed power,

$$P = i\omega \mid E_o \mid^2 \left[\alpha^*(\omega) - \alpha(\omega)\right] .$$

Dividing this value by the radiation flux $c|E_o|^2/(2\pi)$, we obtain

$$\sigma_{abs} = 4\pi \frac{\omega}{c} \operatorname{Im} \alpha(\omega) , \tag{5.20}$$

for the absorption cross section of the particle. Using formula (5.5) for the polarizability of a spherical particle, we obtain

$$\begin{aligned} \sigma_{abs}(\omega) &= \frac{12\pi\omega}{c} \frac{\varepsilon''}{(\varepsilon'+2)^2 + (\varepsilon'')^2} r_o^3 \\ &= \frac{\pi\omega}{c} r_o^3 g_{sph}(\omega) , \quad g_{sph}(\omega) = \frac{12\varepsilon''}{(\varepsilon'+2)^2 + (\varepsilon'')^2} , \end{aligned} \tag{5.21}$$

where the dielectric constant of the particle's material is taken in the form $\varepsilon(\omega) = \varepsilon'(\omega) + i\varepsilon''(\omega)$. As is seen, the absorption cross section is

$$\sigma_{abs} \sim (r_o/\lambda) r_o^2 ,$$

that is, it is small compared with the particle cross section πr_o^2 and is large compared with the cross section (5.19) of wave scattering. The absorption cross section is proportional to the particle's volume r_o^3, or to the number of atoms that constitute the particle.

Problem 5.12

Determine the absorption cross section for a chain aggregate, modeling it by a cylindrical dielectric particle of length $2l$ and radius r_o such that $l \gg r_o$. Assume the wavelength of the radiation to be large compared to the particle's size.

Use formula (5.20) for the absorption cross section of a small particle and formula (5.13) for the polarizabilities of the cylindrical dielectric particle. If the particle's axis forms an angle θ with the direction of the radiation polarization, the particle's polarizability is given by

$$\alpha = \alpha_{\parallel} \sin^2\theta + \alpha_{\perp} \sin^2\theta .$$

Averaging over this angle, we get for the mean absorption cross section on the basis of formula (5.13)

$$\sigma_{abs} = 4\pi \frac{\omega}{c} \operatorname{Im} \alpha(\omega) = \frac{4\pi}{3} \frac{\omega}{c} \operatorname{Im}\left[\alpha_{\parallel}(\omega) + 2\alpha_{\perp}(\omega)\right] = \frac{2\pi}{3} \frac{\omega}{c} r_o^2 l \cdot g_{cyl}(\omega), \tag{5.22}$$

where

$$g_{cyl}(\omega) = \varepsilon'' + \frac{8\varepsilon''}{(\varepsilon'+1)^2 + (\varepsilon'')^2} . \tag{5.23}$$

Problem 5.13

Connect the absorption cross section of a small particle with the area of its surface and the gray coefficient of the surface if the particle size exceeds remarkably the wavelength of the radiation.

We have a spherical particle of large radius $r_o \gg \lambda$, where λ is the wavelength of the radiation, and this particle is located in a cavity surrounded by walls, so the blackbody radiation in thermal equilibrium with the walls is also located in this cavity. If the equilibrium is supported between the particle and the blackbody radiation, the absorbing spectral power is equal to $\pi r_o^2 \hbar\omega\, i(\omega)$, where $i(\omega)$ is the random photon flux of blackbody radiation inside the cavity. The particle emits spectral power $4\pi r_o^2 \hbar\omega \cdot j(\omega)$, where $j(\omega)$ is the radiation flux from the surface of the bulk blackbody. From the equality of radiation powers we obtain

$$j(\omega) = \frac{1}{4} i(\omega) . \tag{5.24}$$

Now let us consider a particle of any form that is located in the space where blackbody radiation propagates. From the equilibrium between radiation and absorption processes we have

$$\int i(\omega)\sigma_{\text{abs}} \frac{d\Omega}{2\pi} = j(\omega) S ,$$

where σ_{abs} is the absorption cross section, $d\Omega$ is the solid angle element that characterizes the photon direction with respect to the particle's surface, and S is the area of the particle's surface. From this it follows that

$$\overline{\sigma_{\text{abs}}} = \int \sigma_{\text{abs}} \frac{d\Omega}{2\pi} = \frac{S}{4} . \tag{5.25}$$

This result for a spherical particle of a blackbody material has the simple form $\overline{\sigma_{\text{abs}}} = S/4 = \pi r_o^2$. The latent assumption in the course of the deduction of this formula is such that the resultant flux of radiation at each surface point is directed perpendicular to the surface. It is valid if the depth of the surface layer that is responsible for emitting and absorbing radiation at the surface is small compared with the radius of curvature of the surface.

Above we considered a particle that absorbs and emits radiation as a blackbody. In the general case let us introduce the gray coefficient $a(\omega)$, which is the mean probability of absorption of a photon that is scattered on the particle's surface. Then formula (5.25) takes the form

$$\overline{\sigma_{\text{abs}}} = \int \sigma_{\text{abs}} d\Omega/(2\pi) = \frac{1}{4} S a(\omega) . \tag{5.26}$$

Problem 5.14

Determine the spectral power of radiation of a small particle of temperature T.

Let us use Kirchhoff's law, which establishes the connection between rates of absorption and emission of radiation. Kirchhoff's law is similar to the principle of detailed balance for rates of emission and absorption of radiation that under thermal equilibrium leads to the equilibrium parameters of radiation near this object. According to the Planck formula [101], this leads to the following formula for the equilibrium random photon flux $i(\omega)$ when this radiation is in equilibrium with the environment having a temperature T [39, 102],

$$i(\omega) = \frac{\hbar\omega^2}{\pi^2 c^2}\left[\exp\left(\frac{\hbar\omega}{T}\right) - 1\right]^{-1} , \tag{5.27}$$

and the random photon flux $i(\omega)/4$ is connected by formula (5.24) with the radiation flux from the surface of a bulk blackbody. In the limit of small and large frequencies this formula is transformed into the Rayleigh–Jeans formula [104–106] and the Wien formula [107], respectively.

We now use Kirchhoff's law. Under equilibrium conditions this leads to the following connection between the spectral power of radiation $p(\omega)$ and the absorption cross section for a small particle $\sigma_{\text{abs}}(\omega)$:

$$p(\omega) = \hbar\omega \cdot i(\omega)\sigma_{\text{abs}}(\omega) .$$

Here $i(\omega)$ is the random photon flux of blackbody radiation inside the space where this radiation propagates. In the equilibrium case we have for the spectral power of radiation of a small particle

$$p(\omega) = \frac{\hbar\omega^3}{\pi^2 c^2}\sigma_{\text{abs}}(\omega)\left[\exp\left(\frac{\hbar\omega}{T}\right) - 1\right]^{-1} . \tag{5.28}$$

Note that formula (5.28) does not require a thermal equilibrium for radiation, which was used as a method for its deduction.

Problem 5.15

Determine the temperature dependence for the power of radiation of a small spherical particle.

The radiation power of a small particle differs from that for a bulk surface because of another relation between the particle size and the wavelength of the radiation [108]. Indeed, we have for the radiation power of a small spherical particle

$$P = \int_0^\infty p(\omega)d\omega = \int_0^\infty \sigma_{\text{abs}}(\omega)\hbar\omega \cdot i(\omega)d\omega$$
$$= 12\pi \int_0^\infty \frac{\omega}{c} r^3 g_{\text{sph}}(\omega)\hbar\omega\, i(\omega)d\omega ,$$

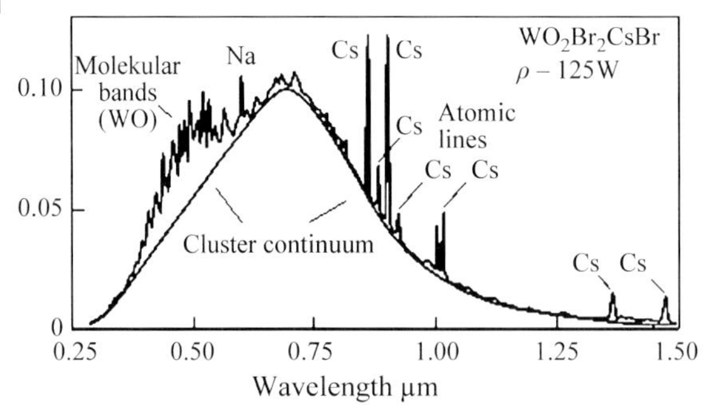

Fig. 5.3 Spectrum of radiation of a cluster plasma resulting from excitation WO_2Br_2 with an admixture of CsBr by microwave discharge of 125 W power [109], so tungsten clusters are responsible for the continuous radiation spectrum of this plasma.

where the parameter $g_{sph}(\omega)$ for the spherical particle is given by formula (5.21). Let us assume this parameter is independent of the radiation wavelength in the wavelength range of the maximum of the radiation. In particular, Figure 5.3 demonstrates the reality of this assumption. This dependence leads to the following expression for the radiation power:

$$P = \frac{\pi}{\hbar c} r_o^3 g_{sph} \sigma T^5 k \,, \quad \frac{T r_o}{\hbar c} \ll 1 \,, \tag{5.29}$$

where σ is the Stefan–Boltzmann constant, and the numerical coefficient k is

$$k = \frac{\int_0^\infty x^4 dx (e^x - 1)^{-1} dx}{\int_0^\infty x^3 dx (e^x - 1)^{-1} dx} = 3.83,$$

so that formula (5.29) takes the form

$$P = \frac{3.8\pi r_o^3 g_{sph} \sigma T^5}{\hbar c} = C g_{sph} r_o^3 T^5 \,, \quad C = \frac{3.8\pi\sigma}{\hbar c} \,, \quad \frac{T r_o}{\hbar c} \ll 1 \,, \tag{5.30}$$

where the numerical coefficient is $C = 3.0 \cdot 10^{-10}\,\mathrm{W/(cm^3K^5)}$. As is seen, the temperature dependence for the radiation power (proportional to T^5) of a small particle differs from the Stefan–Boltzmann law (proportional to T^4), which applies to bulk surfaces.

Problem 5.16

Show that the power of radiation of a hot gas containing small particles does not depend on the size distribution of the particles.

According to formulas (5.20) and (5.21), the absorption cross section is proportional to the particle's volume or to the number of atoms n that the particle contains.

From this it follows that the specific absorption cross section, that is, the cross section per atom, does not depend on the particle size. Therefore, the radiation power of particles per unit volume is proportional to the total number of atoms in the particles that are located in a unit volume, or to the particle mass per unit volume. This value does not depend on the size distribution of the particles if the particles have identical shape. Correspondingly, the total radiation power of a hot gas or plasma containing small particles is proportional to the total mass of the particles in the volume where the radiation is created.

Problem 5.17

Modify Wien's law for the radiation power of a small hot particle in a gas.

One can compare the radiation power P of a small hot particle with $P_{bl} = 4\pi r_o^2 \sigma T^4$ for a bulk blackbody. Taking identical radii of particles in these cases, we have

$$\frac{P}{P_{bl}} = \frac{k g_{sph} r_o T}{\hbar c}$$

with $k = 3.83$. As is seen, a small parameter r_o/λ in the spectral power of radiation transfers into total radiation power in the form of a small factor $r_o T/(\hbar c)$.

Let us include this in the Wien law [107], according to which the maximum spectral power of blackbody radiation corresponds to the wavelength λ_m that satisfies the relation

$$\lambda_m = 0.29 \frac{\text{cm} \cdot \text{K}}{T} .$$

In the case of radiation of a particle that is small compared with the wavelength, its spectral power contains an additional factor r_o/λ. Owing to this, the wavelength λ_m of the maximum of the spectral power for blackbody radiation is 1.2 times lower.

Problem 5.18

Analyze the characteristics of the radiation power of a small hot particle in a gas due to the properties of the particle material.

The characteristics of the optical properties of a radiating particle are included in the quantity g_{sph}, which is defined by formula (5.21). For a hot macroscopic particle the function $g_{sph}(\omega)$ is a continuous function in some range of frequencies. This follows from the example in Figure 5.3 [109], where the spectrum of radiation of tungsten clusters in a hot gas is represented. Let us assume that, similar to the stationary case, the components of the dielectric constant ε', ε'' are positive values. This gives that the maximum value of the quantity g_{sph} corresponds to $\varepsilon' = 0$, $\varepsilon'' = 2$ and is $g_{sph} = 3$.

For definiteness, we will be guided now by soot particles in candle flames where this parameter was measured [110–112] and is $g_{sph} = 0.9 \pm 0.1$ under the assumption that it is independent of the radiation frequency. Then the radiation power P

of small soot particles in a hot gas is proportional to the volume of radiating particles V, and in accordance with formulas (5.30) the above value of g_{sph} gives for the specific power of radiation

$$\frac{P}{V} = \gamma T^5 , \quad \gamma = \frac{3 C g_{sph}}{4\pi} = (6.4 \pm 0.7) \cdot 10^{-11} \frac{\mathrm{W}}{\mathrm{cm^3 K^5}} .$$

Problem 5.19

Determine the total radiation power of a small spherical particle whose radius has an arbitrary relation to the typical wavelength of the radiation.

Let us combine formula (5.30), which corresponds to the radiation power of a particle of small radius, with that of a bulk particle:

$$P_o = \pi r^2 a \sigma T^4 , \tag{5.31}$$

where a is the mean gray coefficient of the particle's material. A general formula can be written in the form

$$P = P_o[1 - \exp(-\delta)] , \quad \delta = \frac{46 g_{sph}}{a} \frac{T r_o}{\hbar c} . \tag{5.32}$$

In the limiting cases this formula is transformed into formulas (5.30) and (5.31).

Problem 5.20

Determine the radiation power of a chain aggregate that is modeled by a cylindrical dielectric particle of radius r_o and length $2l$.

The radiation power of a small cylindrical particle is

$$P = \int_0^\infty \sigma_{abs}(\omega) \hbar\omega\, i(\omega) d\omega = \frac{2\pi}{3} \int_0^\infty \frac{\omega}{c} r^2 l g_{cyl}(\omega) \hbar\omega\, i(\omega) d\omega ,$$

where we use both formula (5.22) for the absorption cross section of the spherical particle σ_{abs} and the function

$$g_{cyl}(\omega) = \varepsilon'' + 8\varepsilon''[(\varepsilon' + 2)^2 + (\varepsilon'')^2]^{-1} .$$

Using formula (5.27),

$$\hbar\omega \cdot i(\omega) = \frac{\hbar\omega^3}{\pi^2 c^3} \left[\exp\left(\frac{\hbar\omega}{T}\right) - 1 \right]^{-1} ,$$

we obtain

$$P = \frac{2\pi}{3} \frac{r_o^2 l}{\hbar c} \sigma T^5 g_{cyl} k = 2.55 \pi r_o^2 l \sigma T^5 g_{cyl}/(\hbar c) , \tag{5.33}$$

for the power of radiation of a small cylindrical particle, assuming that the function $g(\omega)$ does not depend on the frequency of radiation. We use the analogy with a derivation of formula (5.30), so the numerical factor is taken as $k = 3.83$.

Problem 5.21

Determine the mean free path of photons in a gas containing chain aggregates and spherical particles of identical mass per unit gaseous volume if their sizes are small compared with the wavelength of the radiation. Compare these values for the same mass of particles per unit volume.

The mean free path of the radiation is given by

$$L = (N\sigma_{\rm abs})^{-1},$$

where N is the particle number density and we take into account that the particle absorption cross section $\sigma_{\rm abs}$ is larger than the cross section of the radiation scattered, as follows from a comparison of formulas (5.21) and (5.19). Let us introduce the density of particles in a gas $\rho_{\rm o}$ (the particle mass per unit gas volume) and the density of a particle material ρ. Because the mean free path does not depend on particle size, we take the particles to be identical. Let us first consider a system of spherical particles of radius $r_{\rm o}$. Then we have $N = \rho_{\rm o}/M$ for the particle number density, where $M = 4\pi r_{\rm o}^3\rho/3$ is the particle's mass. Using formula (5.20) for the absorption cross section of a small spherical particle, we obtain for the mean free path of the radiation in a gas containing particles

$$L_{\rm sph} = \lambda \cdot \frac{\rho}{\rho_{\rm o}} \frac{1}{18\pi g_{\rm sph}(\omega)}, \tag{5.34}$$

where $\lambda = 2\pi c/\omega$ is the wavelength of the radiation.

In the case of cylindrical particles we get by the same method using the absorption cross section according to formulas (5.22) and (5.23)

$$L_{\rm cyl} = \lambda \cdot \frac{\rho}{\rho_{\rm o}} \frac{3}{4\pi g_{\rm cyl}(\omega)}. \tag{5.35}$$

As is seen, in both cases we get the same dependence of the mean free path of the radiation on the problem parameters $L \sim \lambda \cdot \rho/\rho_{\rm o}$. This value does not depend on the size distribution function of clusters until the clusters are small.

The ratio of the mean free paths for spherical and cylindrical clusters is given by

$$\frac{L_{\rm sph}}{L_{\rm cyl}} = \frac{2g_{\rm cyl}(\omega)}{27g_{\rm sph}(\omega)} = \frac{2}{27}\left[1 + \frac{8}{(\varepsilon'+1)^2 + (\varepsilon'')^2}\right]\cdot[(\varepsilon'+2)^2 + (\varepsilon'')^2]. \tag{5.36}$$

Usually, $\varepsilon'' \ll 1$, so one can write for this ratio

$$\frac{L_{\rm sph}}{L_{\rm cyl}} = \frac{2}{27}\frac{[(\varepsilon')^2 + 2\varepsilon' + 9](\varepsilon'+2)^2}{(\varepsilon'+1)^2}. \tag{5.37}$$

This function has a minimum at $\varepsilon' = 1$, where $L_{\rm sph}/L_{\rm cyl} = 2$. This means that particles of cylindrical shape are stronger absorbers than spherical particles of the same mass.

5.3
Resonance Absorption of Metal Clusters

Problem 5.22

Determine the frequency that corresponds to the maximum of the absorption cross section for a small spherical metal particle within the framework of the plasmon model.

On the basis of the plasmon model [113], we consider cluster electrons to move freely inside the cluster, and then an electromagnetic wave interacts effectively with the collective degree of freedom for electrons. This is valid in the case of high conductivity of a cluster material and is suitable for metal clusters. The dielectric constant of a system of free electrons is

$$\varepsilon'(\omega) = 1 - \frac{\omega_p^2}{\omega^2},$$

as is fulfilled for a plasma, and the plasma frequency is given by [114]

$$\omega_p = \sqrt{\frac{4\pi N_e^2 e^2}{m_e}}.$$

Here N_e is the number density of electrons, e is the electron charge, and m_e is the effective electron mass for this metal. From this and formula (5.20) we have that the maximum cross section follows from the relation $\varepsilon'(\omega_{max}) = -2$ and corresponds to the radiation frequency

$$\omega_{max} = \frac{\omega_p}{\sqrt{3}} = \sqrt{\frac{4\pi N_e^2 e^2}{3m_e}}. \quad (5.38)$$

Note that the plasmon model is guided by the radiation spectrum with one wide resonance as occurs for the lithium and potassium clusters represented in Figures 5.4 and 5.5 [115–117].

Problem 5.23

On the basis of the resonant character of cluster absorption and formula (5.38) for the resonant frequency, analyze the spectral form of the absorption cross section for large metal clusters.

Using the resonant structure of the absorption cross section, we represent it in the form

$$\sigma(\omega) = \sigma_{max}\frac{\Gamma^2}{\hbar^2(\omega - \omega_o)^2 + \Gamma^2}, \quad (5.39)$$

where ω_o is the resonant frequency, $\Gamma = \hbar\omega_o\varepsilon''/6$ is the resonance width for a spherical particle in accordance with formula (5.21), and σ_{max} is the maximum

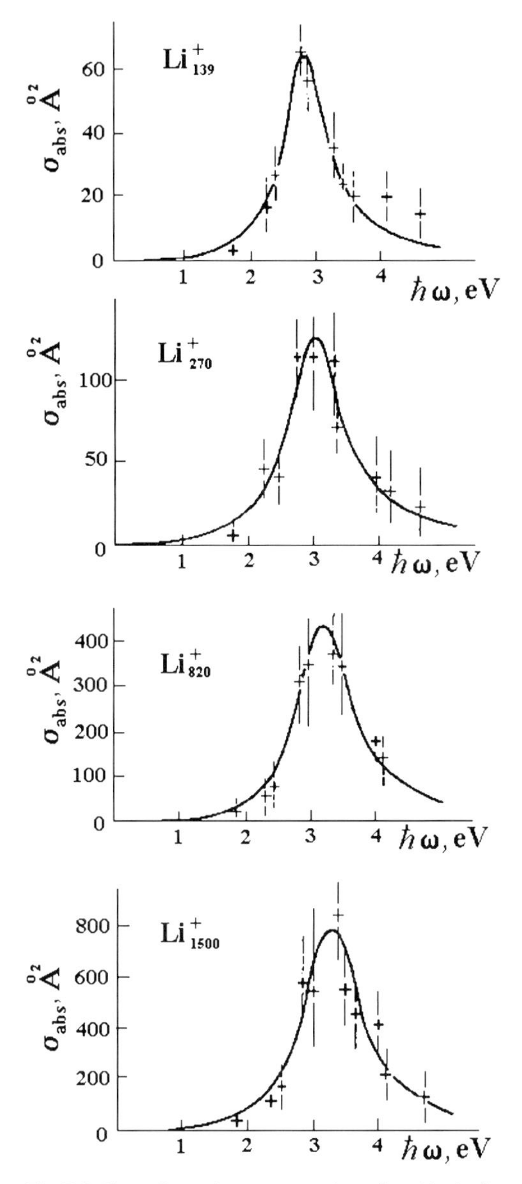

Fig. 5.4 Photoabsorption cross section of positively charged lithium cluster ions [115].

absorption cross section:

$$\sigma_{\max} = \frac{2\pi\hbar\omega_o^2 r_o^3}{\Gamma c} = \frac{2\pi\hbar\omega_o^2 r_{oW}^3}{\Gamma c} n \,, \tag{5.40}$$

where r_W is the Wigner–Seitz radius and n is the number of cluster atoms. The resonance width in formula (5.39) is assumed to be relatively small, corresponding to $\varepsilon'' \ll 1$.

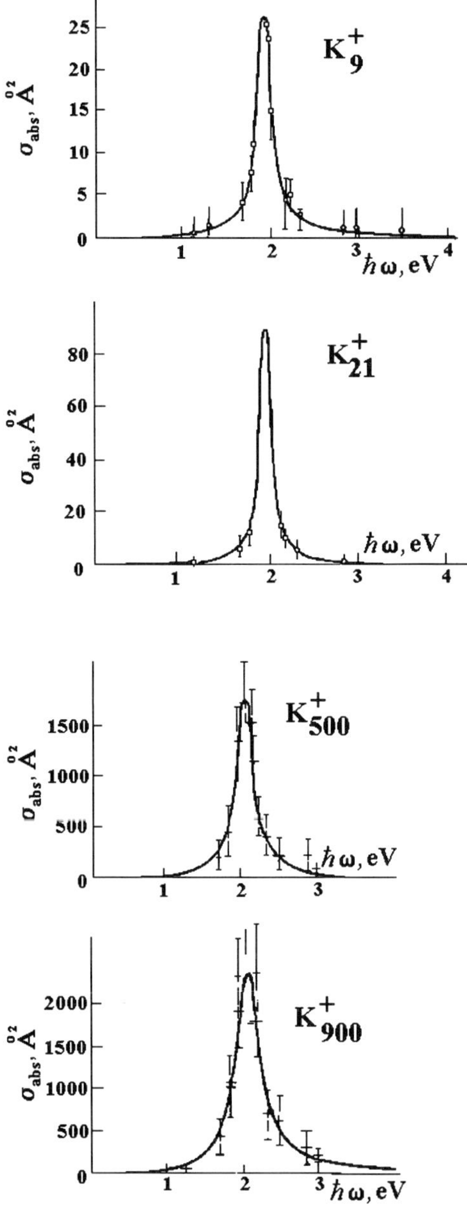

Fig. 5.5 Photoabsorption cross section of positively charged potassium cluster ions [116, 117].

From formula (5.39) the integral relation follows:

$$\int \sigma_{\text{abs}}(\omega)d\omega = \frac{\pi\sigma_{\max}\Gamma}{2\hbar} = \frac{\pi^2\omega_o^2 r_o^3}{c} \,. \tag{5.41}$$

Problem 5.24

Formulate a method for measuring the photoabsorption cross section if the energy of an absorbed photon exceeds the binding energy for a cluster atom.

Because the energy of an absorbed photon exceeds the binding energy of a cluster atom, the process of photon absorption is accompanied by photoinduced dissociation [118], which means that the energy of an absorbed photon is consumed in the breaking of cluster bonds, so that absorption of a photon leads to fragmentation of the cluster. Hence, in measuring the mass spectrum of cluster ions resulting from irradiation of these cluster ions by laser radiation, one can determine the dependence of the current for cluster ions of a given mass on the radiation intensity. This dependence allows one to determine the absorption cross section of a large cluster ion for a given wavelength. One can expect that the absorption cross section of a large single cluster ion does not differ significantly from that of the corresponding neutral cluster because the interaction of the cluster with an electromagnetic wave is determined by its electrons. Therefore, information on the absorption cross sections of large metal cluster ions may be transferred qualitatively to neutral metal clusters as well.

Problem 5.25

Check the validity of the plasmon concept for metal clusters where measurements of the absorption cross section are fulfilled.

Table 5.1 contains parameters of the measured absorption cross section for some metal clusters consisting of atoms with one valence electron for cases where the absorption cross section can be approximated as a function of the photon frequency by the plasmonlike dependence (5.39). The basis of such measurements is the concept of photoinduced evaporation, so the absorption of photons leads to cluster dissociation. Then measurement of the size distribution function of cluster ions as a function of the intensity of an incident laser beam for a given geometry of the experiment allows one to find the absorption cross section.

Because the atoms of the clusters in Table 5.1 have one valence electron, the number of cluster atoms coincides with the number of valence electrons. Hence, according to formula (5.38), the resonant frequency is

$$\omega_{\max} = \sqrt{\frac{4\pi N_e e^2}{3m_e}} = \sqrt{\frac{e^2}{m_e r_W^3}} ,$$

where r_W is the Wigner–Seitz radius. From this formula follows $\hbar\omega_o = 4.5\,\text{eV}$ for large lithium clusters, $\hbar\omega_o = 2.4\,\text{eV}$ for large potassium clusters, and $\hbar\omega_o = 4.9\,\text{eV}$ for large silver clusters if we take the effective electron mass of the metal clusters to be equal to the free electron mass. One can correct the deviation of these values from the data in Table 5.1 by introducing the effective electron mass m_e^*, which does not coincide with the free electron mass m_e and on average is $m_e^* = (0.57 \pm 0.10)\,m_e$.

Table 5.1 Parameters of the absorption cross sections for metal clusters [a].

Cluster	$\hbar\omega_o$, eV	Γ, eV	σ_{max}, Å^2	σ_{max}/n, Å^2	ω_o/ω_{max}	ξ	f
Li_{139}^+	2.92	0.90	62	4.5	0.64	2.8	0.58
Li_{270}^+	3.06	1.15	120	4.4	0.68	3.2	0.73
Li_{440}^+	3.17	1.32	280	6.4	0.70	4.9	1.20
Li_{820}^+	3.21	1.10	440	5.4	0.71	3.3	0.85
Li_{1500}^+	3.25	1.15	830	5.5	0.72	3.5	0.91
Li_n, average	3.1 ± 0.1	1.12 ± 0.12	–	5.2 ± 0.8	0.69 ± 0.03	3.5 ± 0.8	0.85 ± 0.23
K_9^+	1.93	0.22	26	2.9	0.79	2.9	0.91
K_{21}^+	1.98	0.16	88	4.2	0.81	2.9	0.96
K_{500}^+	2.03	0.28	1750	3.5	0.84	4.0	1.40
K_{900}^+	2.05	0.40	2500	2.8	0.84	4.5	1.59
K_n, average	2.00 ± 0.05	0.26 ± 0.10	–	3.4 ± 0.6	0.82 ± 0.02	3.6 ± 0.8	1.2 ± 0.3
Ag_9^+	4.02	0.62	8.84	1.0	0.82	2.6	0.87
Ag_{21}^+	3.82	0.56	16.8	0.9	0.78	2.1	0.64
Ag_n, average	3.9 ± 0.1	0.59 ± 0.03	–	0.9 ± 0.1	0.80 ± 0.02	2.4 ± 0.3	0.76 ± 0.12

a Parameters $\hbar\omega_o$, Γ, and σ_{max} are based on the measurements [115] for Li^+ clusters, [116, 117] for K^+ clusters, and [119] for silver cluster ions. The measured absorption cross section as a function of the photon energy is approximated by formula (5.39).

One more method to check the validity of the plasmon concept for metal clusters uses formula (5.40). Let us introduce the parameter

$$\xi = \frac{\sigma_{\max} \Gamma c}{2\pi \hbar \omega_o^2 r^3} ,$$

which is equal to unity if formula (5.40) is valid. Values of this parameter for metal clusters with the plasmonlike form of the absorption cross section are given in Table 5.1. As is seen, the parameter ξ differs from unity more than might be expected from the accuracy of the used parameters. This means that the assumptions used are not valid. Thus, the expression for the cluster absorption cross section, based on the concept of absorption of an electromagnetic wave by a metal cluster as a result of the interaction of the wave with a bulk plasmon, is invalid for real metal clusters when the spectral form of the absorption cross section corresponds to the plasmon case.

Problem 5.26

Express the effective oscillator strength per electron for a metal cluster through parameters of the absorption cross section in the cases of the plasmonlike form of the cross section.

Let us introduce the effective oscillator strength per electron f such that the sum of oscillator strengths over an active spectral range equals $n_e f$, where n_e is the number of cluster valence electrons. Let us use for a metal cluster a general formula for the absorption cross section of an atomic system

$$\sigma_{\text{abs}}(0 \to k) = \frac{\pi^2 c^2}{\omega^2} \cdot \frac{a_\omega}{\tau_{\text{ok}}} \cdot \frac{g_k}{g_o} = \frac{2\pi^2 e^2}{m_e c} \cdot f_{\text{ok}} g_k a_\omega .$$

Here ω is the frequency of the electron transition between states 0 (lower state) and k (upper state), g_o and g_k are the statistical weights of the transition states, τ_{ok} is the radiative lifetime with respect to this transition, a_ω is the frequency distribution function of radiative photons such that the normalization relation has the form $\int a_\omega d\omega = 1$, and f_{ok} is the oscillator strength for a given transition. The sum rule for radiative dipole transitions of valence electrons over an active spectral range has the form

$$\sum_k f_{\text{ok}} = n_e f ,$$

and this is the definition of the oscillator strength per valence electron for the active range of the cluster spectrum. For definiteness, below we consider clusters with one valence electron, such as the clusters in Table 5.1. Then, assuming that this spectral range includes all resonant dipole transitions of electrons, by integrating over frequencies in the vicinity of each cluster resonant transition and summing over all the resonant transitions, we obtain the following integral relation for the absorption cross section:

$$\int \sigma_{\text{abs}}(\omega)d\omega = \frac{2\pi^2 e^2}{m_e c} n f \, .$$

Comparing this relation with formula (5.41), we obtain for the effective oscillator strength of cluster valence electrons that

$$f = \frac{\sigma_{\max} \Gamma m_e c}{4\pi e^2 n \hbar} \, . \tag{5.42}$$

Table 5.1 contains values of the effective oscillator strength f per valence electron for the metal clusters under consideration. On average, the values of f for each element correspond to the oscillator strength of the lowest resonant transition $^2S_{1/2} \to {}^2P_{1/2}, {}^2P_{3/2}$ of the corresponding atom. These atomic oscillator strengths are 0.74 for lithium, 1.05 for potassium, and 0.77 for silver. The coincidence of the cluster and atom oscillator strengths confirms the concept that the active spectrum of metal clusters is a result of the transformation of resonant atomic spectral lines by means of atomic interaction. Thus, one can consider radiative transitions in clusters as individual radiative transitions for valence electrons in this system of bound atoms, and these transitions are broadened owing to the motion of the nuclei.

Hence, the character of the radiation of metal clusters as systems of bound atoms with interacting valence electrons can be considered in the following way. Let an individual atom have a resonant excited state, so a radiative dipole transition connects this state with the ground state. Usually, the lowest resonant atomic state is characterized by the maximum oscillator strength for the transition from the ground state. Now form a cluster of n atoms. If the positions of the atomic nuclei are fixed, then the resonant atomic spectral line is split into n lines. Owing to the vibrational motion of the nuclei in a solid or liquid cluster, these lines are broadened and partially overlap. Hence, the absorption spectrum of a metal cluster contains several (or one) wide resonances. Within the framework of this description, the effective oscillator strength of a cluster per valence electron does not depend on the cluster size.

Problem 5.27

Compare the radiative transition of an atom between the ground and resonantly excited atomic states and the radiative transition for a metal cluster consisting of these atoms. Assume that the atoms have one valence electron, and the oscillator strength for this atom transition is close to one. Determine the size dependence for the oscillator strength of this transition and follow the variation of the radiation spectrum in the course of the transition from an atom to a cluster.

In this case a system of bound atoms forms a metal cluster, and this occurs for alkali metals and coin metals. Let the state of a valence electron in the atom be characterized by quantum numbers nl, where n is the principal electron quantum number and l is its orbital momentum. We join atomic states with different fine

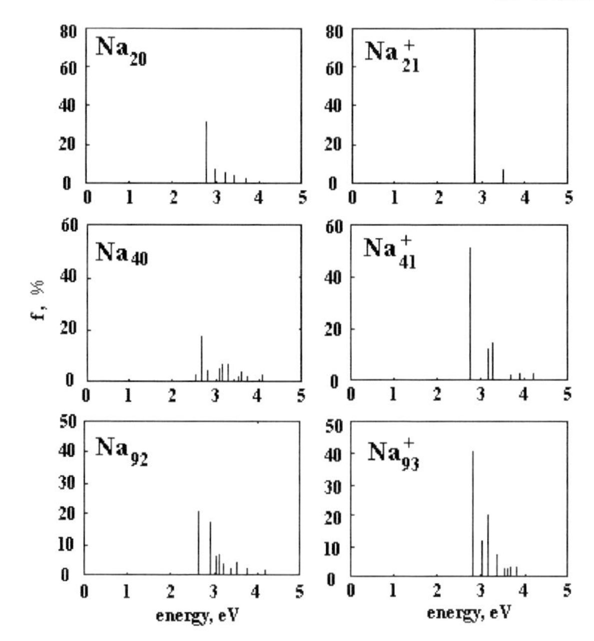

Fig. 5.6 Spectrum of neutral and charged sodium clusters at zero temperature [120]. The oscillator strength for the transition at the energies indicated is given as a percentage with respect to the total oscillator strength.

structures in one state and consider the resonant transition as an electron transition in a self-consistent field of an atomic core. According to the selection rule, the orbital quantum number of the valence electron changes by ± 1 as a result of the radiative dipole transition. If we now take a molecule consisting of two atoms, we obtain two resonantly excited states for zero momentum projection, though the molecule symmetry permits the radiative transition in one state.

In a cluster consisting of n atoms we obtain n resonantly excited states until we ignore splitting due to a certain momentum projection in a certain direction for the ground and resonantly excited states. Thus, the cluster spectrum includes several transition energies that are grouped around the atomic spectrum. But some resonantly excited states may have similar energies, and the cluster spectrum consists of a finite number of spectral lines, as shown in Figure 5.6 [120] for some sodium clusters.

Note that we consider zero temperature assuming that nuclei are motionless. If atoms are moving in clusters, each spectral line is broadened, and neighboring spectral lines overlap. This effect is intensified with a temperature increase, and this fact is demonstrated by the results given in Figure 5.7 [121]. Correspondingly, as a result of broadening and overlapping of neighboring spectral lines, the absorption cross section of metal clusters can have a plasmonlike shape and may include several broadened transitions, as follows from Figure 5.8 [119], where experimental absorption cross sections are given for silver clusters. Figures 5.9

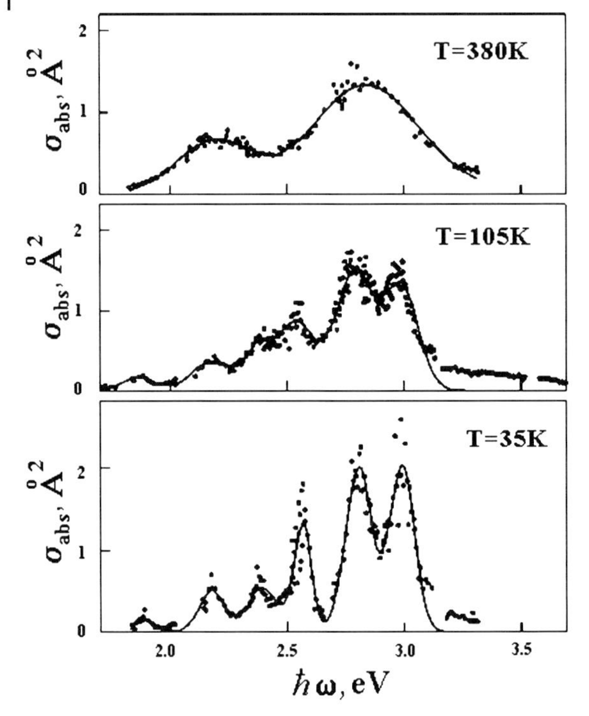

Fig. 5.7 Spectrum of the positively charged sodium cluster Na_{11}^+ at different temperatures [121]. It includes several transitions that are broadened owing to thermal motion of nuclei.

and 5.10 [122, 123] contain the absorption cross sections of sodium clusters, and these cross sections may be composed as a result of the summation of two broadened transitions.

Problem 5.28

Compare limiting cases of resonance transitions in clusters due to electron transitions.

Our goal is to compare various models for electron excitations of metal clusters assuming these excitations to be responsible for radiative dipole transitions in clusters. These models correspond to limiting cases of electron interaction inside metal clusters and include the plasmon model, the jellium model, and the atom model. The plasmon model is based on the collective interaction of electrons. The formation of plasma oscillations proceeds at motionless nuclei, that is, a cluster conserves its shape, but an electron component vibrates. This mechanism of electron excitation leads to a cluster spectrum with one resonance. In addition, the absorption cross section of metal clusters does not depend on the temperature in this mechanism of cluster excitation, because nuclei do not partake in this excitation. In reality all this is realized to a small degree.

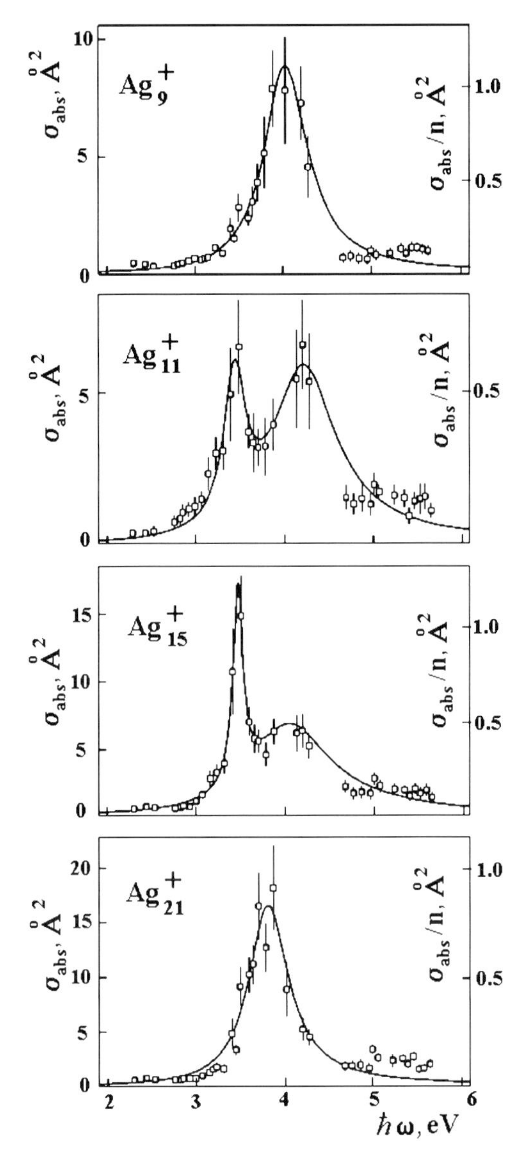

Fig. 5.8 Photoabsorption cross sections of positively charged silver clusters [119].

The jellium model for metal clusters assumes that each cluster electron interacts simultaneously with all nuclei, and cluster electrons have a shell structure. This means that radiative dipole transitions may proceed for electrons of a given shell with quantum numbers nl to a shell with quantum numbers $n', l \pm 1$, if this shell is not occupied. Correspondingly, in contrast to the plasmon model, the radiation spectrum of the jellium model includes many broadened lines. Next, the spectra of a metal cluster and its ion with the same number of atoms coincide more or less.

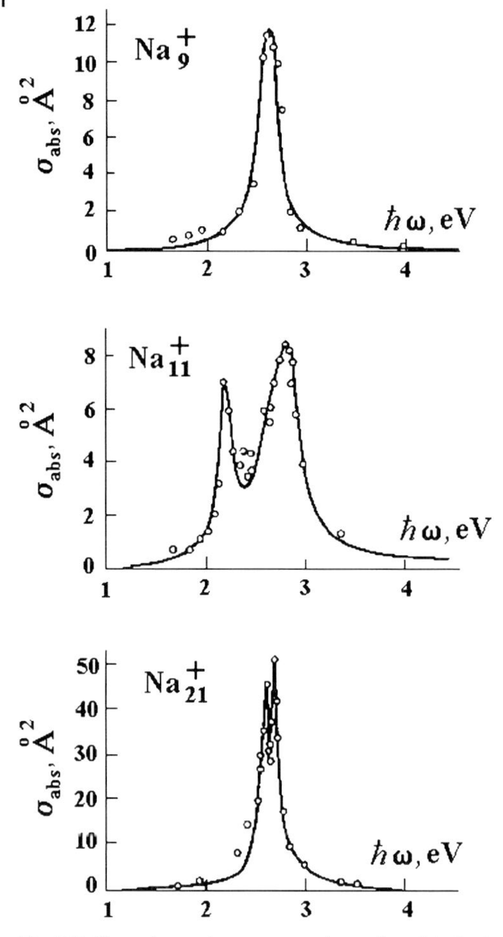

Fig. 5.9 Photoabsorption cross sections of positively charged sodium cluster ions [122].

This contradicts the data in Figure 5.6, where spectra of metal atoms with complete electron shells and their ions are given.

Evidently, a more profitable model considers cluster radiation as radiation of individual atoms interacting with surrounding atoms and electrons. Then the basis of the radiation spectrum is the spectrum of individual atoms that is shifted and broadened owing to interaction with the environment. This corresponds more or less to real spectra. In reality, radiation of individual atoms is a transition of a valence electron in a self-consistent field of the Coulomb nucleus and other electrons. One can apply this concept to a metal cluster, and then we join in this concept the plasmon, jellium, and atom models, which are the limiting cases of this general model. Comparing this general approach with measured data for metal clusters, we find that the atom model is closer to this general concept than the plasmon and jellium models.

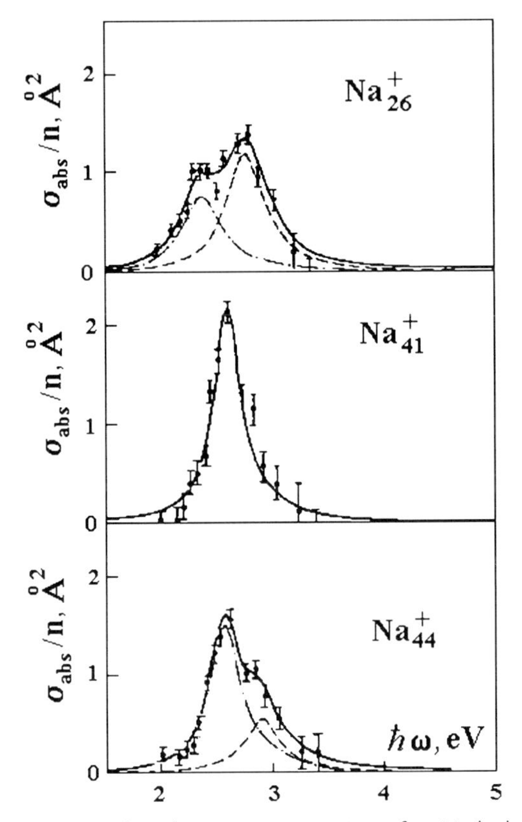

Fig. 5.10 Photoabsorption cross sections of positively charged sodium cluster ions [123], the spectrum of which is composed of two broadened transitions.

Problem 5.29

Determine the ratio η of the total energy of emitted photons by a cluster to the total initial binding energy of the cluster atoms after evaporation. For the parameters of the cross section of cluster absorption use the data of Table 5.1 for lithium, potassium, and silver clusters as a model.

An isolated cluster located in a buffer gas at a certain temperature evaporates and emits radiation during evaporation. In the course of cluster evaporation and a decrease in its size, the energy obtained from a buffer gas is consumed both on atom evaporation and emission of radiation. A size decrease is described by the balance equation using the evaporation rate (4.16)

$$\frac{dn}{dt} = -k_o N_{sat}(T) n^{2/3} \exp\left(\frac{\Delta\varepsilon}{Tn^{1/3}}\right),$$

where k_o is given by formula (2.7), $N_{sat}(T)$ is the atom number density at the saturated vapor pressure, and $\Delta\varepsilon$ is the specific surface energy of the cluster; we neglect the attachment of atoms to the cluster in this equation.

Taking the radiation power of a large cluster in the form $p_o n$, where p_o is the radiation power per atom, we use the data in Table 5.1 as the model data to determine this value. We have for the radiation energy released during cluster evaporation

$$E_{\rm rad} = \int p_o n dt = \frac{p_o}{k_o N_{\rm sat}(T)} \int n^{1/3} dn \exp\left(-\frac{\Delta\varepsilon}{Tn^{1/3}}\right) = \frac{3}{4} \cdot \frac{p_o n_o^{4/3}}{k_o N_{\rm sat}(T)} \exp\left(-\frac{\Delta\varepsilon}{Tn_o^{1/3}}\right),$$

where n_o is the initial number of cluster atoms. For simplicity, we neglect the exponential dependence that corresponds to the relation $\Delta\varepsilon \ll Tn_o^{1/3}$. Introducing the cluster binding energy ε_o per atom and neglecting the contribution of the surface cluster energy to the total binding energy E of cluster atoms, we have $E = \varepsilon_o n_o$. From this we obtain for the efficiency of cluster radiation

$$\eta = \frac{E}{E_{\rm rad}} = \frac{4k_o N_{\rm sat}(T)\varepsilon_o}{3p_o n_o^{1/3}} \exp\left(\frac{\Delta\varepsilon}{Tn_o^{1/3}}\right).$$

Below we determine the radiation power on the basis of the spectral power (5.28) and the cross section in accordance with the data in Table 5.2. The evaporation power $p_{\rm ev}$ in this table is reduced to a tungsten cluster consisting of $n = 10^3$ atoms and is given by the formula

$$p_{\rm ev} = k_o N_{\rm sat}(T) n^{2/3} \exp\left(\frac{\Delta\varepsilon}{Tn^{1/3}}\right).$$

The values of the energy ratio η are given in Table 5.2 at temperatures that are of interest for a light source with clusters. When the parameter η is small, radiation is a more effective channel for transformation of the cluster's energy. As is seen, at low temperatures the process of cluster evaporation is not essential for the cluster energy balance compared with the radiation process. At high temperatures this process becomes significant.

Table 5.2 Parameters of radiation of clusters consisting of 10^3 atoms.

T, K	3000	3500	4000	4500
$p_{\rm ev}$, W	$2.5 \cdot 10^{-15}$	$1.9 \cdot 10^{-13}$	$4.9 \cdot 10^{-12}$	$6.2 \cdot 10^{-11}$
η(Li)	$6 \cdot 10^{-4}$	0.017	0.22	1.5
η(K)	$3 \cdot 10^{-4}$	0.010	0.13	0.9
η(Ag)	0.0016	0.05	0.63	4

5.4
Radiative Processes in the Heat Balance and Relaxation of Clusters

Problem 5.30

Analyze the character of relaxation of hot clusters resulting from their emission.

Hot metal clusters may be created by heating them with a laser beam, and their effective temperatures may be determined by comparison of their emission spectra with spectra of a blackbody at an appropriate temperature. Examples of this type are given in Figure 5.11 [124–126]. In rare gases, emission of hot clusters determines their cooling, and Figure 5.12 [126] gives an example of this when the effective cluster temperature T_{cl} varies as a result of cluster radiation. The effective cluster temperature T_{cl} is introduced as the temperature of a blackbody whose spectral radiation power in a wavelength range near the maximum radiation power coincides with that of a cluster. In particular, Figure 5.12 gives such examples of the determination of the effective cluster temperature T_{cl}.

In considering radiation of hot metal clusters, we assume a weak wavelength dependence for the absorption cross section, which allows us to expand the absorption cross section $\sigma_{abs}(\omega)$ near a frequency ω_o of the maximum radiation power as

$$\sigma_{abs}(\omega) = \sigma_{abs}(\omega_o) + \frac{d\sigma_{abs}}{d\omega}(\omega - \omega_o) \,.$$

The cluster heat balance in the case under consideration is described by the equation

$$C\frac{dT}{dt} = -\int \hbar\omega i(\omega)\sigma_{abs}(\omega)d\omega \,,$$

where C is the cluster heat capacity. Accounting for a weak frequency dependence for the absorption cross section, rewrite this equation for radiation relaxation in the

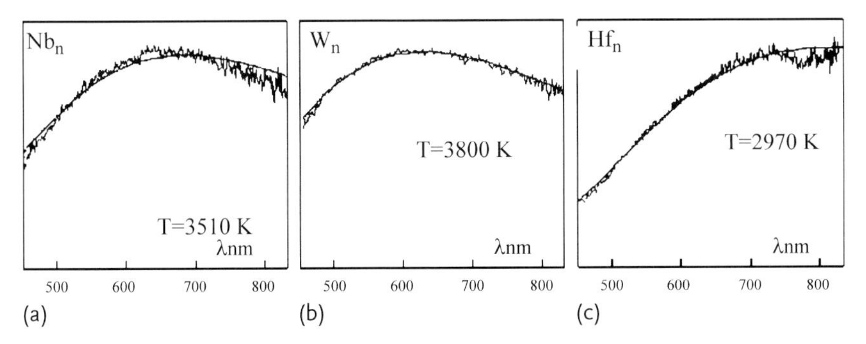

Fig. 5.11 Radiation spectra of niobium (a), tungsten (b), and hafnium (c) clusters heated by a laser beam [124–126]. The effective cluster temperature is determined by a comparison between the spectrum of a cluster and that of a blackbody.

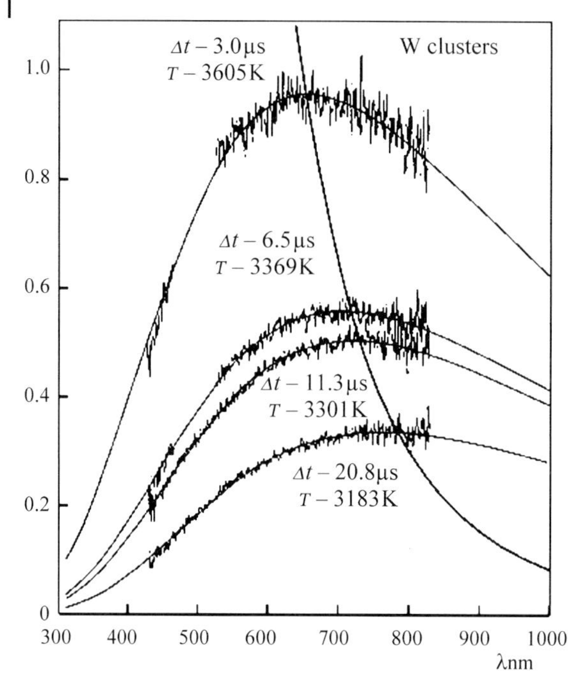

Fig. 5.12 Radiation spectra of tungsten clusters heated by a laser beam depending on the delay time Δt with respect to the laser pulse [126]. The dependence of the relative intensity of cluster emission on the wavelength for the maximum intensity is given by the solid curve.

form

$$C\frac{dT}{dt} = -\frac{\sigma_{\text{abs}}(\omega_o)}{4}\kappa T^4 ,$$

where κ is the Stefan–Boltzmann constant, and the central photon frequency ω_o satisfies the equation

$$\int_0^\infty (\omega - \omega_o) p(\omega) d\omega = 0 .$$

Restricting the exponential temperature dependence in formula (5.27) to $i(\omega) \sim \omega^2 \cdot \exp(-\hbar\omega/T)$ and accounting for $\sigma_{\text{abs}}(\omega) \sim \omega$, we obtain for the optimal photon energy $\hbar\omega_o \approx 5T$. Below we use this relation for the central photon frequency.

Let us use a bulk model for a cluster according to which the cluster heat capacity within the framework of the Dulong–Petit law, that gives $C = 3n$, where n is the number of cluster atoms. Then solving the heat balance equation for a cluster whose temperature decreases owing to its radiation, we obtain

$$\frac{\sigma_{\text{abs}}(\omega_o)}{n} = \frac{4}{\kappa \Delta t}\left(\frac{1}{T_2^3} - \frac{1}{T_1^3}\right) . \tag{5.43}$$

Table 5.3 Absorption parameters of niobium clusters [127] derived on the basis of formula (5.43) using experimental data [124, 126].

ΔT, K	3479–3408	3408–3345	3345–3193	3550–3170
Δt, μs	3.7–5.9	5.9–8.0	8.0–18.7	6.4–31.1
$\sigma_{abs}(\omega_o)/n$, 0.01 Å^2	6.7	6.7	3.6	3.6
λ_o, μ	0.84	0.85	0.88	0.86

Table 5.4 Absorption parameters of tungsten clusters [127] derived on the basis of formula (5.43) using experimental data [124].

ΔT, K	3605–3301	3301–3138	3500–3200	3500–3100	3700–3200
Δt, μs	8.3	9.5	16	17	18
$\sigma_{abs}(\omega_o)/n$, 0.01 Å^2	7.5	4.7	5.8	5.8	5.7
λ_o, μ	0.83	0.89	0.86	0.86	0.83

Here Δt is the time interval during which the effective cluster temperature varies from T_1 to T_2. Note that because the absorption cross section is roughly proportional to the number of cluster atoms n, the rate of cluster relaxation owing to radiation is practically independent of cluster size.

Formula (5.43) may be used to determine the absorption cross section for large metal clusters on the basis of the analysis of their relaxation. Tables 5.3 and 5.4 contain two examples of such clusters. Note that the photoabsorption cross sections for hot niobium and tungsten clusters are less by more than one order of magnitude than those for cold lithium, potassium, and silver clusters.

Problem 5.31

A hot tungsten cluster consisting of $n = 1000$ atoms is located in helium at room temperature at a pressure of 10 Pa. Determine the cluster temperature when the rates of cluster relaxation due to its radiation and collision with helium atoms are identical. Assume that a helium atom after collision acquires the cluster temperature.

Under the given conditions, the heat balance equation for a cluster has the form

$$\frac{dT}{dt} = -\frac{\sigma_{abs}(\omega_o)}{3n}\kappa T^4 - \frac{3T}{2} N_{He} v_T \cdot \frac{\pi r_W^2}{n^{1/3}} .$$

Here T is the cluster temperature, which is high compared with room temperature, $N_{He} = 2.4 \cdot 10^{15}\,\mathrm{cm}^{-3}$ is the number density of helium atoms at the pressure and room temperature indicated, $v_T = 1.3 \cdot 10^5\,\mathrm{cm/s}$ is the thermal velocity of helium atoms at room temperature, and r_W is the Wigner–Seitz radius. We use the collision model for the cluster and helium atom where after collision atoms carry the kinetic energy that corresponds to the cluster temperature. The cross section of atom–

cluster collisions is equal to the geometric cluster cross section $\pi r_o^2 = \pi r_W^2 n^{2/3}$, where r_o is the cluster radius.

We take the photoabsorption cross section per atom according to the data in Table 5.4 to be $\sigma_{abs}(\omega_o)/n = (0.06 \pm 0.01)$ Å^2. Under these conditions we find that the equality of radiative and collision terms in the heat balance equation for the cluster takes place at $T \approx 2500$ K.

Problem 5.32

Determine the contribution of radiation to the relaxation rate of a hot tungsten cluster consisting of $n = 1000$ atoms in a vacuum at cluster temperatures $T = 3500$ K and $T = 4000$ K.

The power of cluster cooling due to the evaporation of atoms is $P_{ev} = \varepsilon_o \nu_{ev}$, where $\varepsilon_o = 8.59$ eV is the atom binding energy in bulk tungsten, and the evaporation rate is given by formula (4.16)

$$\nu_{ev} = k_o N_{sat}(T) \cdot \exp\left(-\frac{2A}{3n^{1/3}T}\right) .$$

Here $A = 4.7$ eV is the specific surface energy of a large tungsten cluster, and the number density at the saturated vapor pressure is $N_{sat} = 2.1 \cdot 10^{13}$ cm^{-3} at $T = 3500$ K, and $N_{sat} = 6.4 \cdot 10^{14}$ cm^{-3} at $T = 4000$ K. From this we have for the cooling power due to cluster evaporation $P_{ev} = 5.2 \cdot 10^{-14}$ W at $T = 3500$ K and $P_{ev} = 2.0 \cdot 10^{-12}$ W at $T = 4000$ K.

The power of cluster cooling due to radiation is

$$P_{rad} = \kappa T^4 \sigma_{abs} ,$$

and for a tungsten cluster with $n = 1000$ atoms the absorption cross section is $\sigma_{abs} = 6 \cdot 10^{-15}$ cm^2. From this formula we have for the radiation power $P_{rad} = 5.1 \cdot 10^{-12}$ W for $T = 3500$ K and $P_{rad} = 8.7 \cdot 10^{-12}$ W for $T = 4000$ K. Correspondingly, the ratio of powers due to the mechanisms under consideration is $P_{ev}/P_{rad} = 0.01$ at $T = 3500$ K and $P_{ev}/P_{rad} = 0.22$ at $T = 4000$ K.

5.5 Hot Clusters as Light Sources

Problem 5.33

Compare the vision function of blackbody radiation with that of small spherical particles located in a flame if the optical depth of the flame is small.

Clusters are effective radiators. This property originates from the resonant radiation of free atoms. For many elements this radiation corresponds to spectral lines located in the optical spectral range. The joining of atoms to molecules and clusters leads to splitting and broadening of these spectral lines, and they are transformed into spectral bands for clusters and bulk systems. In the case of clusters, the absorption spectrum at low temperatures consists of a certain number of broadened spectral lines. An increase in the cluster temperature leads to broadening of these spectral lines owing to the motion of the nuclei and variations in their configurations. As a result, some resonances overlap at high temperatures. Therefore, the absorption spectrum of hot large clusters consists of one or several broad bands. Usually the absorption spectrum of large metal clusters includes the optical spectral range as well as spectral ranges adjoining this one.

The light output is the ratio of the power of radiation perceived by the eye to the total radiation power. Let us assume the flame temperature is relatively small, so the energy of an optical photon is large compared with the thermal energy (proportional to T). Then the radiation power per unit area in the optical region of the spectrum is given by

$$I_\omega d\omega = C\omega^3 \exp(-\hbar\omega/T) ,$$

where C is the normalization constant. The total radiation power per unit area, the energy flux for radiation q, is given by

$$q = \int_0^\infty d\omega \cdot C\omega^3 \left[\exp\left(\frac{\hbar\omega}{T}\right) - 1\right]^{-1} = C\frac{\pi^4}{15}\left(\frac{T}{\hbar}\right)^4 .$$

Let us introduce the vision function $V(\omega)$, which characterizes the luminous properties of radiation and corresponds to the perception of light by the eye. This function has a maximum $V_{max} = 683\,\mathrm{lm/W}$ at wavelength $\lambda_{max} = 554\,\mathrm{nm}$ ($\hbar\omega_{max} = 2.234\,\mathrm{eV}$). Near the maximum it is convenient to approximate this function by the dependence

$$V(\omega) = V_{max} \exp\left[-\frac{(\omega - \omega_{max})^2}{2\Delta\omega^2}\right] , \tag{5.44}$$

where $\Delta\omega = 0.244 \pm 0.016\,\mathrm{eV}$. From this it follows that for the vision function of a blackbody we have

$$V_{bb}(T) = \frac{\int V(\omega) I_\omega d\omega}{\int I_\omega d\omega} = \frac{15}{\pi^4}\left(\frac{\hbar}{T}\right)^4 \int_0^\infty \omega^3 d\omega \exp\left[-\frac{\hbar\omega}{T} - \frac{(\omega - \omega_{max})^2}{2\Delta\omega^2}\right]$$
$$= \frac{V_{max}}{6}\sqrt{\frac{2\pi}{F''(x_o)}} \exp[-F(x_o)] .$$

We evaluate the above integral on the basis of Euler's method. Here $F(x) = x + (x - y)^2/z^2 - 3\ln x$; $y = \hbar\omega_{max}/T$, $z = \hbar\Delta\omega/T$, and the parameter x_o is determined by the equation $F'(x_o) = 0$. Thus, for the vision function of the blackbody we have

$$V_{bb}(T) = \frac{V_{max}}{6}\sqrt{\frac{2\pi}{2/z^2 + 3/x_o^2}} \exp\left[-x_o - \frac{(x_o - y)^2}{z^2}\right] , \tag{5.45}$$

where x_o is given by the equation

$$x_o = \frac{1}{2}(y - z/2) + \frac{1}{2}\sqrt{(y - z/2)^2 + 6z^2}\,. \tag{5.46}$$

The vision function of the blackbody is a strong function of the temperature. In particular, $V_{bb}(3000\,\mathrm{K}) = 22\,\mathrm{lm/W}$ and $V_{bb}(3500\,\mathrm{K}) = 39\,\mathrm{lm/W}$.

Radiation of small particles is characterized by higher values of the vision function than in the case of a blackbody because the spectrum of small particles is shifted to higher frequencies compared with the blackbody spectrum. Let us determine the ratio $\eta = V_p(T)/V_{bb}(T)$, where $V_p(T)$ is the light output of radiation of particles. Taking into account that the functions under the integrals containing vision functions have strong maxima, we find for the vision function of a small particle that

$$V_p(T) = \frac{\int V(\omega) I_\omega \sigma_{abs}(\omega) d\omega}{\int I_\omega \sigma_{abs}(\omega) d\omega} = \frac{\sigma_{abs}(\omega_o) \int V(\omega) I_\omega d\omega}{\sigma_{abs}\left(\frac{3T}{\hbar}\right) \int I_\omega d\omega} = \eta V_{bb}(T)\,,$$

so

$$\eta = \frac{V_p(T)}{V_{bb}(T)} = \frac{\sigma_{abs}(\omega_o)}{\sigma_{abs}(3T/\hbar)}\,. \tag{5.47}$$

Here $\omega_o = T x_o/\hbar$ and $3T/\hbar$ are frequencies corresponding to maxima of the respective functions under integrals. Using formulas (5.21)–(5.23) for spherical or cylindrical particles for the absorption cross sections and assuming functions $g_{sph}(\omega)$ and $g_{cyl}(\omega)$ to be independent of ω, we get

$$\eta = \frac{V_p(T)}{V_{bb}(T)} = \frac{\hbar\omega_o}{3T} = \frac{T_o}{T}\,. \tag{5.48}$$

If we take $\omega_o = \omega_{max}$, which is valid for small temperatures, we obtain $T_o = 6800\,\mathrm{K}$.

Problem 5.34

Using the resonant character of the absorption cross section for metal clusters, compare the light output for radiation of these clusters with that of a blackbody.

Let us use the data for the parameters of the absorption cross section of some metal clusters (Table 5.1) as models for clusters of heat-proof metals at high temperatures. Then one can expect that the light output for radiation of large clusters is higher than that of a blackbody at the same temperature. Indeed, the absorption spectrum of such clusters includes a visible range of the spectrum and cuts off the infrared range of the spectrum. Table 5.5 gives the light output of blackbody and metal clusters at high temperatures if we use the data for silver, lithium, and potassium at these temperatures as models. As is seen, such clusters are more effective radiators than blackbodies.

Table 5.5 The light output η (lm/W) for blackbody and metal clusters whose radiative parameters are taken from Table 5.1.

T, K	3000	3500	4000
blackbody	22	39	57
Ag	51	75	88
Li	51	80	102
K	108	141	165

Problem 5.35

Estimate the absorption cross section of a fractal aggregate of radius R consisting of solid clusters of radius r_o. The fractal dimensionality of the aggregate is β. Compare the specific absorption cross section of a fractal aggregate with that of the solid clusters that constitute the aggregate.

We assume that the aggregate radius R is small compared with the wavelength λ of radiation and also that interaction of this object with radiation proceeds similarly to that with a compact particle of this size [112, 129]. Then the absorption cross section is estimated on the basis of the formula

$$\sigma_{\text{abs}} \sim \frac{R}{\lambda} \cdot \pi R^2 . \tag{5.49}$$

This dependence is valid for a spherical particle whose radius is small compared with the radiation wavelength and was used above for a cluster; [see formula (5.21)]. In this case the mass of the fractal aggregate M is given by [130–133]

$$M = m \cdot \left(\frac{R}{r_o}\right)^{\beta} , \tag{5.50}$$

where m is the mass of an elemental solid cluster, r_o is its radius and β is the aggregate fractal dimensionality, which depends on the mechanism of formation of the aggregate from solid clusters. In particular, if this aggregate results from a consecutive joining of clusters into small aggregates and if, further, these aggregates combine into larger ones, then the fractal dimensionality is $\beta = 1.7$–1.8 if the relative motion of the aggregates is due to their diffusion in a buffer gas.

From this for the specific absorption cross section of a fractal aggregate we have

$$\frac{\sigma_{\text{abs}}}{M} \sim R^{3-\beta} .$$

In the case of an individual small spherical particle or a cluster, this ratio does not depend on the particle's radius. In particular, assuming the numerical coefficient formula (5.49) is identical for clusters and fractal aggregates, we obtain that the absorption cross section of solid clusters of radius r_o increases by a factor of

$(R/r_0)^{3-\beta}$ if they form a fractal aggregate of radius R. Usually, the maximum radius of a fractal aggregate exceeds the radius of the solid particles constituting it by three orders of magnitude. Therefore, the formation of fractal aggregates from solid clusters leads to an increase of absorption by several orders of magnitude for a gaseous system in which clusters are located. Thus, the formation of fractal aggregates in gases and plasmas can significantly influence the radiative parameters of a system.

Part II Cluster Processes in Gases

6
Cluster Transport in Gases and Diffusion-Limited Association of Clusters

The character of the interaction of a cluster with atoms of a surrounding gas depends on the relation between the cluster size and the mean free path of gas atoms. In the kinetic regime, when the cluster size r_0 is small compared with the mean free path λ of atoms in a gas $\lambda \gg r_0$, a cluster at each time may interact strongly with one atom only, and the cluster motion results from separate collisions with gas atoms. In addition, because the cluster mass is large compared with the atomic mass, a remarkable variation in the cluster momentum may result from many collisions with gas atoms. Therefore, if a cluster moves in a gaseous flow and the velocity of this flow varies sharply in value or direction, the cluster conserves the initial velocity and can be removed from the gaseous flow in this manner.

In the diffusion regime of cluster motion in a gas, when the mean free path of gas atoms is small compared with the cluster size ($\lambda \ll r_0$), a cluster collides with many atoms simultaneously, and the hydrodynamic laws describe the cluster motion in this case. In both regimes, the diffusion coefficient for cluster motion in a gas is large compared with that of atoms, which leads to a low displacement of clusters in a gas. This influences the character of association of clusters in a gas where the approach of clusters is determined by their motion in the gas, and the growth of a system of bound clusters may lose a pairwise character.

6.1
Transport of Large Clusters in Gases

Problem 6.1

On the basis of the hard sphere model for collisions between atoms and clusters, determine the force that acts on a spherical cluster moving in a gas with a small velocity. Consider the kinetic regime of the cluster behavior when the cluster radius is small compared with the mean free path of atoms but exceeds a region dimension of action of atomic forces.

As a result of collisions between atoms and clusters, momentum is transferred to a cluster if the average cluster velocity differs from the atom velocity. The force acting on a cluster results from collisions of atoms with the cluster. The momen-

Cluster Processes in Gases and Plasmas. Boris M. Smirnov

ISBN: 978-3-527-40943-3

tum $m\mathbf{g}(1-\cos\vartheta)$ is transferred to the cluster in the collision of an individual atom with the cluster when we consider the process under conditions in which the cluster is motionless, and ϑ is the scattering angle. Here m is the atom mass, and the relative atom–cluster velocity is $\mathbf{g} = \mathbf{v} - \mathbf{w}$, where $\mathbf{v}$ is the atom velocity and $\mathbf{w}$ is the cluster velocity. Clusters do not influence the Maxwell distribution function of atoms $\varphi(v)$, and because the cluster mass M is considerably larger than the atomic mass, we obtain the change in the cluster momentum per unit time due to collisions with atoms or the force $\mathbf{F}$ that acts on a cluster as a result of collisions with atoms:

$$\mathbf{F} = \int m\mathbf{g}(1-\cos\vartheta)\varphi(v) g\, d\sigma d\mathbf{v}\,.$$

Here $d\sigma$ is the differential cross section of elastic atom–cluster collision, and the velocity distribution function of atoms $\varphi(v)$ is normalized to the number density N_a of atoms, that is, $\int \varphi(v) d\mathbf{v} = N_a$. Since clusters do not influence the distribution of atoms, we have for the Maxwell distribution function of atoms $\varphi(v)$ accounting for its normalization

$$\varphi(v) = N_a \left(\frac{m}{2\pi T}\right)^{3/2} \exp\left(-\frac{mv^2}{2T}\right)\,.$$

According to formulas (2.2) and (2.3), we reduce the above expression for the friction force to

$$\mathbf{F} = m\sigma_o \int \mathbf{g}\varphi(v) g d\mathbf{v} = \frac{m}{\lambda}\langle \mathbf{g}g\rangle\,, \quad \lambda = \frac{1}{N_a \sigma_o}\,, \tag{6.1}$$

where $\sigma_o = \pi r_o^2$ and r_o is the cluster radius.

In evaluating this value, we take into account that the force $\mathbf{F}$ is directed along the cluster drift velocity $\mathbf{w}$. Taking this direction as z and accounting for $w \ll v$, we have for this integral

$$\langle \mathbf{g}g\rangle = \frac{1}{N_a}\int \mathbf{g}g\varphi(v) d\mathbf{v} = \mathbf{k}\frac{1}{N_a}\int g_z g w_z \frac{\partial\varphi(\mathbf{g})}{\partial g_z} d\mathbf{g}$$
$$= -\mathbf{w}\left\langle \frac{m g_z^2 g}{T}\right\rangle = -\mathbf{w}\frac{m}{3T}\left\langle g^3\right\rangle\,,$$

where we take $\mathbf{v} = \mathbf{g} + \mathbf{w}$ and expand $\varphi(v)$ over a small parameter that is proportional to w, and $\mathbf{k}$ is the unit vector along $\mathbf{w}$. As a result, we obtain for the resistance force (6.1) that acts on a cluster moving in an atomic gas [134]

$$\mathbf{F} = -\mathbf{w}\cdot\frac{m^2}{3T\lambda}\langle v^3\rangle = -\mathbf{w}\cdot\frac{8\sqrt{2\pi m T}}{3}\cdot N_a r_o^2\,. \tag{6.2}$$

The resistance force $\mathbf{F}$ acting on a cluster moving in a gas is obtained under the assumption that a cluster does not influence the spatial distribution of atoms. This is valid in the kinetic regime of atom–cluster collisions

$$\lambda \gg r_o\,,$$

Under this assumption, violation the spatial distribution of atoms owing to collisions with a cluster near the atom is restored as a result of collisions between atoms in a range of distances λ around the cluster.

Problem 6.2

Determine the mobility of a small charged cluster in a gas if the cluster radius is small compared to the mean free path of atoms in a gas.

According to the definition, the mobility of a charged cluster K is given by

$$\mathbf{w} = K\mathbf{E},$$

where $\mathbf{w}$ is the cluster drift velocity and $\mathbf{E}$ is the electric field strength. On the basis of formula (6.2) for the resistance force acting on a cluster and taking into account that the resistance force in this case is equal to $e\mathbf{E}$, we get for the cluster mobility in the kinetic regime of atom–cluster collisions, that is, the Knudsen number is small ($\mathrm{Kn} = r_o/\lambda \ll 1$) [86]

$$K = \frac{3e}{8\sqrt{2\pi m T} N_a r_o^2} = \frac{K_o}{n^{2/3}}, \quad K_o = \frac{3e}{8\sqrt{2\pi m T} N_a r_W^2}. \tag{6.3}$$

Note that this expression for the cluster mobility in a gas corresponds to the first Chapman–Enskog approximation [135, 136], where the cluster cross section is taken as the average cross section of atom–cluster elastic scattering, and the average atom velocity is taken as the average relative collision velocity.

Problem 6.3

Obtain the Einstein relation between the diffusion coefficient and the mobility of a charged cluster moving in a gas.

The Einstein relation applies to the limit of a small electric field strength when the electric field does not disturb the thermodynamic equilibrium in a system including a gas with clusters, which leads to the Maxwell velocity distribution function of atoms. Therefore the cluster flux created by the action of an external electric field is equalized by the flux due to a gradient in the cluster number density, and therefore we have for the total cluster flux

$$\mathbf{j} = N_{cl}\mathbf{w} - D\nabla N_{cl} = 0,$$

where N_{cl} is the number density of clusters, $\mathbf{w}$ is the cluster drift velocity, and D is the diffusion coefficient for clusters in a gas. From this we determine the diffusion coefficient of clusters as $D = N_{cl}\mathbf{w}/\nabla N_{cl}$. Because of the thermodynamic equilibrium of clusters in an external field, the cluster number density is given by the Boltzmann equation

$$N_{cl} = N_o \exp\left(-\frac{U}{T}\right),$$

where N_o is a parameter and U is the potential of an external field. From this we have

$$\nabla N_{cl} = -N_{cl}\frac{\nabla U}{T} = N_{cl}\frac{\mathbf{F}}{T},$$

where $\mathbf{F} = -\nabla U$ is the force that acts on the cluster from an external field. Thus, we have $D = \mathbf{w}T/\mathbf{F}$. In this case we have $\mathbf{F} = e\mathbf{E}$, and on the basis of the definition of cluster mobility, we obtain the following relation between the cluster mobility and the diffusion coefficient:

$$K = \frac{eD}{T}. \tag{6.4}$$

This is the Einstein relation between the mobility and the diffusion coefficient for any charged particle [137–139], though this relation was first obtained by Nernst [140] and Townsend and Bailey [141, 142] (see also [143, 144]).

On the basis of formulas (6.3) and (6.4), we have for the diffusion coefficient of a cluster in a buffer gas within the framework of the hard sphere model for clusters [86]

$$D = \frac{3\sqrt{T}}{8\sqrt{2\pi m}N_a r_o^2} = \frac{D_o}{n^{2/3}}, \quad D_o = \frac{3\sqrt{T}}{8\sqrt{2\pi m}N_a r_W^2}, \quad \lambda \gg r_o. \tag{6.5}$$

This formula is valid when the cluster radius r_o is small compared with the mean free path λ of atoms in the gas. This formula may be represented in the form [145]

$$D = D_* \left(\frac{b}{r_o}\right)^2.$$

In particular, taking in this formula $b = 1\,\text{Å}$ for a large cluster moving in air, we have $D_* N_a = 4.4 \cdot 10^{19}\,\text{cm}^{-1}\text{s}^{-1}$ in the kinetic regime of atom–cluster collisions.

Problem 6.4

Determine the force that acts on a spherical cluster moving in a gas in the diffusion regime of atom–cluster collisions when the cluster radius r_o greatly exceeds the mean free path λ of gas atoms

$$r_o \gg \lambda. \tag{6.6}$$

In the diffusion regime, the resistance force occurs because a gas stream flows around the cluster, so the relative velocity of the gas and cluster is zero on the cluster surface and equal to $\mathbf{v}$ far from the cluster. Therefore, the gas velocity varies in a region near the cluster from zero to the gas flow velocity. Displacement of gas atoms in this region leads to transport of the momentum and creates a frictional force. Because this process is connected with the gas viscosity, the frictional force may be expressed through the gas viscosity coefficient.

By definition, the viscosity force F per unit surface is determined by the equation

$$\frac{F}{S} = \eta\frac{\partial v_\tau}{\partial R},$$

where η is the viscosity coefficient of the gas, S is the area of the frictional surface, v_τ is the tangential component of the velocity with respect to the flow, and the coordinate R characterizes the normal direction to this flow. From this one can estimate the value of the friction force. On the basis of estimations $S \sim r_o^2$, $R \sim r_o$, $v_\tau \sim v$, where v is the cluster velocity in a gas, we obtain the following estimation:

$$F \sim \eta r_o v \, .$$

For a spherical particle, the numerical coefficient that follows from the accurate solution of this problem yields the Stokes equation [146, 147]:

$$F = 6\pi \eta r_o v \, , \quad \lambda \ll r_o \, . \tag{6.7}$$

From this we have the following expression for the cluster mobility:

$$K = \frac{\mathbf{v}}{\mathbf{E}} = \frac{e}{6\pi r_o \eta} = \frac{K_o}{n^{1/3}} \, , \quad K = \frac{e}{6\pi r_W \eta} \, . \tag{6.8}$$

Using the Einstein relation (6.4), we then have the diffusion coefficient of the cluster [145]

$$D = \frac{KT}{e} = \frac{T}{6\pi r_o \eta} = \frac{D_o}{n^{1/3}} \, , \quad D_o = \frac{T}{6\pi r_W \eta} \, , \quad \lambda \ll r_o \, . \tag{6.9}$$

Note that according to formula (6.9) the diffusion coefficient of clusters does not depend on the cluster material. The only cluster parameter in the above expression for the diffusion coefficient is the cluster radius.

Problem 6.5

Determine the cluster diffusion coefficient in gases in an intermediate case by combining the kinetic and diffusion regimes.

The limiting cases of the kinetic and diffusion regimes of collisions between atoms and clusters are given by formulas (6.5) and (6.9), respectively. To combine these cases, we express the gas viscosity η through the mean free path λ of atoms in a gas, that is, $\lambda = (N_a \sigma_g)^{-1}$, where N_a is the number density of atoms and $\sigma_g = \pi \rho_o^2$ is the gas-kinetic cross section. Then we apply to the transport of atoms the hard sphere model, and the cross section σ_g of the collision of two atoms does not depend on the collision velocity. Within the framework of this model, the viscosity of a gas is given by [135, 136]

$$\eta = \frac{5\sqrt{\pi T m}}{24\sigma_g} = \frac{5\sqrt{Tm}}{24\sqrt{\pi}\rho_o^2} \, , \tag{6.10}$$

where m is the atom mass. The ratio of the diffusion coefficients in the kinetic D_{kin} and diffusion D_{dif} regimes of collisions between atoms and clusters, which are given by formulas (6.5) and (6.9), respectively, is

$$\frac{D_{kin}}{D_{dif}} = \frac{15\pi}{32\sqrt{2}} \cdot \frac{\lambda}{r_o} = 0.96 \mathrm{Kn} \, , \tag{6.11}$$

where $\mathrm{Kn} = \lambda / r_o$ is the Knudsen number, and the mean free path of atoms in a gas is $\lambda = (N_a \sigma_g)^{-1}$.

Combining the kinetic and diffusion regimes of atom–cluster collisions, we represent the cluster diffusion coefficient in a gas as

$$D = D_{\mathrm{kin}} + D_{\mathrm{dif}} ,$$

and in the limiting cases the diffusion coefficient is given by formulas (6.5) and (6.9). As a result, the cluster diffusion coefficient in a gas may be represented in the form

$$D = \frac{T}{6\pi r_o \eta}(1 + 0.96\mathrm{Kn}) , \qquad (6.12)$$

and this equation transforms to formulas (6.5) and (6.9) in the limits of small and large Knudsen numbers.

Let us apply this formula to the motion of water clusters in air at a temperature of $T = 300$ K, when formula (6.12) takes the form

$$D = \frac{k_*}{r_o}\left(1 + \frac{1}{N_a r_o s_o}\right) ,$$

where N_a is the number density of air molecules, $k_* = 1.2 \cdot 10^{-11}\,\mathrm{cm}^3/\mathrm{s}$, and $s_o = 2.7 \cdot 10^{-15}\,\mathrm{cm}^2$.

Problem 6.6

Compare the diffusion coefficients of atoms and clusters in a buffer gas in the kinetic regime.

It is convenient to use the hard sphere model for collisions between atoms, as well as for collisions of atoms with clusters. Within the framework of this model, the mean free path of atoms is independent of the collision velocity, which allows us to model colliding atoms by hard balls. Then the diffusion coefficient of atoms D_a is given by formula (6.5), where the radius of action of atomic forces is used instead of the cluster radius. Therefore, the ratio of the diffusion coefficients of clusters D and atoms D_a in a gas is given by [17, 146]

$$\frac{D}{D_a} = \left(\frac{\rho_o}{r_o}\right)^2 ,$$

where ρ_o is the effective interaction distance for atoms, so $\sigma_g = \pi \rho_o^2$ is the gas-kinetic cross section for atom–atom collisions in a gas, and r_o is the cluster radius. The interaction radius of atoms ρ_o according to the model used is assumed to be independent of the collision velocity. In particular, for atmospheric air at the temperature $T = 300$ K the parameters of this equation are $\rho_o = 3.6\,\text{Å}$, $D_a N_a = 4.8 \cdot 10^{18}\,\mathrm{cm}^{-1}\mathrm{s}^{-1}$, $D N_a r_o^2 = 4.4 \cdot 10^3\,\mathrm{cm/s}$, where N_a is the number density of air molecules.

Problem 6.7

Determine the free-fall velocity for a spherical cluster in a gas under the action of the gravitational force. The cluster radius greatly exceeds the mean free path of gas molecules (the diffusion regime), and the Reynolds number of the moving cluster is small.

The cluster equilibrium velocity **w** follows from the equality of the gravitational and resistance forces, and this relation has the form

$$\frac{4\pi}{3}\rho \mathbf{g} r_o^3 = 6\pi\eta r_o \mathbf{w},$$

where ρ is the density of the cluster material and g is the free-fall acceleration. From this we obtain for the cluster free-fall velocity

$$w = \frac{2\rho g r_o^2}{9\eta}. \tag{6.13}$$

This formula is valid at small values of the Reynolds numbers $\mathrm{Re} = w r_o/\nu \ll 1$, where $\nu = \eta/\rho_g$ is the kinematic viscosity of the gas and ρ_g is the gas density. For the cluster radius, this criterion gives

$$r_o \ll \frac{\eta^{2/3}}{(\rho\rho_g g)^{1/3}}. \tag{6.14}$$

In particular, for large water clusters moving in atmospheric air this criterion gives $r_o \ll 30\ \mu\mathrm{m}$.

Problem 6.8

Determine the diffusion coefficient of clusters in an atomic gas as a function of the number of cluster atoms within the framework of the hard sphere model for atom–cluster collisions for the kinetic and diffusion regimes of collisions.

The cluster diffusion coefficient in a gas is given by formula (6.5) for the kinetic regime of atom–cluster collisions and by formula (6.9) for the diffusion regime of their collisions. Express these equations through the number of cluster atoms on the basis of formula (2.5) for the cluster radius r_o, that is, $r_o = r_W n^{1/3}$, where r_W is the Wigner–Seitz radius for a cluster material and n is the number of cluster atoms. Then we represent formula (6.5) for the cluster diffusion coefficient in the kinetic regime of collisions in the form

$$D_n = \frac{D_o}{n^{2/3}}, \quad D_o = \frac{3\sqrt{T}}{8\sqrt{2\pi m} N_a r_W^2}. \tag{6.15}$$

Here N_a is the number density of buffer gas atoms, T is the gas temperature, and m is the atomic mass. The kinetic regime of collisions is realized when atoms collide with the cluster separately, that is, each time a strong cluster interaction is possible with one atom only. Therefore, the mean free path of buffer gas atoms λ is large compared with the cluster radius r_o.

Table 6.1 The reduced diffusion coefficient of metal clusters in argon D_o according to formula (6.15) at the argon temperature $T = 1000$ K [36]. The reduced diffusion coefficient is given for the normal number density of atoms ($N_a = 2.69 \cdot 10^{19}$ cm^{-3}).

Cluster	D_o, cm^2/s
Ti	0.91
V	1.05
Fe	1.17
Co	1.20
Ni	1.22
Zr	0.74
Nb	0.90
Mo	0.98
Rh	1.05
Pd	1.01
Ta	0.90
W	0.98
Re	1.01
Os	1.05
Ir	1.01
Pt	0.98
Au	0.93
U	0.81

Note that the cluster parameters are expressed in formula (6.15) through r_W only and depend weakly on the cluster material. We give in Table 6.1 the values for the reduced diffusion coefficients of metal clusters in argon, and these data evidence a weak dependence on the cluster material. Note that the basic temperature dependence for the cluster diffusion coefficient is $D_n \sim \sqrt{T}$ for the kinetic regime of atom–cluster collisions.

Problem 6.9

Determine the diffusion coefficient of clusters in an atomic gas as a function of the number of cluster atoms for the diffusion regime of atom–cluster collisions.

According to formula (6.9) the cluster diffusion coefficient in a gas for the diffusion regime of atom–cluster collisions is given by

$$D_n = \frac{T}{6\pi r_o \eta} = \frac{d_o}{n^{1/3}}, \quad d_o = \frac{T}{6\pi r_W \eta}, \quad \lambda \ll r_o . \tag{6.16}$$

The reduced diffusion coefficient d_o of clusters in a gas depends on the cluster material weaker than the reduced diffusion coefficient D_o given in Table 6.1. We give in Table 6.2 values of the reduced diffusion coefficient for titanium clusters in inert

Table 6.2 The reduced diffusion coefficient d_o of titanium clusters in inert gases and diatomic molecular gases according to formula (6.12) at the gas temperature $T = 1000$ K. The reduced diffusion coefficient is given for the normal number density of atoms ($N_a = 2.69 \cdot 10^{19}$ cm^{-3}).

Gas	He	Ne	Ar	Kr	Xe	H_2	N_2	O_2
d_o, 10^{-4} cm^2/s	9.8	6.2	8.0	6.8	6.8	22	11	9.2

gases in the case of the diffusion regime of atom–cluster collisions. Note that the cluster diffusion coefficient in the diffusion regime of atom–cluster collisions is independent of the number density of gas atoms or molecules N_a and is proportional to the gas temperature T.

Problem 6.10

Derive the criterion of validity of the kinetic and diffusion regimes for cluster motion in a gas.

For this it is convenient to use formula (6.12), according to which the kinetic regime of cluster drift takes place for large Knudsen numbers $\mathrm{Kn} \gg 1$, and the diffusion regime of atom–cluster collisions corresponds to the opposite criterion. Hence, the criterion of the kinetic regime is $\lambda \gg r_o$, where λ is the mean free path of atoms in a gas, and r_o is the cluster radius. We represent this criterion in the form

$$n^{1/3} N_a \ll (r_W \sigma_g)^{-1}, \qquad (6.17)$$

where σ_g is the gas-kinetic cross section that follows from formula (6.10). Note that the basic dependence in criterion (6.17) is determined by the number density N_a of gas atoms. The values of the parameter σ_g obtained from formula (6.10) using the measurable viscosity values [148] are given in Table 6.3.

Problem 6.11

Determine the mobility of large clusters in an atomic gas in the kinetic regime of atom–cluster collisions.

The cluster mobility in the kinetic regime of atom–cluster collisions is given by formula (6.3), which may be rewritten in the form

$$K_n = \frac{K_o}{n^{2/3}}, \quad K_o = \frac{e D_o}{T} = \frac{3e}{8\sqrt{2\pi m T} N_a r_W^2}. \qquad (6.18)$$

If a cluster of charge Z is moving in an electric field, the force acting on the cluster from the electric field $\mathbf{F} = Ze\mathbf{E}$ is equalized by the friction force (6.2). Here $\mathbf{E}$ is the electric field strength, and defining the cluster mobility K from the relation

Table 6.3 The gas-kinetic cross sections (Å^2) and obtained on the basis of formula (6.10).

T, K	100	200	300	400	600	800	1000
He	17	15	15	14	13	12	12
Ne	25	22	20	19	18	18	17
Ar	64	47	40	37	33	32	30
Kr	–	–	52	46	41	40	38
Xe	–	–	71	62	54	50	47
H_2	49	25	23	22	20	19	19
N_2	64	49	43	40	37	36	35
O_2	62	45	40	37	34	32	31
Air	63	48	42	39	36	34	34
CO	–	49	43	41	38	37	36

$\mathbf{w} = K\mathbf{E}$, we obtain for the cluster mobility

$$K_n = \frac{3Ze}{8\sqrt{2\pi m T} N r^2} = \frac{3Ze}{8\sqrt{2\pi m T} N r_W^2 n^{2/3}} , \quad \lambda \gg r_o , \tag{6.19}$$

This formula is valid if the cluster radius is small compared with the mean free path of atoms in a gas.

Formula (6.19) also follows from the Einstein relation (6.4) and expression (6.9) for the cluster diffusion coefficient, giving

$$K_n = \frac{Ze D_n}{T} = \frac{K_o}{n^{2/3}} , \quad K_o = \frac{3Ze}{8\sqrt{2\pi m T} N_a r_W^2} , \tag{6.20}$$

where we use formula (6.15) for the cluster diffusion coefficient D_n in a gas.

6.2
Dynamics of Cluster Motion in Gases

Problem 6.12

Clusters are injected into a gas flow. Analyze the character of relaxation of cluster motion to the motion of a gas in the diffusion regime of atom–cluster collisions.

The evolution of clusters in a gas flow is determined by the motion equation for the cluster in a gas:

$$M \frac{d\mathbf{w}}{dt} = 6\pi\eta r_o (\mathbf{v}_o - \mathbf{w}) ,$$

where M is the particle mass, $\mathbf{w}$ is the cluster velocity, and $\mathbf{v}_o$ is the flow velocity or the average velocity of atoms. From the solution of this equation we have

$$\mathbf{w} = \mathbf{v}_o \left[1 - \exp\left(-\frac{t}{\tau} \right) \right] ,$$

where

$$\tau = \frac{M}{6\pi r_o \eta} .$$

The relaxation time may be represented in the form

$$\tau = \tau_o \cdot \left(\frac{r_o}{r_W}\right)^2 = \tau_o n^{2/3} , \quad \tau_o = \frac{m_a}{6\pi r_W \eta} , \tag{6.21}$$

where n is the number of cluster atoms, η is the gas viscosity, and we use formula (2.5) for the cluster radius. In particular, for large water clusters moving in atmospheric air this equation gives $\tau_o = 4.5 \cdot 10^{-13}$ s.

Problem 6.13

Determine a typical time of relaxation for clusters injected into a gas flow in the kinetic regime of atom–cluster interactions.

Using formula (6.2) for the force that returns the cluster velocity to the gas velocity, we have the relaxation equation for the cluster velocity $\mathbf{w}$:

$$M\frac{d\mathbf{w}}{dt} = \mathbf{F} = \frac{8\sqrt{2\pi}}{3}\sqrt{mT} \cdot N_a r_o^2 (\mathbf{v}_o - \mathbf{w}) .$$

The solution of this equation has the usual form

$$\mathbf{w} = \mathbf{v}_o \left[1 - \exp\left(-\frac{t}{\tau}\right)\right] ,$$

with the relaxation time

$$\tau = \frac{3M}{8\sqrt{2\pi m T} N_a r_o^2} = \frac{n^{1/3}}{k_{ef} N_a} , \quad k_{ef} = \frac{8\sqrt{2\pi m T} r_W^2}{3m_a} . \tag{6.22}$$

Here M is the cluster mass, m_a is the mass of a cluster atom, m is the mass of a gas atom, and N_a is the number density of atoms in a gas. In particular, for large water clusters in air at temperature $T = 300$ K formula (6.22) gives $k_{ef} = 1.2 \cdot 10^{-10}$ cm^3/s, and for large copper clusters in argon $k_{ef} = 1.9 \cdot 10^{-11}$ cm^3/s.

Problem 6.14

Analyze the motion of a spherical cluster in a gas in the field of a propagating sound wave.

Propagation of a sound wave is accompanied by compression and rarefaction of a gas, that is, by a macroscopic motion of the gas. Let us take the velocity of a given point in the gas in the form $v_o \cos\omega t$, where ω is the frequency of the sound wave and v_o is its amplitude. Then the equation of motion of a cluster in the field of the sound wave has the form

$$\frac{dw}{dt} = \frac{1}{\tau}(v_o \cos\omega t - w) ,$$

where τ is the relaxation time of cluster motion. The stationary solution of this equation is

$$w = \frac{v_o}{1+\omega^2\tau^2} \cdot (\cos \omega t + \omega\tau \sin \omega t) . \qquad (6.23)$$

As follows from this formula, at small values of the parameter $\omega\tau \ll 1$ the cluster is moving together with the gas, whereas at large values of this parameter $\omega\tau \gg 1$ the amplitude of cluster vibrations is $\omega^2\tau^2$ times less than that for the gas, and the cluster vibration phase is shifted with respect to the sound wave by $\pi/2$.

Problem 6.15

In considering the motion of a spherical cluster in a gas in the field of a propagating sound wave, take into account nonuniformities resulting from the action of an acoustic wave on the gas and analyze the influence of this effect on the cluster motion.

Sound vibrations cause both gas motion and variations of the gas density. Let us represent the velocity of the gas in the form

$$v = v_o \cos(kx - \omega t) ,$$

where k is the wave vector and x is the direction of wave propagation. The wave frequency ω and the wave vector k are connected by the dispersion relation $\omega = c_s k$, where c_s is the sound velocity. We now take into account the spatial distribution of the gas velocity in a sound wave. The gas density is

$$\rho = \rho_o + \rho_1 \cos(kx - \omega t) ,$$

where ρ_o is the stationary gas density and ρ_1 corresponds to gaseous vibrations and is relatively small, that is, $\rho_1 \ll \rho_o$. The amplitude of the variation of the gas density is connected with the amplitude of the variations of the gas velocity owing to the continuity equation for gas motion:

$$\frac{\partial \rho}{\partial t} + \frac{\partial}{\partial x}(\rho v) = 0 .$$

From this it follows that $\rho_1 = \rho_o(kv_o/\omega) = \rho_o v_o/c_s$, and since $v_o \ll c_s$, we have $\rho_1 \ll \rho_o$.

Let us determine the force that acts on a cluster and arises owing to a nonuniformity of a gas in the field of a sound wave. This force is given by $F = \int p \cos\theta \, dS$, where the area of a surface cluster element is $dS = 2\pi r_o^2 d\cos\theta$, and θ is the angle between the motion direction and the cluster surface direction. In addition, the gas pressure is $p = p_o + p_1 \cos(kx - \omega t)$. Since $\partial p/\partial\rho = c_s^2$ in a sound wave, we have $p_1 = c_s^2 \rho_1$. To find the force that acts on a cluster ($kr_o \ll 1$), we replace x in the above equation with $x = r_o \cos\theta$ and expand it over a small cluster radius r_o.

This gives

$$F = \int_{-1}^{1} \rho_1 c_s^2 \cos(kx - \omega t) \cdot \cos\theta \cdot 2\pi r_o^2 d\cos\theta$$
$$= -\frac{4\pi}{3} r_o^3 \rho_1 c_s^2 \sin(kx - \omega t) = -\frac{4\pi}{3} r_o^3 \rho_o \omega v_o \sin(kx - \omega t)\,.$$

Then the equation of cluster motion takes the form

$$\frac{\partial w}{\partial t} = \frac{1}{\tau}\left[v_o \cos(kx - \omega t) - w\right] - \beta\omega v_o \sin(kx - \omega t)\,,$$

where $\beta = \frac{4\pi}{3} r_o^3 \rho_o / M = \rho_o/\rho \ll 1$ is a small parameter that is the ratio of the mass of the gas located in the cluster volume to the cluster mass, M is the cluster mass, ρ is the density of the cluster material, and ρ_o is the average gas density. Solving the above equation, we get

$$w = \frac{v_o}{1 + \omega^2\tau^2} \cdot \left[(1 + \beta\omega^2\tau^2)\cos\omega t + \omega\tau(1 - \beta)\sin\omega t\right]\,. \tag{6.24}$$

This formula differs from expression (6.23) by terms containing a small parameter β. These terms take into account the gaseous nonuniformity due to propagation of the sound wave. As a matter of fact, formula (6.24) corresponds to expansion over a small parameter $\beta = \rho_o/\rho$. One can see that the influence of the gas nonuniformity is essential at large values of $\omega\tau$. If $\omega^2\tau^2 \gg 1/\beta$, then cluster motion is determined mostly by the force owing to the nonuniformity of the gas. Then the vibration phase of cluster oscillations coincides with that of gas oscillations, but the amplitude of cluster oscillations is much less than the amplitude of gas oscillations.

Problem 6.16

Clusters are injected into a motionless gas with a velocity w. On the basis of the kinetic equation for clusters, analyze the character of cluster braking and the parameters of this process.

In considering a small cluster concentration, we ignore interactions of clusters with each other. Therefore we account in the Boltzmann kinetic equation for only elastic collisions between clusters and gas atoms. In addition, clusters do not influence the Maxwell distribution function $\varphi(\mathbf{v})$ of atoms, where $\mathbf{v}$ is the atom velocity. Therefore, the kinetic equation for the velocity distribution function $f(\mathbf{v})$ of clusters has the form

$$\frac{\partial f(\mathbf{v}_{cl})}{\partial t} = \int \left[f(\mathbf{v}'_{cl})\varphi(\mathbf{v}') - f(\mathbf{v}_{cl})\varphi(\mathbf{v})\right] g\, d\sigma\, d\mathbf{v}\,,$$

where $\mathbf{v}_{cl}$, $\mathbf{v}$ are the cluster and atom velocities before collision, $\mathbf{v}'_{cl}$, $\mathbf{v}'$ are the cluster and atom velocities after collision, $\mathbf{g} = \mathbf{v} - \mathbf{v}_{cl}$ is the relative velocity of atom–cluster collisions, and $d\sigma$ is the differential cross section of elastic atom–cluster collisions that leads to a given variation of velocities.

Multiplying this equation by the cluster momentum $M\mathbf{v}_{cl}$, where M is the cluster mass, and integrating over the cluster velocities $\mathbf{v}_{cl}$, we obtain the following equation for the average cluster momentum $\mathbf{P} = \int M\mathbf{v}_{cl} f(\mathbf{v}_{cl}) d\mathbf{v}_{cl}/N_{cl}$:

$$N_{cl}\frac{d\mathbf{P}}{dt} = \int \mu \mathbf{g} f(\mathbf{v}_{cl})\varphi(\mathbf{v})\sigma^*(g) d\mathbf{v}_{cl} d\mathbf{v} . \tag{6.25}$$

Here N_{cl} is the number density of clusters (the distribution functions are normalized by the number densities of corresponding particles), $\mu = Mm/(M+m)$ is the reduced atom–cluster mass (m is the atom mass), and $\sigma^*(g) = \int(1-\cos\vartheta)d\sigma$ is the transport cross section for elastic atom–cluster scattering that according to formula (2.3) is $\sigma^* = \pi r_o^2$, where r_o is the cluster radius.

In accounting for peculiarities of atom–cluster collisions, we take the cluster mass to be large ($M \gg m$), that gives $\mu = m$. Next, the cluster distribution function is relatively narrow and can be written as

$$f(\mathbf{v}_{cl}) = N_{cl}\delta(\mathbf{v}_{cl} - \mathbf{w})$$

where $\mathbf{w}$ is the cluster drift velocity. Hence, equation (6.25) takes the form

$$M\frac{d\mathbf{w}}{dt} = \mathbf{F} = -\mathbf{w}\cdot\frac{8\sqrt{2\pi m T}}{3}\cdot N_a r_o^2 ,$$

where we use formula (6.2) for the force acting on a moving cluster in a gas. It is convenient to represent this equation in the form

$$\frac{d\mathbf{w}}{dt} = -\nu\mathbf{w} , \quad \nu = \frac{8\sqrt{2\pi m T}}{3nm_a}\cdot N_a r_o^2 = \frac{8N_a r_W^2\sqrt{2\pi m T}}{3m_a}\cdot n^{-1/3} = \frac{N_a k_{ef}}{n^{1/3}} . \tag{6.26}$$

Here m is the mass of a gas atom, m_a is the mass of a cluster atom, n is the number of cluster atoms, and k_{ef} is given by formula (6.22). As is seen, $\nu = 1/\tau$, where τ is the relaxation time in accordance with formula (6.22). The solution of this equation gives

$$\mathbf{w} = \mathbf{w}_o\left[1 - \exp(-\nu t)\right] ,$$

where $\mathbf{w}_o$ is the initial cluster drift velocity. As follows from this formula, a typical time of a cluster stopping proceeds through of order of $M/m \sim n$ cluster collisions with gas atoms and depends as $n^{1/3}$ on cluster size.

Problem 6.17

Determine a typical cluster size in the kinetic regime of atom–cluster collisions if the drift cluster velocity differs from the gas flow velocity v and the gas flow velocity varies remarkably at a distance variation of order of l.

Assuming the kinetic regime of atom–cluster collisions to be valid, which corresponds to criterion (6.17), we require that the relaxation time (6.22) be relatively large. This corresponds to the criterion $\tau \gg l/v$ or

$$\frac{N_a}{n^{1/3}} \ll \frac{l}{vk_{ef}} , \tag{6.27}$$

where we use expression (6.22) for the relaxation time. Here N_a is the number density of atoms and n is the number of cluster atoms.

6.3
Cluster Motion in Gas Flows

Problem 6.18

A gas flow with clusters passes through the cylinder tube with a conic exit (Figure 6.1) to a vacuum. The angle α of the cone with respect to the tube axis is small, the cylinder radius is R_o, and the orifice radius is ρ_o. Assuming the temperature and pressure inside the tube to be constant, and the gas flows through the orifice with the speed of sound c_s, find the tube radius R in the conic part at which the cluster velocity differs from the flow velocity.

We consider the dynamics of a gas flow for a large orifice radius $\rho_o \gg \lambda$ in the kinetic regime of atom–cluster collisions, so the following criterion holds true:

$$\rho_o \gg \lambda \gg r_o\,, \tag{6.28}$$

where λ is the mean free path for gas atoms. Because of the small angle α, we model gas motion in a conic tube to be identical to that in a cylindrical tube, and from the conservation of gas consumption in each cross section we have

$$\rho_o^2 c_s = R^2 v(R)\,, \tag{6.29}$$

where $v(R)$ is the velocity in a cross section of radius R.

We take the cylinder frame of reference with the origin at the orifice, so the distance z from the origin is connected with the radius R of the corresponding cross section by the formula

$$z = \frac{R - \rho_o}{\tan\alpha}\,.$$

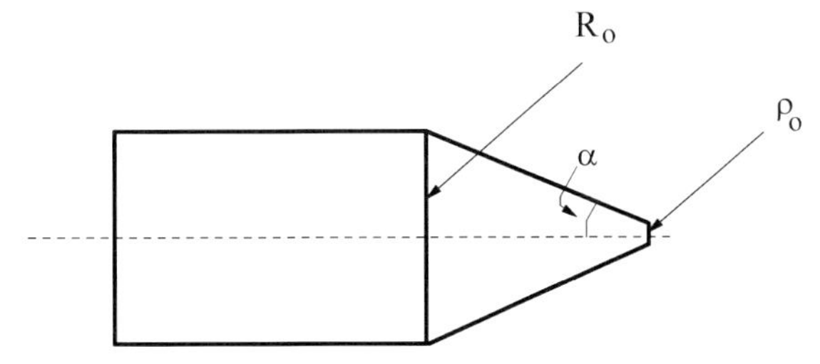

Fig. 6.1 The geometry of a cylindrical tube with a conic exit through which a gas with clusters flows. R_o is the radius of the cylindrical tube, ρ_o is the orifice radius, α is the conic angle, so $(R_o - \rho_o)/\cos\alpha$ is the length of the conic tube part.

From this we have for the flow velocity at the axis for a distance z from the origin

$$v(z) = \frac{c_s \rho_o^2}{(\rho_o + z \tan \alpha)^2} . \tag{6.30}$$

Introduce the parameter

$$\xi = 1 + \frac{z \tan \alpha}{\rho_o} . \tag{6.31}$$

Accounting for $v(z) = dz/dt$, we obtain from the solution of equation (6.30)

$$v(z) = -\frac{dz}{dt} = \frac{c_s}{\xi^2} ,$$

and $\xi(t)$ may be taken in the form

$$\xi^3 = \frac{t_o - t}{\tau_{or}} , \quad \tau_{or} = \frac{\rho_o}{3c_s \tan \alpha} . \tag{6.32}$$

The flow velocity $v(t)$ may be represented as

$$v(t) = c_s \left(\frac{\tau_{or}}{t_o - t} \right)^{2/3} . \tag{6.33}$$

Here t varies from zero, when $\xi = \xi_o$, up to $t_o - \tau_{or}$, when $\xi = 1$. These equations describe gas motion without cluster location in a tube.

We now analyze the behavior of clusters in a gas flow with a varied velocity. Note that if criterion (6.27) holds true, the relaxation of the cluster velocity to the average flow velocity is absent. This connection between the gas and cluster velocities is violated when the relaxation time (6.22) for a cluster coincides with the typical time of a flow $(dv/dz)^{-1}$. Here the flow velocity $v(R)$ depends on the tube radius R by formula (6.29), and the latter is connected to the distance from the orifice z by the relation $R = z \tan \alpha$ $(R \gg \rho_o)$. From this we have

$$\frac{dv}{dz} \sim \frac{\rho_o^2 c_s \tan \alpha}{R^3} .$$

Equating a time of variation in the flow velocity with the relaxation time (6.22), we find the radius R of the cross section of the conic tube part where the relaxation time is comparable to the time of flow velocity variation.

We now use the criterion according to which a typical time of variation of the flow velocity proceeds faster than a time of variation of the cluster velocity as a result of collisions with gas atoms. From this it follows that

$$R \sim \left(\frac{n^{1/3} c_s \rho_o^2 \tan \alpha}{N_a k_{ef}} \right)^{1/3} ,$$

where we assume $\rho_o < R < R_o$.

Problem 6.19

A gas flow with clusters passes through the cylinder tube with a conic exit (Figure 6.1) and then expands into a vacuum. Near the orifice of radius ρ_o, criterion (6.27) is fulfilled, that is, the drift velocity of clusters near the orifice is below the gas drift velocity. Determine the drift velocity of clusters.

The cluster velocity w_z along the tube axis is given by equation (6.26), which, accounting for the motion of a gas, has the form

$$\frac{dw_z}{dt} = \nu(v_z - w_z),$$

where the velocity of the gas flow v_z is determined by equation (6.29). We assume that at the initial time $t = 0$ the velocities of the gas flow and clusters coincide. Using formula (6.33) for the flow velocity, we represent this equation in the form

$$\frac{d}{dt}\left[w_z \exp(\nu t)\right] = \nu v e^{\nu t} = \nu c_s \exp(\nu t)\left(\frac{\tau_{or}}{t_o - t}\right)^{2/3},$$

where t_o is a parameter, so a test gas point goes through the exit orifice at $t = t_o - \tau_{or}$. At the beginning, $t = 0$, we have

$$w(0) = v(0) = \frac{c_s \tau_{or}^{2/3}}{t_o^{2/3}},$$

where τ_{or} is determined by formula (6.32).

Solving the above equation with the indicated initial condition, we obtain

$$w(t) = \nu c_s \tau_{or}^{2/3} e^{-\nu t} \int_0^t \frac{e^{\nu t'} dt'}{(t_o - t')^{2/3}} = \nu c_s \tau_{or}^{2/3} \int_0^t d\tau \frac{e^{-\nu\tau}}{(\tau + \tau_{or})^{2/3}},$$

where we use a new variable $\tau = t - t'$. The condition of small initial drift velocities of a cluster and a gas flow compared with their final values near the orifice leads to a small parameter $\nu\tau_{or} \ll 1$. Then we have for the drift velocity of clusters at the orifice based on the solution of this equation [149]

$$w_o = c_s \left(\nu\tau_{or}\right)^{2/3} \Gamma\left(\frac{1}{3}\right) = 2.68 c_s \left(\nu\tau_{or}\right)^{2/3}. \tag{6.34}$$

Thus, in the limit $\nu\tau_{or} \ll 1$ the equilibrium between the drift velocity of clusters and buffer gas atoms is violated, and the drift velocity of clusters near the orifice is less than the flow velocity.

Problem 6.20

A gas flow with clusters passes through a rectangular tube whose width does not vary. This tube finishes with a gap through which a gas with clusters flows in a vacuum. Near the gap criterion (6.27) is fulfilled, and the tube width is relatively large when the gas flow and cluster velocities become different. Determine the drift velocity of clusters near the gap assuming the gas drift velocity there to be equal to the speed of sound c_s.

We now solve the equation for relaxation of the cluster drift velocity in a gas flow with a varied velocity in the same way as was done in the derivation of formula (6.34) for a tube with axial symmetry. Denote by l the distance between inclined tube planes, and the gas flow velocity $v(l)$ is given by analogy with formula (6.29) from the conservation of gas consumption:

$$l v(l) = \Delta c_s ,$$

where Δ is the gap width at the tube exit. Let us introduce a new variable by analogy with formula (6.32):

$$\xi = 1 + \frac{2z \tan \alpha}{l} ,$$

where $z = (l - \Delta)/2 \tan \alpha$ is the distance from the exit gap. Then we have for this geometry

$$v(z) = -\frac{dz}{dt} = \frac{c_s}{\xi^2} ,$$

and solving this equation for $\xi(t)$, we get

$$\xi^2 = \frac{t_o - t}{\tau_{or}} , \quad \tau_{or} = \frac{l}{4 c_s \tan \alpha} ,$$

which gives for the gas flow velocity

$$v(t) = c_s \left(\frac{\tau_{or}}{t_o - t} \right)^{1/2} .$$

Next, solving the motion equation for the cluster drift velocity, we obtain for this value at the gap [149]

$$w_o = c_s \left(\nu \tau_{or} \right)^{1/2} \Gamma \left(\frac{1}{2} \right) = c_s \sqrt{\pi \nu \tau_{or}} . \tag{6.35}$$

As is seen, for this geometry the final cluster drift velocity is small compared to the flow velocity owing to the large value for the relaxation time, but the dependence on a small parameter is different for this tube geometry than that according to formula (6.34) for a cylindrical tube.

Problem 6.21

The goal of an impactor [150, 151] is to extract particles in a gas flow from a gas, and a simple example of an impactor is a nondirect tube through which a gas with small particles flows. Consider a gas flow with a velocity u through a cylindrical tube whose continuation is a torus with radius R. Estimate the radius of small particles in a flow when these particles leave the gas flow in a toroidal region through holes of walls and may be collected there.

When a gas moves along a ring, each atom presses onto an external wall with a force $F = mu^2/R$, that creates a pressure

$$p_u \sim \frac{mu^2}{R} N_a \lambda ,$$

where m is the atomic mass, u is the flow velocity, R is the tube radius, N_a is the number density of gas atoms, λ is the mean free path of atoms, and $\lambda \ll R$. Note that the condition of smallness of this pressure compared with the gas pressure $p = N_a T$ is valid if the criterion

$$v_T \gg u\sqrt{\frac{\lambda}{R}}$$

holds true, where $v_T = \sqrt{8T/(\pi m)}$ is the average velocity of atoms. As is seen, this criterion is valid if the flow velocity is small compared with the sound speed.

The force acting on a small particle (or a cluster) located in a flow is Mu^2/R, where $M = nm_a$ is the particle mass, n is the number of cluster atoms, and m_a is the mass of an individual particle atom. Then the drift velocity of particles toward the walls in the kinetic regime is, according to formula (6.3),

$$w = \frac{K}{e}F = \frac{3Mu^2}{8\sqrt{2\pi mT}N_a r_o^2 R} = \frac{3m_a u^2 n^{1/3}}{8\sqrt{2\pi mT}N_a r_W^2 R} ,$$

where r_W is the Wigner–Seitz radius defined by relation (2.6). In the diffusion regime of atom–particle interaction ($r_o \gg \lambda$), the drift velocity of particles toward the walls according to formula (6.8) is

$$w = \frac{Mu^2}{6\pi r_o \eta R} = \frac{m_a u^2 n^{2/3}}{6\pi r_W \eta R} .$$

The time of particle drift from the center to the walls of the torus tube is R/w, whereas the time for a particle to pass in a gas flow through a torus is $2\pi R/u$. From this we obtain the criterion of particle collision with torus walls

$$w > \frac{u}{2\pi} .$$

In particular, in the diffusion regime of atom–particle interaction this equation gives for the number n of particle atoms

$$n > \left(\frac{3r_W \eta R}{m_a u}\right)^{3/2} . \tag{6.36}$$

Note that the measure of particle drift to the walls of a crooked tube is the Stokes number St [152, 153]:

$$\mathrm{St} = \frac{u\tau}{R} , \tag{6.37}$$

where u is the flow velocity, τ is the drift time of a particle to the walls, and R is the curvature radius. Since $\tau = R/w$, where w is the drift velocity of a small particle to the walls, the Stokes number is $\mathrm{St} = u/w$. Hence, the Stokes number compares the flow velocity and the drift velocity of particles to the walls. Therefore, it characterizes the possibility for particles to reach the walls.

Problem 6.22

In considering the gas flow through an orifice in the gas dynamic regime, we assume the radius ρ_o of the orifice through which the flow leaves a chamber to be large compared to the mean free path of gas atoms. Assuming the velocity of the flow through the orifice to be equal to the sound speed and the gas dynamic regime of argon flowing to be realized at room temperature, find a range of argon consumptions S.

The character of the flow of a buffer gas with clusters is represented in Figure 6.2. The buffer gas flow far from the orifice proceeds along straightforward lines, and clusters move together with the gas and may propagate in a transverse direction owing to diffusion in the gas. Because the orifice size is large compared to the mean free path of atoms, the gas-dynamic regime of flow is realized. This means that the flow remains continuous also after the orifice. We use a simple model for passage of the flow through the orifice in this regime assuming the gas drift velocity at the orifice to be equal to the sound speed at any point of the orifice cross section. This follows from experimental results and is valid with an accuracy of up to 20%.

The sound speed in argon at room temperature is $c_s = \sqrt{\gamma T/m} = 3.2 \cdot 10^4$ cm/s, where $\gamma = c_p/c_V = 5/3$, the argon temperature is taken to be $T = 293$ K, and m

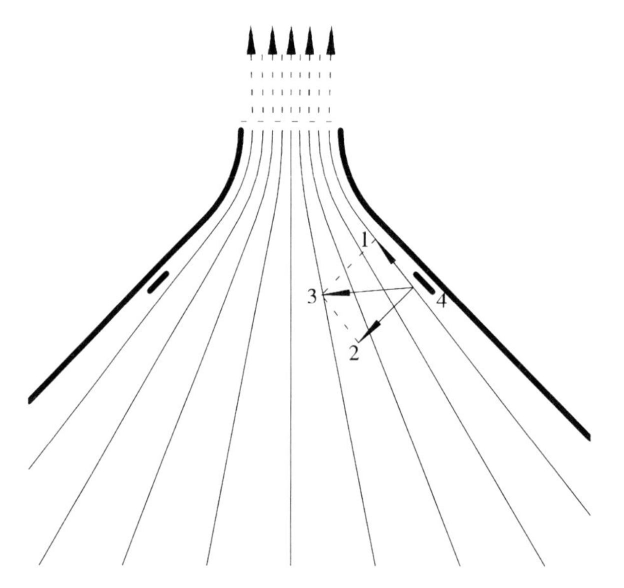

Fig. 6.2 Character of gas flow through a cylindrical tube with a conic exit.

is the argon atomic mass. The consumption of a gas with an admixture of clusters S under the assumption that the gas flows through an orifice of radius ρ_o with the sound speed is

$$S = \pi \rho_o^2 N c_s \,.$$

We define the consumption S as the number of atoms that crossed the orifice per unit time, and N is the number density of argon atoms in the chamber from which argon flows in a vacuum. One can see that consumption is proportional to

$$S = C p \rho_o^2 \,, \tag{6.38}$$

where p is the argon pressure. Let us measure the consumption S in units of *sccm* (standard cubic centimeter per minute) which corresponds to the number of atoms located in 1 cm^3 per minute under normal conditions or $2.69 \cdot 10^{19}/\mathrm{min} = 4.48 \cdot 10^{17}\,\mathrm{s}^{-1}$. If the argon pressure is expressed in *Torr* and the orifice radius ρ_o in *mm*, we have for argon at room temperature $C = 74$.

Taking the gas-kinetic cross section for collision of two argon atoms at room temperature to be $\sigma_g = 3.7 \cdot 10^{-15}\,\mathrm{cm}^2$, we find that the condition of the gas-kinetic regime of the flow, if the mean free path of atom $\lambda = 1/(N\sigma_g)$, is small compared to the initial radius under the assumption

$$p\rho_o \gg 8 \cdot 10^{-4} \,,$$

where p is measured in *Torr* and the orifice radius is measured in *mm*. Correspondingly, we have the following criterion for the gas-dynamic regime of argon passing through an orifice:

$$\frac{S}{\rho_o} \gg 10^{-3} \,,$$

where the argon consumption is measured in *sccm* and the orifice radius is measured in *mm*.

Note that the argon consumption is conserved in each cross section, that allows us to connect this value with the velocity in each cross section. In particular, let us determine the flow velocity in the cylindrical part of the tube. Taking the gas flow to be laminar in the cylindrical tube, we have for the flow velocity u inside the tube

$$u(\rho) = u_{max}\left(1 - \frac{\rho^2}{R_o^2}\right) \,,$$

where ρ is the distance from the axis, R_o is the radius of the cylindrical tube, and u_{max} is the flow velocity at the flow center. Assuming that the flow passes through an orifice of radius ρ_o with sound speed c_s and the number density of buffer gas atoms is identical at the orifice and in the cylindrical part, we find from the conservation of the flow rate in the cylinder part of the tube

$$u_{max} = 2c_s \frac{\rho_o^2}{R_o^2} \,. \tag{6.39}$$

Problem 6.23

Argon with an admixture of silver clusters containing $n = 10^4$ atoms on average flows through a chamber with a conic exit (Figure 6.1) with angle $\alpha = 30°$ through an orifice of a radius of $\rho_o = 2\,\text{mm}$. Determine the drift velocity of clusters at the orifice if the argon pressure is $p = 0.1\,\text{Torr}$. Assuming that clusters do not attach to walls, find the increase in the cluster concentration near the walls compared with a region of the cylindrical tube part.

The drift velocity of clusters near the orifice is given by formula (6.34). We have for the parameters of this formula $\tau_{or} = 3.6 \cdot 10^{-6}\,\text{s}$ according to formula (6.32) and $\nu = 2.6 \cdot 10^3\,\text{s}^{-1}$ according to formula (6.26). As a result, we obtain ($c_s = 3.2 \cdot 10^4\,\text{cm/s}$) $w_o = 3.8 \cdot 10^3\,\text{cm/s}$.

If clusters do not attach to walls and their growth finishes in the cylindrical part of the chamber (Figure 6.1), we find an increase in the cluster concentration near the orifice:

$$\frac{c_s}{w_o} = 8.5\,.$$

Problem 6.24

Argon with an admixture of silver clusters flows through a cylindrical chamber having a conic exit (Figure 6.1). Taking the parameters of the previous problem for flowing argon as a buffer gas ($p = 0.1\,\text{Torr}$), the chamber parameters ($\alpha = 30°, \rho_o = 2\,\text{mm}$), and the cluster size ($n = 10^4$ atoms on average), determine the radius of the cross section of the conic part of the chamber R_* so that starting from this cross section, the drift velocity of the clusters becomes lower than the argon drift velocity. The number density of the clusters is very low.

When clusters are moving in a buffer gas flow in a chamber, as shown in Figure 6.1, and at the beginning have the drift velocity of the flow, they have the flow drift velocity in the conic part of the tube, if a time for establishment of this velocity is small enough. After this the cluster drift velocity w_o does not vary over the course of the transition to the exit. Hence, the cross-section radius R_* where the equilibrium is violated is given by

$$w_a(R_*) = w_o\,,$$

where w_a is the flow drift velocity. Accounting for the fact that the argon velocity at the orifice is the sound speed c_s, we find from the conservation of the flow rate for the drift velocity w_a in the cross section of radius R

$$w_a = c_s \left(\frac{\rho_o}{R}\right)^2\,,$$

where ρ_o is the orifice radius. From this we find

$$R_* = \rho_o \left(\frac{c_s}{w_o}\right)^{1/2}\,.$$

In particular, under the given conditions we have $c_s/w_o = 8.5$, and $R_* \approx 6\,\text{mm}$.

Problem 6.25

Argon with an admixture of silver clusters flows through a chamber shaped as in Figure 6.1. Under the conditions of Problem 6.23, determine a radius of the conic tube part starting from which a part of the clusters located in the neutral gas flow attach to walls.

For this goal we compare a typical time $\tau_{\rm dif} = \rho^2/(4D_{\rm cl})$ for attachment to walls due to cluster diffusion in buffer gas, where ρ is a current tube radius, and a typical drift time $\tau_{\rm dr} = \rho/(w_{\rm o}\tan\alpha)$ on a distance ρ. We assume that in this region the drift velocity of the clusters differs from that of the buffer gas. From the relation $\tau_{\rm dif}(\rho_*) = \tau_{\rm dr}(\rho_*)$ we find a typical tube radius ρ_*, so the attachment of clusters to walls takes place at this cross section:

$$\rho_* \sim \frac{4D_{\rm cl}}{w_{\rm o}\tan\alpha}\,.$$

Under the conditions of Problem 6.23 ($p = 0.1$ Torr, $\alpha = 30°$, $\rho_{\rm o} = 2$ mm, $n = 10^4$) we have $w_{\rm o} = 3.8\cdot10^3$ cm/s. Next, on the basis of formula (6.5), we have $D_{\rm cl} = 190\,{\rm cm^2/s}$ under the given parameters ($D_{\rm o} = 4.1\cdot10^3\,{\rm cm^2/s}$). This gives $\rho_* = 0.35$ cm, that is, under the given conditions the attachment of clusters to walls starts close to the orifice.

Problem 6.26

Under the conditions of Problem 6.23 determine the portion of clusters that attach to walls while argon is flowing through an orifice.

We use a simple dependence for the number density of clusters $N_{\rm cl}$ in a conic tube:

$$N_{\rm cl} = C(z)\left(R_{\rm o}^2 - \rho^2\right)\,,$$

where the coordinate z is directed along the flow, ρ is the distance from the tube axis, and $R_{\rm o}$ is the tube radius. Assuming the angle α between the walls and the axis to be small, we use an approximation of cylindrical walls, and the cluster flux to the walls is

$$j = -D_{\rm cl}\frac{\partial N_{\rm cl}}{\partial\rho} = 2R_{\rm o}C(z)\,.$$

Let us introduce the total number of clusters per unit length of a tube Q,

$$Q(z) = \int_0^{R_{\rm o}} N_{\rm cl}(\rho)\cdot 2\pi\rho\,d\rho = \frac{\pi}{2}R_{\rm o}^4 C(z)\,,$$

and the evolution of this quantity along a tube is determined by the balance equation

$$\frac{dQ}{dt} = w_{\rm o}\frac{dQ}{dz} = -2\pi R_{\rm o}j = -\frac{4D_{\rm cl}}{R_{\rm o}^2}Q\,,$$

where $w_o = dR_o/dt$ is the cluster drift velocity that does not change in the orifice region. Using the connection $\tan\alpha dz = dR_o$ (R_o is the tube radius) and solving this equation, we obtain for the total number of clusters in a given cross section

$$Q = Q_o \exp(-\xi)\,, \quad \xi = \frac{4D_{cl}}{w_o \rho_o \tan\alpha}\,. \tag{6.40}$$

This formula shows the character of a decrease in the number of clusters in the course of their drift along a tube. In particular, under the conditions of Problem 6.23 ($\rho_o = 2\,\text{mm}, \alpha = 30°, w_o = 3.8 \cdot 10^3\,\text{cm/s}, D_{cl} = 180\,\text{cm}^2/\text{s}$) we have the passed portion of clusters $\xi = 1.6$, $Q/Q_o = 20\%$.

Problem 6.27

Determine the condition of cluster attachment to walls if clusters are located in a laminar flow of a buffer gas in the chamber shown in Figure 6.1 consisting of cylindrical and conic parts. Apply the results to the parameters of Problem 6.23.

In ignoring the diffusion of clusters, we have that clusters move together with a buffer gas along the current lines. But because of cluster inertia, they shift from the current lines if they turn. In the case of the chamber of Figure 6.1, current lines turn by an angle β at the transition from the cylindrical chamber part to the conic one, and this angle is given by

$$\tan\beta = \frac{\rho}{R_o}\tan\alpha\,,$$

where α is the turn angle for the tube boundary, R_o is the radius of the cylindrical tube part, and ρ is the distance from the center. For the laminar character of motion, the velocity of the buffer gas flow u inside the tube is

$$u(\rho) = u_{max}\left(1 - \frac{\rho^2}{R_o^2}\right)\,,$$

where u_{max} is the flow velocity in the flow center.

When a current line turns, a cluster located in a flow obtains a component $u(\rho)\sin\beta$ that is perpendicular to the current line. This component will be damped according to formula (6.26), and the solution of this equation gives a cluster shift Δx with respect to its current line that is

$$\Delta x = \frac{u(\rho)\sin\beta}{\nu}\,,$$

where the parameter ν is determined by formula (6.26). In particular, near chamber walls $\Delta\rho = R_o - \rho \ll R_o$ and $\alpha \approx \beta$ this gives

$$\Delta x = \Delta\rho\frac{2u_{max}\sin\alpha}{\nu R_o}\,.$$

From this it follows that cluster attachment to the walls due to a turn of the current lines is absent if the following criterion holds:

$$u_{max} < \frac{\nu R_o}{2 \sin \alpha} .$$

In particular, for the parameters of Problem 6.23 ($R_o = 3$ cm, $\alpha = 30°$, $\nu = 2.6 \cdot 10^3$ cm/s) this criterion is fulfilled if $u_{max} < 8 \cdot 10^3$ cm/s. Note that under the conditions of Problem 6.23 ($\rho_o = 2$ mm), according to formula (6.39), $u_{max} = 290$ cm/s.

6.4 Pairwise Association of Clusters Limited by Motion in a Gas

Problem 6.28

Determine the mean free path for a cluster in a gas in the kinetic regime of atom–cluster collisions.

If a cluster moves with a certain velocity v, the direction of its motion varies after many collisions with atoms. Indeed, a typical cluster momentum is $P \sim \sqrt{TM}$, where T is a thermal energy of atoms and clusters, $M = n m_a$ is the cluster mass, where n is the number of cluster atoms and m_a is the mass of the cluster atoms, which is assumed to be of the same order of magnitude as the mass of gas atoms m. The cluster momentum variation as a result of collision with a gas atom is $\Delta P \sim \sqrt{Tm} \sim n^{1/2} P$, which is relatively small, that is, relaxation of the cluster velocity proceeds after many collisions, and a typical time of cluster relaxation is given by formula (6.22) for the kinetic regime of atom–cluster collisions.

If a cluster has a velocity $\mathbf{v}_o$ at the initial time that we denote as $t = 0$, the average velocity at time t is given by

$$\mathbf{v}(t) = \mathbf{v}_o \exp(-t/\tau) ,$$

owing to relaxation with a relaxation time τ. From this we obtain for a path Λ during relaxation

$$\Lambda = \int v(t) dt = v_o \tau .$$

On the basis of formula (6.22) for the relaxation time in the kinetic regime of atom–cluster collisions we obtain the mean free path Λ of clusters with respect to variation in the initial direction of the cluster velocity

$$\Lambda \sim \frac{v n^{1/3}}{k_{ef} N_a} , \tag{6.41}$$

where v is a typical cluster velocity.

Problem 6.29

Determine the rate constant of association of two neutral clusters in a dense buffer gas assuming the cluster number density to be relatively small.

Under these conditions, in the course of association each cluster changes the direction of its motion many times, which corresponds to the criterion

$$N_{cl}\Lambda^3 \ll 1 . \qquad (6.42)$$

Hence, the approach of neutral clusters results from their diffusion in a gas, and the rate of this process is determined by the Smoluchowski formula (4.22). Using the liquid model for the cluster structure, we have that the contact of two clusters leads to their joining, so the rate constant of cluster association limited by their diffusion is given by

$$k_{dif} = 4\pi D(r_1 + r_2) ,$$

where the diffusion of clusters in a gas is responsible for their approach, and D is the diffusion coefficient of clusters in a gas. Above we took into account that the association radius is the sum of radii r_1 and r_2 of joining clusters.

In the determination of the diffusion coefficient of clusters approaching in the course of their association, we take into account that the motion of each cluster is independent of that of the others. Then, according to the nature of diffusional motion, we have for the relative distance between clusters at time t

$$\overline{\mathbf{R}^2} = \overline{(\mathbf{R}_1 - \mathbf{R}_2)^2} = 6Dt ,$$

where $\mathbf{R}_1, \mathbf{R}_2$ are coordinates of joined clusters and a bar means averaging over time. By definition, we have for the diffusion coefficients D_1, D_2 of each cluster

$$\overline{\mathbf{R}_1^2} = 6D_1 t , \quad \overline{\mathbf{R}_2^2} = 6D_2 t .$$

We have

$$\overline{\mathbf{R}^2} = \overline{(\mathbf{R}_1 - \mathbf{R}_2)^2} = \overline{\mathbf{R}_1^2} + \overline{\mathbf{R}_2^2} = 6(D_1 + D_2)t ,$$

and taking into account that the motion of each cluster is independent, we get that $\overline{\mathbf{R}_1\mathbf{R}_2} = \overline{\mathbf{R}_1} \cdot \overline{\mathbf{R}_2} = 0$. From this it follows for the rate constant of association of two clusters that

$$k_{dif} = 4\pi(D_1 + D_2)(r_1 + r_2) , \qquad (6.43)$$

and the diffusion coefficient of clusters in a gas is given by formula (6.15) for the kinetic regime of atom–cluster collisions, and by formula (6.16) for the diffusion regime of atom–cluster collisions. In particular, in the diffusion regime of atom–cluster collisions we have, on the basis of formula (6.16) for the cluster diffusion coefficient,

$$k_{dif} = \frac{2T}{3\eta}\left(\frac{1}{r_1} + \frac{1}{r_2}\right)(r_1 + r_2) . \qquad (6.44)$$

Problem 6.30

Determine the rate constant of association limited by the cluster motion in a gas for positively and negatively charged clusters in a dense buffer gas.

Assuming for simplicity the cluster charge to be $\pm e$, we find the rate of cluster recombination that consists in the approach of clusters to their contact when clusters are joined. Hence, the recombination process in a dense gas is restricted by a cluster approach that is slowed by frictional forces. Denoting by v_+ and v_- velocities of positively and negatively charged clusters, we have that these clusters approach toward each other owing to an electric field created by another charged cluster. This electric field strength is $E = e/R^2$, when the distance between clusters is R, and the velocities of clusters moving toward each other are $v_+ = EK_+$, $v_- = EK_-$, where K_+, K_- are the mobilities of the clusters. Thus, the rate for negatively charged clusters to intersect a sphere of radius R around a positively charged cluster is $J_- = 4\pi R^2(v_+ + v_-)N_- = 4\pi e(K_+ + K_-)N_-$, where N_- is the number density of negative clusters. Define the recombination coefficient k_{rec} of charged clusters according to the balance equation

$$\frac{dN_+}{dt} = -k_{rec}N_+N_- = -J_-N_+ \,,$$

where N_+ is the number density of positively charged clusters. As a result, we get the Langevin formula [87] for the recombination coefficient of positive and negative cluster ions in a dense gas:

$$k_{rec} = 4\pi e(K_+ + K_-) \,. \tag{6.45}$$

The recombination coefficient is multiplied by Z_1Z_2 if the charges of the clusters Z_1e and Z_2e differ from the electron charge e.

Using formula (6.20) for the cluster mobility in a dense gas for the kinetic regime of atom–cluster collisions, or formula (6.16) for the cluster mobility in the diffusion regime of atom–cluster collisions, one can find the recombination coefficient of clusters on the basis of the Langevin formula (6.45). In particular, in the kinetic regime of atom–cluster collisions we have

$$k_{rec} = \frac{3\sqrt{\pi}Z_1Z_2e^2}{2\sqrt{2mT}N_a}\left(\frac{1}{r_+^2} + \frac{1}{r_-^2}\right) \,, \quad \lambda \gg r_+, r_- \,, \tag{6.46}$$

where r_+, r_- are the radii of positively and negatively charged clusters. As is seen, the dependence of the recombination coefficient of charged clusters on cluster size is $k_{rec} \sim 1/r_o^2$, where $r_o \sim r_+, r_-$. In particular, taking the cluster radii to be identical $r_+ = r_- = r_o$ and their charges to be $Z_1 = Z_2 = 1$, we obtain at temperature $T = 300$ K for clusters in atmospheric air $k_{rec}r_o^2N_a = 0.79\,\mathrm{cm^2/s}$.

Problem 6.31

Compare the rate constants for association of two neutral liquid clusters and two oppositely charged clusters in a dense buffer gas.

Taking the cluster radii to be $r_1 = r_2 = r_o$ and their charge to be Z, we obtain for the rate constant of their joining due to diffusion in a gas on the basis of formula (6.43)

$$k_{dif} = 16\pi D r_o ,$$

where D is the diffusion coefficient of each cluster in a gas. For the association rate of these clusters as a result of their approach due to Coulomb attraction we have from formula (6.45)

$$k_{rec} = 8\pi Z^2 e K ,$$

where K is the mobility of each cluster in a gas. Using the Einstein relation (6.4), we find for the ratio of these rate constants

$$\frac{k_{rec}}{k_{dif}} = \frac{Z^2 e^2}{2 T r_o} . \tag{6.47}$$

As is seen, this ratio is independent of both gas and the cluster material. In particular, if $Z = 1$ and $T = 300$ K, this ratio is 1 at $r_o = 0.03$ μm. The mechanism of cluster association due to the interaction of the charges is preferable at small cluster sizes.

Problem 6.32

Determine the rate constant of association of charged and neutral clusters in a dense gas as a result of a polarization interaction between them.

Under this condition, the approach of charged and neutral clusters in a buffer gas proceeds owing to the polarization interaction between them, and the interaction potential of clusters at a distance R between them is given by

$$U(R) = -\frac{\alpha Z^2 e^2}{2 R^4} ,$$

where Z is the cluster charge and α is the polarizability of a neutral cluster. The velocity of motion v of clusters toward each other is $v = F(K_1 + K_2)/e$, where $F = -\partial U/\partial R$ is the attraction force due to cluster interaction and K_1, K_2 are the mobilities of clusters. We have for the approach velocity

$$v = \frac{2\alpha Z^2 e}{R^5}(K_1 + K_2) .$$

Solving the equation $v = dR/dt$, we obtain the association time for clusters τ if initially they are located at a distance R:

$$\frac{1}{\tau} = \frac{12\alpha Z^2 e(K_1 + K_2)}{R^6} . \tag{6.48}$$

Let us determine the average time of association for a charged cluster that is located in a gas of neutral clusters, assuming that at the beginning p neutral clusters are located in a spherical volume V_o with a charged cluster at the center. Let us introduce the probability dP that the nearest cluster is located at a distance R from the center in a volume range dV. We have

$$dP = p\frac{dV}{V_o}\cdot\left(1-\frac{V}{V_o}\right)^p ,$$

where the volume V occupies the space inside a sphere with radius R. Since the number density of neutral clusters is $N = p/V_o$, taking $p \gg 1$, we obtain

$$dP = NdVe^{-NV} ,$$

where N is the number density of neutral clusters. This distribution function gives for the average time of joining of a neutral cluster with the nearest neutral one $(\overline{V^2} = \overline{(4\pi R^3/3)^2} = 2/N^2, \overline{R^6} = 9/(8\pi^2 N^2))$

$$\overline{\tau} = \int \tau V dP = \frac{\overline{R^6}}{12\alpha Z^2 e(K_1 + K_2)} = \frac{3}{32\pi^2 \alpha Z^2 e(K_1 + K_2)N^2} . \tag{6.49}$$

From this we have for the effective rate constant of the association process defined as $k_{\text{pol}} = 1/(N\overline{\tau})$

$$k_{\text{pol}} = \frac{32\pi^2 \alpha Z^2 e N(K_1 + K_2)}{3} . \tag{6.50}$$

Note that the effective rate constant of this pairwise process depends on the cluster number density because of the nonlinear character of cluster approach in a dense gas.

Problem 6.33

Compare the rate constants of association of charged and neutral clusters if their approach toward each other proceeds owing to the polarization interaction or to the diffusion motion of clusters in a gas.

Let us take the cluster radii to be $r_1 = r_2 = r_o$ and the mobilities in a buffer gas to be $K_1 = K_2 = K$. Formula (6.43) gives $k_{\text{as}} = 16\pi D r_o$ for the association rate constant of clusters due to their diffusion in a gas. According to formula (6.50), the rate constant of cluster association due to the polarization interaction between clusters is $k_{\text{pol}} = 64\pi^2 \alpha Z^2 e N K/3$. The ratio of these rate constants with accounting for the Einstein relation (6.4) gives

$$\frac{k_{\text{pol}}}{k_{\text{as}}} = \frac{4\pi}{3}\frac{\alpha Z^2 e^2 N}{r_o T} . \tag{6.51}$$

As is seen, the polarization mechanism of cluster association is important at a high number density N of clusters.

$$N \gg \frac{r_o T}{Z^2 e^2 \alpha} .$$

Though we are based on formula (6.43) for the association rate, which relates to a dense buffer gas ($\lambda \ll r_o$), this estimate is also suitable for a rare buffer gas. For a metal cluster whose polarizability is $\alpha = r_o^3$ according to formula (5.3), we have the above criterion in the form

$$N r_o^2 \gg \frac{T}{Z^2 e^2} \, .$$

This criterion may be rewritten in the form

$$\lambda_{cl} \ll \frac{Z^2 e^2}{T} \, ,$$

where λ_{cl} is the mean free path of a cluster with respect to collisions with buffer gas atoms.

7
Charging of Clusters in Ionized Gas

Attachment of electrons and ions to clusters in a weakly ionized gas is an effective process, and clusters may be a sink for charged atomic particles in an ionized gas. A cluster acquires an equilibrium charge in an ionized gas, and the cluster field influences the character of the charging process. There are various regimes of the charging process depending on the cluster size and the charge density in an ionized gas. As for the nature of the charging process for a dielectric cluster, ions (or electrons) of an ionized gas find active centers on the cluster surface and form a bound state with the cluster, and ions of the opposite charge recombine with these bound ions. Thus, the processes on the cluster surface in the course of cluster charging are similar to those during chemical reactions in a gas where a cluster plays the role of a catalyst. Usually, there are relatively few occupied active centers on the cluster surface, and the rate of cluster charging is determined by the processes in the gas volume rather than those on the cluster surface.

7.1
Attachment of Ions to Clusters in Dense Gas

Problem 7.1

Determine the current of positive ions of a dense ionized gas on the cluster surface if electrons and ions perturb weakly the field of a negatively charged cluster in a region that is responsible for cluster charging.

In considering the behavior of clusters in a weakly ionized gas, we assume that the collision of electrons and ions with the cluster surface leads to transfer of their charges to the cluster. Since the ionization degree is small, positive ions and electrons transfer their charge to the cluster independently as a result of their contact, and because of a high mobility of electrons, clusters are charged negatively. If an ionized gas contains positive and negative ions, the sign of the cluster charge depends on the ion types. For example, aerosols, small particles of the Earth atmosphere, are charged depending on their sorts, which in turn depend on air admixtures at local atmosphere points. Charged aerosols fall down in the atmosphere under the action of gravity that creates an electric charge and the electric fields

Cluster Processes in Gases and Plasmas. Boris M. Smirnov

ISBN: 978-3-527-40943-3

of the Earth's atmosphere. As a result, electrical phenomena in the atmosphere near the Earth's surface are determined by small water aerosols located in the atmosphere. Below we consider processes of cluster charging in an ionized gas as a result of transport phenomena for electrons and ions of the ionized gas near the cluster.

We model a cluster by a spherical particle with a radius r_o, and this radius is large compared with the radius of action of atomic forces. Then chemical interaction of a cluster with atomic particles (atoms, electrons, and ions) in the vicinity of a cluster surface may be reduced to the hard sphere model, and the cross section of collisions of a charged atomic particle with a cluster is given by formula (2.2), that is, a strong interaction occurs between them at a distance r_o from the cluster center. Below we take the probability of charge transfer as a result of contact between a charged atomic particle and the cluster surface to be one. Under these model assumptions we evaluate below the ion currents to the cluster surface.

We deal with a dense gas that surrounds the cluster under consideration. The following criterion holds true:

$$r_o \gg \lambda ,$$

where λ is the mean free path of electrons or ions in a gas. Charging of the cluster results from attachment of electrons or ions to its surface and is hampered by their motion toward the cluster. We now determine the currents of plasma electrons and ions toward the cluster surface if the motion in a gas of atomic particles is determined by their diffusion and drift under the action of a cluster electric field. Then we have for the current I of positive ions toward the cluster surface at a distance r from it

$$I = 4\pi r^2 \left(-D_+ \frac{dN}{dr} + K_+ E N \right) e .$$

Here N is the current number density of ions, D_+, K_+ are the diffusion coefficient and the mobility of positive ions, e is the ion charge, that is equal to the electron charge, $E = Ze/r^2$ is the electric field strength acting on an ion from the charged cluster, and Z is the cluster charge. The first term of this expression corresponds to diffusion motion, the second term relates to drift motion, and we take into account the absence of recombination involving positive ions outside the cluster.

This relation may be considered as the equation for the ion number density in a space. Assuming that a positive ion loses its charge after contact with the cluster surface, we have the boundary condition $N(r_o) = 0$. Another boundary condition far from the cluster is $N(\infty) = N_o$, where N_o is the equilibrium number density of atomic ions in an ionized gas. Assuming the cluster electric field to be relatively small, we use the Einstein relation (6.4) between the mobility K_+ and the diffusion coefficient D_+ of an ion that has the form

$$K_+ = \frac{eD_+}{T} ,$$

where T is the ion temperature. This gives for the positive ion current toward the cluster surface

$$I = -4\pi r^2 D_+ e \left(\frac{dN}{dr} - \frac{Ze^2 N}{Tr^2} \right) .$$

Assuming the absence of ion recombination in a space and hence the ion current to be independent of r, we solve this equation with the above boundary conditions. This solution gives, using the boundary condition $N(r_o) = 0$ for the ion number density,

$$N(r) = \frac{I}{4\pi D_+ e} \int_{r_o}^{r} \frac{dr'}{(r')^2} \exp\left(\frac{Ze^2}{Tr'} - \frac{Ze^2}{Tr} \right)$$
$$= \frac{IT}{4\pi D_+ Ze^3} \left[\exp\left(\frac{Ze^2}{Tr_o} - \frac{Ze^2}{Tr} \right) - 1 \right] . \quad (7.1)$$

Because of the boundary condition $N(\infty) = N_o$ we find from this the Fuks formula [154] for the ion current on the cluster surface:

$$I = \frac{4\pi D_+ N_+ Ze^3}{T\{\exp[Ze^2/(Tr_o)] - 1\}} . \quad (7.2)$$

Problem 7.2

Consider the limiting case of the Fuks formula (7.2) for the ion current on the surface of a neutral cluster for the diffusion regime of ion–cluster collisions.

In the limiting case $Z \to 0$ the Fuks formula (7.2) is transformed into the Smoluchowski formula (4.22) for the diffusion flux J_o of neutral particles to the surface of an absorbed sphere of radius r_o [89]:

$$J_o = \frac{I}{e} = 4\pi D_+ N_+ r_o . \quad (7.3)$$

This formula is valid if the cluster radius is large compared to the mean free path of atoms, which allows us to consider the diffusion character of atomic motion near the cluster surface.

Problem 7.3

Consider the limiting case of the Fuks formula (7.2) for the current of negative ions on the surface of a positively charged cluster for the diffusion regime of ion–cluster collisions.

The Fuks formula (7.2) describes the positive ion current if the cluster charge has the same sign. To obtain the expression for the negative ion current toward a positively charged cluster, it is necessary to substitute in this formula $Z \to -Z$, and the parameters of positive ions must be replaced by the parameters of negative ions. Then we have for the negative ion current toward the positively charged cluster

$$I_- = -\frac{4\pi D_- N_- Z e^3}{T \cdot \left[1 - \exp\left(-\frac{Ze^2}{Tr_o}\right)\right]} . \tag{7.4}$$

In the limit $Ze^2/(r_o T) \gg 1$ this formula is transformed into the Langevin formula [87]

$$I_- = \frac{4\pi Z e^3 D_-}{T} = 4\pi Z e^2 K_- , \tag{7.5}$$

from which formula (6.45) follows for the rate constant of join of oppositely charged particles in a dense gas. Note that the flux of negatively charged ions on the cluster surface J_- and the current I_- of ions are connected by the relation $I_- = eJ_-$.

Problem 7.4

Generalize the formula for the current of positive and negative ions on the surface of a positively charged spherical cluster for the diffusion regime of ion–cluster collisions.

Introducing the reduced variable $x = |Z|e^2/(r_o T)$, one can represent the Fuks formula (7.2) in the form [36]

$$I_+ = \begin{cases} eJ_o x/(e^x - 1) , & Ze^2 > 0 , \\ eJ_o x/(1 - e^{-x}) , & Ze^2 < 0 , \end{cases} \quad \lambda \ll r . \tag{7.6}$$

Here J_o is the diffusion flux of neutral atomic particles on the surface of an absorbed sphere with radius r_o according to the Smoluchowski formula (7.3); $J_o = 4\pi D N r_o$, where N is the number density of atoms and D is the diffusion coefficient of atoms in a gas. In the limiting case $x \gg 1$ this formula is transformed into the Langevin formula (7.5).

Problem 7.5

Determine the average charge of a spherical cluster located in a quasineutral plasma if its radius is large in comparison with the mean free path of gas atoms.

In a quasineutral plasma $N_+ = N_-$ the currents of positive and negative ions given by formula (7.2) and (7.4) are identical. This yields for the cluster equilibrium charge

$$Z = \frac{r_o T}{e^2} \ln \frac{D_+}{D_-} . \tag{7.7}$$

From this it follows that the cluster has a positive charge if $D_+ > D_-$, that is, positive ions have greater mobility than negative ones. Note that formula (7.7) is valid under the condition

$$r_o \gg \frac{e^2}{T} .$$

If this criterion holds true, an individual ion captured by the cluster does not vary essentially the cluster electric potential. An additional criterion of validity of the above expressions is $r_o \gg \lambda$. At room temperature, the first criterion gives $r_o > 0.06\mu$, and according to the second criterion for atmospheric air we have $\lambda = 0.1\mu$.

Problem 7.6

Find the charge of a large spherical cluster located in a nonquasineutral plasma if its radius is large compared to the mean free path of gas atoms.

Taking the number densities of positive N_+ and negative ions N_- far from a charged cluster to be different, we repeat the derivation of the previous problem for ion currents and the average cluster charge Z. As a result, instead of formula (7.7) we obtain

$$Z = \frac{r_o T}{e^2} \ln \frac{D_+ N_+}{D_- N_-} . \tag{7.8}$$

Problem 7.7

Determine the average charge of a small spherical particle located in a plasma of glow gas discharge if the particle radius is large compared to the mean free path of gas atoms.

A plasma of the positive column of glow gas discharge includes electrons and ions located in an electric field, so that the velocity distribution function of electrons differs from the Maxwell distribution. The Fuks formula (7.2), without using the Einstein relation between the mobility and diffusion coefficient of attached charged particles, has the following form for the current I_e of electrons to a cluster:

$$I_e = \frac{4\pi K_e N_e Z e^3}{1 - \exp[-Ze^2/(T_{ef} r_o)]} .$$

Here we introduce the effective electron temperature T_{ef} as $T_{ef} = eD_e/K_e$ instead of the electron temperature T_e, and K_e, D_e are the mobility and diffusion coefficients of electrons in a gas. We take the cluster charge to be $-Z$ and use that a typical electron energy eD_e/K_e in a gas discharge plasma is large in comparison to a typical ion energy. Then according to the Langevin formula (7.5) the ion current I_+ on the cluster surface is

$$I_+ = 4\pi Z e^3 K_+ N_+ ,$$

where N_+ is the number density of positive ions far from the cluster and K_+ is the mobility of positive ions. Equalizing the fluxes of charged particles, taking into account the plasma quasineutrality $N_e = N_+$, we get for the cluster charge

$$Z = -r_o \frac{D_e}{e K_e} \ln \frac{K_e}{K_+} \,. \tag{7.9}$$

According to this formula, we have $Ze^2/(r_o T) \gg 1$, where T corresponds to the temperature of both electrons and ions.

Problem 7.8

A large spherical cluster is injected into a weakly ionized quasineutral gas where the diffusion coefficients for positive and negative ions are nearby. Consider the character of time evolution of the cluster charge.

Let us introduce the mean diffusion coefficient of ions $D = (D_+ + D_-)/2$ and their difference $\Delta D = |D_+ - D_-|$. As a result of charging, the cluster obtains the charge

$$Z_o = \frac{r_o T}{e^2} \cdot \frac{\Delta D}{D} \,, \tag{7.10}$$

where we account for $\Delta D \ll D$. The equation of charging of the cluster is

$$\frac{dZ}{dt} = \frac{I_- - I_+}{e} \,.$$

Using the initial condition $Z(0) = 0$ and the Fuks formula (7.4) for the charging current, we have

$$\frac{dZ}{dt} = \frac{4\pi N_o e^2 Z}{T} \cdot \left[D_- - D_+ \exp\left(\frac{Ze^2}{Tr_o}\right)\right] \cdot \left[\exp\left(\frac{Ze^2}{Tr_o}\right) - 1\right]^{-1} ,$$

where N_o is the average number density of positive and negative ions of the ionized gas.

From the equation for Z_o we have $Z_o e^2/(Tr_o) = \Delta D/D \ll 1$, and since $Z \le Z_o$ in the course of cluster charging, one can expand the above equation over a small parameter $Ze^2/(Tr_o)$. Then the above equation of cluster charging takes the form

$$\frac{dZ}{dt} = \frac{Z_o - Z}{\tau} \,,$$

where

$$\frac{1}{\tau} = \frac{4\pi N_o e^2 D}{T} = 2\pi\Sigma \,. \tag{7.11}$$

Here $\Sigma = N_o e(K_+ + K_-) = 2N_o e^2 D/T$ is the conductivity coefficient of the ionized gas.

The solution of the above equation at the initial condition $Z(0) = 0$ is

$$Z = Z_o(1 - e^{-t/\tau}) \,.$$

As is seen, a typical time for establishment of the equilibrium cluster charge does not depend on the cluster radius.

Problem 7.9

Determine the current of cluster charging in a plasma for an arbitrary cluster shape if the cluster size greatly exceeds the mean free path of plasma ions. Compare the result with the case of a spherical particle assuming the diffusion coefficients of positive and negative ions to be close.

Let us repeat the deduction of the Fuks formula (7.2) for a cluster of an arbitrary shape. For definiteness, we assume that we have a unipolar plasma containing only positive ions. Then the ion flux toward the cluster surface is equal to

$$\mathbf{j} = -D_+ \nabla N + K_+ \mathbf{E} N .$$

Using the Einstein relation (6.4) $K_+ = eD_+/T$ and introducing the cluster electric potential φ from the relation $\mathbf{E} = -\nabla\varphi$, we get

$$\mathbf{j} = -D_+ \left(\nabla N + \frac{eN}{T} \nabla \varphi \right) = -D_+ e^{-e\varphi/T} \nabla \left(N e^{e\varphi/T} \right) .$$

From this we obtain for the rate of attachment of ions to the cluster surface

$$J = -\mathbf{jS} = eD_+ e^{-e\varphi/T} \mathbf{S} \nabla (N e^{e\varphi/T}) ,$$

where $\mathbf{S}$ is the area of the cluster surface and the surface electric potential of the cluster is $\varphi = \text{const}$. Because ions are not lost in space, the ion current is identical for any equipotential surface. This condition and the boundary condition on the cluster surface, where the ion number density is zero, allow us to determine the distribution of the ion number density in space.

Take the equation of the cluster surface as $\xi = \text{const}$, and because the equation for the cluster potential in space is $\Delta\varphi = 0$, the spatial coordinates may be separated, and ξ is one of these variables. Then the equation for the cluster potential takes the form

$$\frac{d}{d\xi} \left(S_\xi \frac{d\varphi}{d\xi} \right) = 0 ,$$

where S_ξ is the area of an equipotential surface with $\xi = \text{const}$. Taking the cluster electric potential to be zero far from the cluster, we obtain

$$\varphi(\xi) = a \int_\xi^\infty \frac{d\xi}{S_\xi} ,$$

where a is the integration constant. Evidently, the integration constant is proportional to the cluster charge and we find this value by taking a spherical cluster when $\xi = r, \varphi = Ze/r,\ S_\xi = 4\pi r^2$ (r is the distance from the cluster center, and the cluster charge Z is expressed in electron charges). The above relation is fulfilled if $a = 4\pi Ze$.

Thus, we have

$$d\varphi = -\frac{4\pi Zed\xi}{S_\xi}\,.$$

Let us use this in the equation for the number density of ions by using the electric potential as the variable. This gives for the ion current I toward the cluster surface

$$I = -ej_\xi S_\xi = eD_+ e^{-e\varphi/T} S_\xi \frac{d}{d\xi}\left(Ne^{e\varphi/T}\right)$$
$$= 4\pi Ze^2 D_+ e^{-e\varphi/T} \frac{d}{d\varphi}\left(Ne^{e\varphi/T}\right).$$

Solving this equation with the above boundary condition, we obtain, similar to the Fuks formula (7.2),

$$I_+ = \frac{4\pi D_+ N_o Ze^3}{T[\exp(e\varphi_o/T) - 1]}\,,$$

where N_o is the ion number density far from the particle and φ_o is the electric potential of the particle on its surface if it is zero far from the particle. Using the cluster's electric capacity $C = Ze/\varphi_o$, it is convenient rewrite this relation in the form

$$I_+ = \frac{4\pi D_+ N_o Ze^3}{T\cdot\left(\exp\frac{Ze^2}{TC} - 1\right)}\,. \tag{7.12}$$

In the case of a spherical particle we take $C = r_o$, and formula (7.12) is transformed into formula (7.2) for the current on the surface of a spherical particle. Note that the Smoluchowski formula (4.22) for the total flux of neutral atomic particles on the surface of a small cluster of any form is

$$J = \frac{I}{e} = 4\pi DN_o C\,.$$

This formula is transformed into formulas (4.22) and (7.3) for a spherical cluster.

7.2
Field of a Charged Cluster in Dense Ionized Gas

Problem 7.10

Determine the electric potential of a cluster in a quasineutral plasma with respect to the plasma if the cluster size is large compared with the mean free path of plasma ions in a gas. Assume that the average cluster charge is large, the cluster shape may differ from spherical one, and screening of the cluster field by the plasma charge is relatively small.

We find the charge of a cluster of an arbitrary shape in a quasineutral plasma by equalizing the currents of positive and negative ions on the cluster surface. This gives

$$Z = \frac{CT}{e^2} \ln \frac{D_-}{D_+} \tag{7.13}$$

instead of formula (7.7). In this formula we assume the cluster charge to be negative, which corresponds to $D_- > D_+$. This formula coincides with formula (7.7) for a spherical cluster, where $C = r_o$ is the cluster radius.

From this it follows for the cluster electric potential that

$$\varphi_o = \frac{Ze}{C} = \frac{T}{e} \ln \frac{D_-}{D_+} . \tag{7.14}$$

As is seen, the potential energy $e\varphi_o$ corresponding to ions on the cluster surface is on the order of an ion thermal energy T. This follows from the nature of the process, which requires the equality of currents of positive and negative ions on the cluster's surface.

Problem 7.11

Summarize the connection between the character of cluster charging in a weakly ionized plasma and the parameters of this process.

We have three parameters that have the size dimensionality r_o, e^2/T, λ and determine the character of charging of a cluster in a plasma. If $r_o \gg e^2/T$, the change of the cluster electric potential resulting from the attachment of one ion to the cluster is low compared with T/e; if $r_o \ll e^2/T$, the cluster can have only a single charge, and the ion interaction potentials with charged and neutral clusters differ strongly. If $r_o \gg \lambda$, the diffusion character of ion motion in a gas determines their attachment of ions to the cluster; if $r_o \ll \lambda$, cluster charging results from successive collisions with ions in the kinetic regime. Thus, the character of charging of a cluster depends on the relation among the above parameters of the size dimensionality.

Problem 7.12

Analyze the evolution of charging of a small spherical cluster in a unipolar dense plasma.

The equation of cluster charging in time has the form

$$\frac{dZ}{dt} = \frac{I}{e} .$$

On the basis of formula (7.6) for the ion current on the cluster surface, we transform this equation into

$$\frac{dx}{d\tau} = \frac{x}{e^x - 1} ,$$

where $x = Ze^2/Tr_o$, $\tau = t\cdot(e/Tr_o)\cdot(1/I_> + 1/I_<)^{-1}$. The solution of this equation in the limiting cases has the form

$$x = \begin{cases} \tau\,, & \tau \ll 1 \\ \ln(1+\tau)\,, & \tau \gg 1 \end{cases}\,.$$

The accurate solution of this equation, if at the beginning the cluster is neutral, $x(0) = 0$, is as follows:

$$\tau = Ei(x) - \ln x\,.$$

Problem 7.13

Analyze the screening of the field of a spherical charged cluster that is located in a weakly ionized dense gas containing electrons and ions. The cluster charge is large $Z \gg 1$.

According to the Fuks formula (7.1), the number densities of electrons and ions in the field of a negatively charged cluster are

$$N_e(r) = N_o\frac{e^{-x} - e^{-x_o}}{1 - e^{-x_o}}\,, \quad N_i(r) = N_o\frac{1 - \exp(x - x_o)}{1 - e^{-x_o}}\,. \tag{7.15}$$

Here r is the distance from the cluster, N_e and N_i are the number densities of electrons and singly charged positive ions located in a gas, N_o is their number density far from the cluster, $x = Ze^2/Tr$, and $x_o = Ze^2/Tr_o$; the ion temperature T is equal to the electron temperature. According to the Fuks formula (7.15) the number densities of both electrons and ions are equal to zero at the cluster surface ($x = x_o$) and tend to the equilibrium number density N_o of electrons and ions far from the charged cluster. From this we have for the difference between the ion and electron number densities

$$\Delta N \equiv N_i(r) - N_e(r) = N_o\cdot\frac{1 - \exp(x - x_o) + \exp(-x_o) - \exp(-x_o)}{1 - e^{-x_o}}\,.$$

Note the symmetry of ΔN with respect to the transformation $x \to x_o - x$. Next, this number density difference has a maximum at $x = x_o/2$, that is, $r = 2r_o$, where this quantity is

$$\Delta N = N_o\cdot\frac{1 - e^{-x_o/2}}{1 + e^{-x_o/2}}\,.$$

The cluster electric potential φ is the solution of the Poisson equation

$$\Delta\varphi \equiv \frac{1}{r}\frac{d^2(r\varphi)}{dr^2} = 4\pi e\Delta N\,. \tag{7.16}$$

Note that the Fuks approach is valid if the shielding of the cluster charge by a surrounding plasma is small. Roughly, this criterion has the form

$$N_o r_o^3 \ll 1\,. \tag{7.17}$$

Problem 7.14

Ascertain the conditions when the Debye screening of the cluster charge takes place.

In the limit $x \ll 1$ the difference between the ion and electron number densities is $\Delta N = x = Ze^2/Tr$, and the Poisson equation (7.16) takes the form

$$\frac{1}{r}\frac{d^2(r\varphi)}{dr^2} = \frac{4\pi Z e^2}{T} .$$

The solution of this equation far from the cluster ($x \ll 1$ or $r \gg Ze^2/T$) is given by

$$\varphi(r) = \frac{Ce}{r}\exp\left(-\frac{r}{r_D}\right) , \quad r_D = \sqrt{\frac{T}{4\pi Z N_o e^2}} , \tag{7.18}$$

where r_D is the Debye–Hückel radius [155]. The constant C coincides with the cluster charge Z if criterion (7.17) holds true; in other cases this is the cluster charge taking into account the cluster screening, and formula (7.18) with $C = Z$ holds true only in the cluster vicinity.

Problem 7.15

Give the criterion of a weak screening of the cluster charge in a surrounding weakly ionized gas.

We have for a charge Q of a layer with a depth Δr due to a difference in the ion and electron number densities ΔN according to the Gauss formula [156, 159]

$$Q(\Delta r) = \int\limits_{r_o}^{r_o+\Delta r} 4\pi r^2 dr \Delta N ,$$

and $Q \ll Z$. Since according to formula (7.15) we have $\Delta N = x_o - x$ near the cluster surface $x_o - x \ll x_o$, the charge in a layer of thickness Δr is

$$Q(\Delta r) = 2\pi N_o (\Delta r)^2 \frac{Ze^2}{T} .$$

Taking the mean free path of electrons and ions λ in a gas as the thickness of a charged layer ($\Delta r = \lambda$), we obtain the above criterion ($Q \ll Z$) in the form

$$2\pi N_o \lambda^2 \frac{e^2}{T} \ll 1 , \tag{7.19}$$

instead of criterion (7.17). In addition, we used the criterion $r_o \gg \lambda$ of the diffusion regime of ion–cluster scattering.

Problem 7.16

For the diffusion regime of ion–cluster collisions generalize the Fuks formula (7.2) to the case where criterion (7.19) is violated and screening of the cluster field is strong.

Repeating the derivation of the Fuks formula (7.2), we start from the equation for the current I_+ that goes through a sphere of a radius of r, where the center is the cluster center:

$$I_+ = 4\pi r^2 \left(-D_+ \frac{dN}{dr} + w_+ N\right) e\,.$$

Here N is the current number density of ions, D_+ is the ion diffusion coefficient, and w_+ is the ion drift velocity. We have for the drift velocity of ions

$$w = K_+ \frac{F}{e}\,,$$

where K_+ is the ion mobility, $F = dU/dr$ is the force that acts on an ion due to a self-consistent field, and $U(r)$ is the potential of this field that is created by the Coulomb field of the charged cluster and also by the induced field in a surrounding plasma. Introducing the effective ion temperature on the basis of the Einstein relation (6.4) $T = eD_+/K_+$, we obtain for the ion current

$$I_+ = 4\pi r^2 e K_+ \left(-T\frac{dN}{dr} - N\frac{dN}{dr}\right) = -4\pi r^2 e K_+ \frac{d}{dr}\left[N \exp(U/T)\right]\,.$$

As a result, instead of the Fuks formula (7.2) we have

$$I_+ = \frac{4\pi D_+ N r_o U(r_o) e}{T\{\exp[U(r_o)/(T)] - 1\}}\,. \tag{7.20}$$

Of course, this formula is transformed into the Fuks formula (7.2) if we take $U(r_o) = Ze^2/r_o$.

Basing on the derivation of the Fuks formula (7.2), we use the assumptions

$$K_+(r) = \text{const}\,, \quad T(r) = \text{const}$$

in formula (7.20). As a matter, these assumptions require the above values to be independent of the electric field strength and are not fulfilled in reality. Indeed, these assumptions mean that the mobility and the diffusion coefficient of ions are independent of the electric field strength at large fields when the average ion energy significantly exceeds a thermal energy.

One can avoid the assumption $T(r) = \text{const}$ in formula (7.20). Then to solve the above differential equation we replace the value $U(r_o)/T$ by

$$\chi(r_o) = \int_{r_o}^{\infty} \frac{\frac{dU}{dr} dr}{T(r)}\,.$$

As a result, we have

$$I_+ = \frac{4\pi D_+ N r_o \chi(r_o) e}{T\{\exp[\chi(r_o)] - 1\}}, \quad \chi(r_o) = \int_{r_o}^{\infty} \frac{\frac{dU}{dr} dr}{T(r)}. \tag{7.21}$$

This formula is based on the assumption $K_+(r) = \text{const}$ and allows us to express the ion current on the surface of a charged cluster or particle under conditions where a self-consistent distribution of a plasma is established in the vicinity of a charged particle located in a plasma.

Problem 7.17

In a dense ionized gas the Debye–Hückel screening results from thermodynamic equilibrium for plasma electrons and ions. Near a spherical particle or cluster of radius r_o and charge Z this screening may result from the interaction of the charged particle with an ionized gas. Determine the criterion when this takes place.

For Debye–Hückel screening, the interaction potential $U(r)$ of the charged cluster of charge Z_o with singly charged atomic particles is

$$U(r) = -\frac{Z_o e^2}{r} \exp\left(-\frac{r}{r_D}\right), \tag{7.22}$$

where the Debye–Hückel radius is determined by the relation

$$\frac{1}{r_D^2} = 4\pi N_o e^2 \left(\frac{1}{T_e} + \frac{1}{T_i}\right). \tag{7.23}$$

Here T_e, T_i are the electron and ion temperatures, respectively, and N_o is the number density of plasma electrons or ions far from the particle.

Evidently, this formula holds true if the thermodynamic equilibrium for the spatial distribution of electrons and ions is established near the particle surface, which requires that the following criterion be valid for the mean free path λ of electrons and ions in a surrounding gas:

$$\lambda \ll r_o .$$

This corresponds to the diffusion regime of electron and ion collisions with a particle surface.

Another criterion requires a small interaction potential of electrons or ions with the particle field compared to their thermal energy. This gives

$$U(r) \ll T_e, T_i .$$

From this it follows that the Debye–Hückel screening takes place near the particle surface if its charge is restricted, that is,

$$Z_o \ll \frac{r_o T}{e^2}, \tag{7.24}$$

where $T \equiv T_e, T_i$.

Problem 7.18

Prove that the Debye–Hückel screening takes place for an ionized gas surrounding a charged particle of a large charge Z_o at large distances from the particle.

Below we consider the electric potential of an isolated charged particle in a plasma, accounting for a self-consistent character of this interaction potential, that is, the particle electric field establishes the distribution of electrons and ions, that in turn influences the electric potential in a plasma. Because of the spherical symmetry of the problem, the electric potential created by a charged particle does not depend on angles but only on the distance r from a particle.

We start from the Poisson equation for the electric potential $U(r)/e$ that is created by the charged particle and a surrounding weakly ionized gas:

$$-\frac{1}{r^2}\frac{d}{dr}\left[r^2\frac{dU(r)}{dr}\right] = -\frac{1}{r}\frac{d^2}{dr^2}(rU(r)) = 4\pi e^2[N_i(r) - N_e(r)], \qquad (7.25)$$

where $N_e(r)$, $N_i(r)$ are, respectively, the number densities of electrons and ions. Integrating the Poisson equation (7.25), one can reduce it to the form

$$U(r) = -\frac{Z_o e^2}{r} + e^2\int_{r_o}^{\infty} dr'\cdot(r')^2\int_0^{\pi} d\vartheta' \sin\vartheta' \cdot\int_0^{2\pi} d\varphi' \frac{\Delta N(r')}{\sqrt{r^2 - 2rr'\cos\Theta + (r')^2}}, \qquad (7.26)$$

where we denote $\Delta N(r) = N_i(r) - N_e(r)$, and Θ is the angle between the radius vector of a given point and the location of an electron or ion. Use the relation

$$\cos\Theta = \cos\vartheta\cos\vartheta' + \sin\vartheta\sin\vartheta'\cos(\varphi - \varphi'),$$

where ϑ, φ and ϑ', φ' are polar and azimutal angles of vectors $\mathbf{r}$ and $\mathbf{r}'$. Note formula (7.26) which follows from the Poisson equation includes the difference $\Delta N(r)$ for the number densities of electrons and ions, that in turn depends on the electric potential $U(r)/e$. Hence, an additional equation is necessary for the number densities of electrons and ions that that involves this potential. Thus, determination of the electric potential of a charged particle in a plasma is a self-consistent problem that is not exhausted by the Poison equation. Nevertheless, it may be used in a region where the influence of the particle field on a plasma is weak.

Indeed, expanding the denominator of formula (7.26) over the Legendre polynomials and integrating over angles, taking into account the symmetry of the spatial charge distribution, one can reduce it to the form [156, 159]

$$U(r) = -\frac{Z_o e^2}{r} + \frac{4\pi e^2}{r}\int_{r_o}^{r} dr'\cdot(r')^2\Delta N(r') + 4\pi e^2\int_r^{\infty} r'dr'\Delta N(r'). \qquad (7.27)$$

This also gives the following expression of the electric field strength:

$$e E(r) = -\frac{Z_o e^2}{r^2} + \frac{4\pi e^2}{r^2} \int_{r_o}^{r} (r')^2 dr' \varDelta N(r')$$
$$= -\frac{Z_o e^2}{r^2} + \frac{eQ}{r^2} = -\frac{Z_o e^2 - eQ}{r^2}, \qquad (7.28)$$

where the charge of a plasma in a given region is determined by the relation

$$Q(r) = 4\pi e \int_{r_o}^{r} (r')^2 dr' \varDelta N(r'), \qquad (7.29)$$

and relation (7.28) is the Gauss formula. Now

$$Z(r) = Z_o - \frac{Q(r)}{e}$$

is the noncompensated charge of the particle and plasma at a given distance from the particle.

We now consider a range of large distances r from a charged particle when the interaction potential for electrons and ions is small compared with a thermal energy, so that

$$N_i(r) = N_o \exp\left[-\frac{U(r)}{T_i}\right] = N_o\left[1 - \frac{U(r)}{T_i}\right], \quad N_e(r) = N_o\left[1 - \frac{U(r)}{T_e}\right].$$

This gives

$$\varDelta N = N_o U(r)\left[\frac{1}{T_e} + \frac{1}{T_i}\right] = \frac{U(r)}{4\pi r_D^2},$$

and at large distances from the particle $r \gg r_D$ the Debye electric potential (7.22) is the solution of the Poisson equation (7.25).

7.3 Attachment of Ions to Clusters in Rare Gas

Problem 7.19

Find the ion current on the cluster surface in a rare buffer gas.

The case of a rare gas that surrounds a cluster corresponds to the kinetic regime in ion–cluster collisions. In this case a charged cluster does not violate the spatial distribution of atomic particles in its vicinity, and criterion (4.25) is required:

$$r_o \ll \lambda .$$

Assuming that each contact of a colliding ion or electron with the cluster surface leads to the transfer of its charge to the cluster and the classical character of collisions, we express below currents of ions and electrons moving to the cluster surface through classical cross sections of their attachment to a charged cluster. Because the distance of closest approach $r_{\min}$ of classical particles is connected with the impact parameter of their collision ρ by the relation $1-\rho^2/r_{\min}^2 = Ze^2/(r_{\min}\varepsilon)$, where ε is the collision energy in the center-of-mass frame of reference, we have that the cross section of ion collision with the cluster surface under these conditions is

$$\sigma = \pi r_o^2 \left(1 - \frac{Ze^2}{r_o \varepsilon}\right) , \quad \varepsilon \geq \frac{Ze^2}{r_o} ,$$

where r_o is the cluster radius. When $Z > 0$, that is, the charges of the cluster and ions have the same sign, it is necessary to account for the fact that if $\varepsilon \leq Ze^2/r_o$, then the cross section is zero, because in this case the potential energy of repulsion of the cluster and colliding ion exceeds their kinetic energy near the cluster surface. If a charge sign is different for the colliding ion and cluster, the cross section of their contact collision is

$$\sigma = \pi r_o^2 \left(1 + \frac{|Z|e^2}{r_o \varepsilon}\right) , \quad \varepsilon \geq 0 .$$

Averaging over the Maxwell velocity distribution of ions the rate constants obtained on the basis of these formulas, we find for an identical charge sign of a colliding ion and cluster for the averaged rate constant

$$k = \langle v\sigma \rangle = k_o \int_x^\infty t\,dt \exp(-t) \left(1 - \frac{x}{t}\right) = k_o e^{-x} ,$$
$$t = \frac{\varepsilon}{T} , \quad x = \frac{|Z|e^2}{r_o T} , \quad k_o = \sqrt{\frac{8T}{\pi M}} \pi r_o^2 , \tag{7.30}$$

where k_o is the rate constant of ion collision with a neutral cluster when this collision leads to their contact and M is the ion mass. In the case of attraction of a cluster and an ion, the average rate constant of this collision is given by

$$k = k_o \int_0^\infty t\,dt \exp(-t) \left(1 + \frac{x}{t}\right) = k_o (1 + x) . \tag{7.31}$$

Using the reduced parameter $x = |Z|e^2/(r_o T)$ and the probability ξ that the ion will transfer its charge to the cluster as a result of their contact, we combine the above relations for the rate of ion attachment to the cluster and obtain

$$J_< = \xi k_o N_i \cdot (1 + x) , \quad Z < 0 ;$$
$$J_> = \xi k_o N_i \cdot \exp(-x) , \quad Z > 0 . \tag{7.32}$$

Problem 7.20

Determine the cluster charge in a rare ionized gas with identical temperatures of electrons and ions.

This regime of cluster charging in an equilibrium plasma with identical temperatures of electrons and ions takes place in an afterglow plasma. The flux of attaching electrons in this regime according to formula (7.30) is

$$j_{at} = N_e \sqrt{\frac{T}{2\pi m_e}} \exp\left(-\frac{|Z|e^2}{r_o T}\right) , \tag{7.33}$$

where m_e is the electron mass, Z is the cluster charge expressed in electron charges, r_o is the cluster radius, and we assume that each contact of an electron with a cluster surface leads to electron attachment. Assuming the criterion

$$|Z|e^2/(r_o T) \gg 1$$

to be fulfilled, we have for the ion flux $j_+ = J_>$ to the cluster surface

$$j_+ = \sqrt{\frac{T}{2\pi M}} N_i \frac{|Z|e^2}{r_o T} , \tag{7.34}$$

where M is the ion mass, N_i is the number density of positive ions, and $N_i = N_e$ for the quasineutral plasma. Equalizing these current densities, we get for a cluster charge under this regime of cluster charging

$$x = \ln\left(\frac{1}{x}\sqrt{\frac{M}{m_e}}\right) , \quad |Z| = x \cdot \frac{r_o T}{e^2} = x \cdot \frac{r_W n^{1/3} T}{e^2} , \tag{7.35}$$

where r_W is the Wigner–Seitz radius. Table 7.1 gives the solution of this equation for an inert buffer gas with positive atomic ions and for the case where nitrogen is a buffer gas and its plasma contains molecular nitrogen ions.

Problem 7.21

Combine the formulas for the ion flux on a cluster surface within the limits of rare and dense ionized gas.

One can combine the above results when criterion (6.6) is valid or the opposite relation is fulfilled. For this goal we consider first the limiting case $\lambda \gg r_o$ with

Table 7.1 The solution of (7.35) [36].

Buffer gas	He	Ne	Ar	Kr	Xe	N_2
x	3.26	3.90	4.17	4.47	4.65	4.03

a general boundary condition on a cluster's surface $N(r) = N_1 \neq 0$. Using expressions (7.30) and (7.31) for the ion flux toward the cluster surface, we repeat the derivation of the Fuks formula (7.2). Then the ion number density varies from N_1 at the cluster surface up to N_i, the value far from the cluster. Under these boundary condition we get the ion number density in an intermediate region instead of formula (7.1):

$$N(r) - N_1 = \frac{I}{4\pi e D_\text{i}} \int_{r_o}^{r} \frac{dr'}{(r')^2} \exp\left(\frac{Ze^2}{Tr'} - \frac{Ze^2}{Tr}\right)$$
$$= \frac{IT}{4\pi D_\text{i} Ze^3}\left[\exp\left(\frac{Ze^2}{Tr_o} - \frac{Ze^2}{Tr}\right) - 1\right].$$

Using the second boundary condition $N(\infty) = N_\text{i}$, we obtain for the ion current

$$I = \frac{4\pi D_\text{i}(N_\text{i} - N_1)Ze^3}{T\left[\exp\left(\frac{Ze^2}{Tr_o}\right) - 1\right]}.$$

Taking the boundary value $N = N_1$ of the ion number density at the cluster surface, we get the following expression for the ion current $I = eJ$:

$$I = \left(\frac{1}{I_>} + \frac{1}{I_<}\right)^{-1}, \tag{7.36}$$

where the ion current $I_>$ corresponds to the case $\lambda \gg r_o$ and is given by formula (7.32) for $J_>$ ($I_> = eJ_>$), whereas the ion current $I_<$ corresponds to the opposite relation between the free path length of ions λ in a gas and the cluster radius r_o, and the expression for the attachment rate is given by formula (7.6). Formula (7.36) is transformed into formula (7.2) in the limit $I_> \gg I_<$, and into formula (7.5) for the opposite relation between these currents. Thus, formula (7.36) includes any relations between the problem parameters. Note that the ratio of these currents is estimated as

$$\frac{I_<}{I_>} \sim \frac{\xi r_o}{\lambda}.$$

Problem 7.22

A spherical cluster whose radius is small compared to the mean free path of ions in a gas is charged by attachment of electrons and positive ions. Find the cluster charge if the electron and ion temperatures are identical.

The charge of clusters in a weakly ionized buffer gas is negative because of the higher electron mobility. We assume the Maxwell distribution function of electrons, and each contact of an electron and an ion with the cluster surface leads to the transferring of their charges to the cluster. We have the following rate of electron attachment to the cluster surface with a radius of $r_o \ll \lambda$ (λ is the mean free path of gas atoms):

$$J_e = \frac{2}{\sqrt{\pi}} \int_x^\infty z^{1/2} e^{-z} dz \sqrt{\frac{2\varepsilon}{m_e}} \pi r_o^2 = N_e \sqrt{\frac{8T}{\pi m_e}} \pi r_o^2 e^{-x} .$$

Here ε is the electron energy, T is the electron temperature, N_e is the number density of electrons, m_e is the electron mass, $z = \varepsilon/T$, $x = |Z|e^2/(r_o T)$, Z is the negative cluster charge, and we take into account that the electron contact with the cluster surface is possible if the electron energy ε exceeds the repulsion energy of charge interaction $|Z|e^2/r_o$. The rate constant of ion contact with the surface of a charged cluster is given by formula (7.30), which leads to the following ion current to the cluster surface:

$$I_i = e(1 + x) \sqrt{\frac{8T}{\pi M}} N_i \pi r_o^2 ,$$

where N_i is the ion number density and M is the ion mass.

Equalizing the electron and ion currents ($I_i = eJ_e$) to the surface of a charged cluster and assuming the plasma to be quasineutral $N_e = N_i$, we find for the cluster charge

$$|Z| = \frac{r_o T}{2e^2} \left[\ln \frac{M}{m_e} - 2 \ln \left(1 + \frac{|Z|e^2}{r_o T} \right) \right] . \qquad (7.37)$$

Problem 7.23

Evaluate the number density of charged particles of both signs in a quasineutral plasma if the particle radius is small, $r_o \ll e^2/T$.

In this case most of the particles are neutral, and the number of particles with charge $2e$ or more is exponentially small because the Coulomb barrier due to the interaction of a charged particle with an ion of a same charge greatly exceeds a typical thermal energy of ions. Introduce the rate constant $k_{m,m+1}$ by which a particle of charge m increases its charge by one as a result of attachment of positive ions, and use the same notation for other rate constants. According to the Smoluchowski formula (4.22) we have

$$k_{0,1} = 4\pi D_+ r , \quad k_{0,-1} = 4\pi D_- r ,$$

and the Langevin formula (6.45) for the other processes gives

$$k_{1,0} = 4\pi e K_- = 4\pi D_- e^2/T , \quad k_{-1,0} = 4\pi e K_+ = 4\pi D_+ e^2/T .$$

From this we obtain for the number densities of neutral N_0, positive N_1, and negative N_{-1} small particles, using the balance equations of charging (for example, $k_{0,1} N_0 N_+ = k_{1,0} N_1 N_-$),

$$N_1 = N_0 \frac{r_o T}{e^2} \frac{D_+}{D_-} , \quad N_{-1} = N_0 \frac{r_o T}{e^2} \frac{D_-}{D_+} ; \quad \lambda \ll r_o \ll \frac{e^2}{T} . \qquad (7.38)$$

Note that the number density of doubly charged clusters is exponentially small, (proportional to $\exp[-e^2/(r_o T)]$). Furthermore, formula (7.38) corresponds to the relation $\lambda \ll r_o \ll e^2/T$.

In the case $r_o \ll \lambda, r_o \ll e^2/T$, we use formulas (7.30) and (7.31) for the rate constants of a cluster charge variation

$$k_{0,1} = k_o^+ = \pi r_o^2 \sqrt{\frac{8T}{\pi m_+}} , \quad k_{0,-1} = k_o^- = \pi r_o^2 \sqrt{\frac{8T}{\pi m_-}} ,$$

$$k_{1,0} = k_o^+ x , \quad k_{-1,0} = k_o^- x ,$$

where m_+, m_- are the masses of positive and negative ions, and $x = e^2/(r_o T)$. For simplicity, we assume the probability of attachment of a charge to particles as a result of contact between an ion and a particle to be $\xi = 1$. From this we obtain, on the basis of the above operations,

$$N_1 = N_0 \frac{r_o T}{e^2} \sqrt{\frac{m_+}{m_-}} , \quad N_{-1} = N_0 \frac{r_o T}{e^2} \sqrt{\frac{m_-}{m_+}} ; \quad r_o \ll e^2/T , \quad r_o \ll \lambda .$$

In this case the average cluster charge is

$$Z = \frac{N_1 - N_{-1}}{N_0} = \frac{r_o T}{e^2} \cdot \frac{m_- - m_+}{\sqrt{m_- m_+}} . \tag{7.39}$$

Taking $m_- \sim m_+$, we have from this $Z \ll 1$.

Problem 7.24

Find the charge of a spherical cluster if its radius r_o is small compared to the mean free path of electrons and ions and the cluster is located in a weakly ionized gas with different electron T_e and ion T_i temperatures. Assume the cluster radius to be small in comparison with e^2/T_e.

Because the parameter $e^2/(r_o T_e)$ is small, attachment of an individual electron or ion to a particle changes its field weakly, and the charging process is continuous. We use formulas (7.30) and (7.31) for the rate constants of electron and ion attachment to a cluster and substitute into these equations different electron and ion temperatures. Equalizing electron and ion currents on the particle surface or equalizing the rate constants (7.30) and (7.31), we obtain for a cluster charge $Z = -|Z|$ in a quasineutral plasma

$$|Z| = \frac{r_o T_e}{e^2} \ln \left[\sqrt{\frac{M T_e}{m_e T}} \left(1 + \frac{|Z| e^2}{r_o T} \right)^{-1} \right] , \tag{7.40}$$

where T_e is the effective electron temperature, T is the gas and ion temperature, and m_e and M are the electron and ion masses, respectively.

This is the equation for the cluster charge $|Z|$ and may be rewritten as the equation for the reduced parameter (7.30) in the form

$$x = \left[\ln \sqrt{\left(\frac{M T_e}{m_e T} \right)} - \ln \left(1 + x \frac{T_e}{T} \right) \right] , \quad |Z| = x \frac{r_o T_e}{e^2} . \tag{7.41}$$

Problem 7.25

Find the cluster charge in the limit of large sizes in atmospheric air.

When a cluster is large, collisions of ions and electrons with the cluster surface correspond to the diffusion regime. Then the cluster diffusion coefficient in a gas is given by formula (6.16), and $D \backsim 1/r_o$. According to formula (7.7), the cluster charge $Z \backsim r_o$, so the value ZD is independent of the cluster size.

The mobility of a large cluster in a gas is expressed through its diffusion coefficient in the gas according to the Einstein relation (6.4)

$$K = \frac{ZeD}{T} ,$$

and this quantity is independent of the cluster size. In the case of motion of a large cluster in atmospheric air, we take the ratio of the ion diffusion coefficients $D_+/D_- \approx 0.8$, which corresponds to a typical humidity of atmospheric air. Then using formula (7.7) for the cluster charge and formula (6.16) for the cluster diffusion coefficient in a dense gas, we obtain the limiting value of the cluster mobility in atmospheric air with a typical humidity:

$$K = 1.8 \cdot 10^{-5} \frac{\text{cm}^2}{\text{V} \cdot \text{s}} .$$

Problem 7.26

Find the ratio of the number densities of positively and negatively charged small clusters in a rare ionized gas where the electrons and ions are characterized by the temperatures T_e and T_i, respectively.

If clusters are located in a weakly ionized gas, their charge results from attachment of electrons and ions. This charge is negative because of the higher electron mobility, and below we determine this charge and a typical time of its establishment using the Maxwell distribution function of electrons. We assume that a weakly ionized plasma is quasineutral and each contact of an electron and an ion with the cluster surface leads to the transfer of their charges to the cluster. We have the rate of electron attachment to the cluster surface with a radius of $r_o \ll \lambda$ (λ is the mean free path of gas atoms):

$$J_e = \frac{2}{\sqrt{\pi}} \int_{z_o}^{\infty} x^{1/2} e^{-x} dx \sqrt{\frac{2\varepsilon}{m_e}} \pi r_o^2 = (1 + z_o) e^{-z_o} \sqrt{\frac{8T_e}{\pi m_e}} N_e \pi r_o^2 .$$

Here ε is the electron energy, T_e is the electron temperature, N_e is the number density of electrons, m_e is the electron mass, $x = \varepsilon/T_e$, $z_o = |Z|e^2/(r_o T_e)$, Z is the negative cluster charge, and we take into account that electron attachment to the cluster surface is possible if the electron energy ε exceeds the repulsion energy of charge interaction $|Z|e^2/r_o$. The cross section of the ion contact with the surface of a charged cluster is

$$\sigma = \pi r_o^2 \left(1 + \frac{|Z|e^2}{r_o \varepsilon}\right) ,$$

and the ion current to the cluster surface is

$$J_i = (1 + z_o)\sqrt{\frac{8T_i}{\pi m_i}} N_i \pi r_o^2 ,$$

where N_i is the ion number density, T_i is the ion temperature, and m_i is the ion mass. Equalizing the electron and ion currents on the surface of a charged cluster and assuming the plasma to be quasineutral, $N_e = N_i$, we find for the cluster charge

$$|Z| = \frac{r_o T_e}{2e^2} \ln \frac{m_i T_e}{m_e T_i} ,$$

under the assumption $|Z| \gg 1$, that is, attachment of one electron or ion does not change the interaction between a charged cluster and electrons or ions.

Under given conditions this criterion is violated, and hence the cluster charge may be 0, +1, or −1. Therefore, the following processes of cluster charging determine the cluster charge:

$$\begin{aligned} &e + M_n \to M_N^- , \quad e + M_n^+ \to M_N , \\ &A^+ + M_n \to M_N^+ + A , \quad A^+ + M_n \to M_N^+ + A . \end{aligned} \tag{7.42}$$

This leads to the following set of balance equations for the number density of neutral clusters N_0, singly negatively charged clusters N_-, and singly positively charged clusters N_+:

$$\begin{aligned} \frac{dN_0}{dt} &= -k_e N_e N_0 - k_i N_i N_0 + k_i \left(1 + \frac{e^2}{r_o T_i}\right) N_i N_- + k_e \left(1 + \frac{e^2}{r_o T_e}\right) N_e N_e ; \\ \frac{dN_+}{dt} &= k_i N_i N_0 - k_e \left(1 + \frac{e^2}{r_o T_e}\right) N_e N_e ; \\ \frac{dN_-}{dt} &= k_e N_e N_0 - k_i \left(1 + \frac{e^2}{r_o T_i}\right) N_i N_- . \end{aligned}$$

Here we use the notation $k_e = k_{0,-1}$, $k_i = k_{0,+1}$, and

$$k_{-1,0} = k_i \left(1 + \frac{e^2}{r_o T_i}\right) ; \quad k_{+1,0} = k_e \left(1 + \frac{e^2}{r_o T_e}\right) ,$$

so that $k_{Z,Z\pm1}$ is the rate constant of a cluster charge from Z to $Z \pm 1$.

For simplicity we restrict ourselves to the stationary case. The ratios of the equilibrium number densities of clusters are

$$\frac{N_-}{N_0} = \frac{k_e N_e}{k_i N_i \left(1 + \frac{e^2}{r_o T_i}\right)} , \quad \frac{N_+}{N_0} = \frac{k_i N_i}{k_e N_e \left(1 + \frac{e^2}{r_o T_e}\right)} .$$

From this we obtain the following equation for the ratio of positively and negatively charged clusters:

$$\frac{N_+}{N_-} = \left(\frac{k_e N_e}{k_i N_i}\right)^2 \frac{\left(1 + \frac{e^2}{r_o T_e}\right)}{\left(1 + \frac{e^2}{r_o T_i}\right)} .$$

Problem 7.27

A spherical cluster or particle of a radius of r_o is located in a rare ionized gas, so that the mean free path of atoms in a gas λ exceeds the cluster size $\lambda \gg r_o$. Determine the number density of ions near the charged particle if its charge is large, $Z \gg 1$.

We are based on the above character of interaction of electrons and ions with a charged particle. Far from the particle we have a quasineutral plasma, so that the number densities of electrons and ions are equal $N_e = N_i = N_o$. Near the particle, electrons and ions move toward the particle and charge it negatively because of the higher velocity of electrons. Since the particle field is repulsive with respect to electrons, only a small number of incident electrons located near the particle attach to it, and the number density of electrons in the particle field is given by the Boltzmann equation:

$$N_e = N_o \exp\left(-\frac{e\varphi}{T_e}\right) ,$$

where T_e is the electron temperature and $\varphi(r)$ is the particle electric potential near the particle at a distance r from its center. If the number density of the surrounding plasma is small and plasma ions do not screen the Coulomb field of the charged particle, we have

$$\varphi(r) = \frac{Ze}{r} ,$$

and we take into account the negative value of the particle's charge. Usually

$$x = \frac{Ze^2}{r_o T_e} > 1 ,$$

which gives $N_e \ll N_o$ for the electron number density near the particle and allows us to neglect an electron charge near the particle.

In contrast to electrons, for ions the particle field is attractive and they are accelerated toward the particle surface without collisions. Hence, the probability dP_i for an ion to be located at distances ranging between r and $r + dr$ is

$$dP_i \backsim dt = \frac{dr}{v_r} = \frac{dr}{v\sqrt{1 - \rho^2/r^2 - U(r)/\varepsilon}} ,$$

where dt is the time during which an ion is found in this distance range, v_r is the normal component of the ion velocity, v is the ion velocity far from the particle, ε is the ion energy far from the particle, ρ is the impact parameter for ion motion with respect to the particle, and $\varepsilon = m_i v^2/2$ is the ion energy far from the particle (m_i is the ion mass). From this we have for the number density of ions

$$N_i \backsim \frac{\int \rho d\rho d P_i}{4\pi r^2 d r} .$$

In integrating this expression over the impact parameter ρ, we take into account that ions pass a given distance range between r and $r + dr$ both during their approach to the particle and during removal from it. Taking the probability that the ion will stick to the particle upon contact to be unity, we obtain for the return trajectories $\rho \geq \rho_c$, where the capture impact parameter ρ_c corresponds to the distance of closest approach for a colliding ion and particle to be r_o, that is,

$$\rho_c^2 = r_o^2 \left[1 - \frac{U(r_o)}{\varepsilon}\right] .$$

We obtain after integration

$$N_i(r) = \frac{N_o}{2}\left[\sqrt{1 - \frac{U(r)}{\varepsilon}} + \sqrt{1 - \frac{\rho_c^2}{r^2} - \frac{U(r)}{\varepsilon}}\right] ,$$

where we assume $U(\lambda) \ll \varepsilon$ and $\rho_c \ll \lambda$.

We now average the number density of ions over the kinetic ion energy ε, which corresponds to large distances from the particle – the Coulomb center. A term containing the ion energy is essential near the particle at distances where $U(r) \gg \varepsilon$. Averaging over ion energies in this range of distances with the Maxwell distribution function, we obtain finally

$$N_i(r) = \frac{N_o}{2}\left[\sqrt{1 - \frac{4\,U(r)}{\pi T_i}} + \sqrt{1 - \frac{\rho_c^2}{r^2} - \frac{4\,U(r)}{\pi T_i}}\right] , \tag{7.43}$$

where T_i is the ion temperature.

Problem 7.28

Give the criterion for when ions attracted to a spherical charged particle with a radius r_o and a charge Z do not screen the Coulomb field of the particle.

The total charge of ions that partake in screening the particle charge is

$$q = \int_{r_o}^{R_o} 4\pi r^2 dr \cdot N_i(r) ,$$

where the upper limit is determined by the relation

$$U(R_o) = T_e , \quad \lambda \geq R_o . \tag{7.44}$$

Using formula (7.43) for the number density of ions, one finds that in the limit $\lambda \gg R_o$ the main contribution to the screening charge gives a range of $r \gg r_o$, though the number density of ions near the particle is higher than that far from it.

Thus, the criterion $q \ll Z$ that atomic ions near a charged particle do not screen its charge has the form

$$\frac{4\pi}{3} R_o^3 N_o \ll Z ,$$

because large distances provide the main contribution to cluster charge screening.

Problem 7.29

Estimate the parameters of interaction between a particle and a gas discharge plasma under typical conditions of a dusty plasma when the particle radius is $r_o \sim 1\,\mu\text{m}$, the number density of atoms is $N \sim 10^{17}\,\text{cm}^{-3}$ (a pressure of several Torr), and the number density of electrons and ions far from the particle equals to $N_o \sim 10^9\,\text{cm}^{-3}$, the electron temperature is $T_e = 1\,\text{eV}$, and the ion temperature is close to the gas temperature and is 500 K.

Taking a typical cross section of electron–atom collisions to be $\sigma_{ea} \sim 10^{-15}\,\text{cm}^2$, we find for the mean free path of electrons $\lambda \sim 10^{-2}\,\text{cm}$, which corresponds to the limit of a rare gas. Next, we have

$$N_o \lambda^3 \sim 10^3 ,$$

and according to formula (7.41) we now obtain

$$x = \frac{|Z|e^2}{r_o T_e} \approx 3 , \quad |Z| = 2 \cdot 10^3 .$$

Next, the electric potential φ on the particle surface and the electric field strength $E = \varphi / r_o$ are

$$\varphi = \frac{|Z|e}{r_o} = 3\,\text{V} , \quad E = 30\,\text{kV/cm} .$$

Problem 7.30

Analyze the character of formation of an ionic coat around a charged particle in a rare ionized gas as a result of a resonant charge exchange process involving an atomic ion and an atom.

Ions from an ionized gas move in the field of a charged particle along hyperbolic trajectories, and we assume that the mean free path of ions with respect to the resonant charge exchange process is relatively large, that is,

$$N_a \sigma_{res} R_o \ll 1 , \tag{7.45}$$

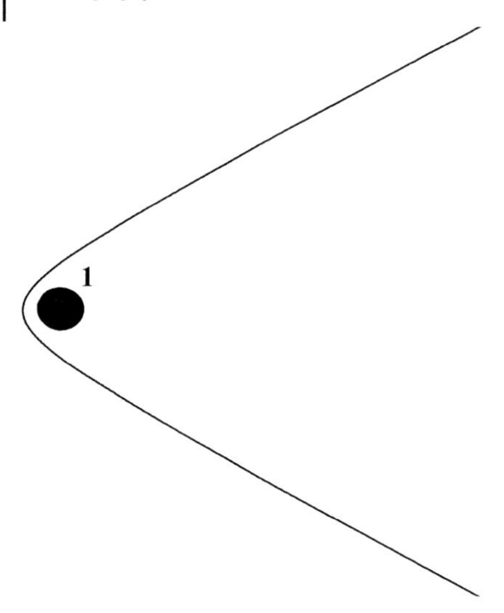

Fig. 7.1 Trajectories of a free positive ion in the field of a negatively charged particle (1) in the ion motion plane.

Here N_a is the number density of atoms, σ_{res} is the cross section of resonant charge exchange, which is independent of the collision velocity [157, 158], R_o is a radius of action of a particle field, which is defined by formula (7.44), and $U(r)$ is the potential of ion–particle interaction at a distance r from the particle that accounts for screening under the action of ions near the particle, and ε is the ion energy.

We have the following character of ion motion in a particle field. If a test ion remains free while it is moving in a particle field, its trajectory has a hyperbolic form for the Coulomb field, or close to it if the particle Coulomb field is screened (Figure 7.1). If resonant charge exchange involving a test ion and a gas atom proceeds in the field of the negatively charged particle, the formed ion may be captured by this field and moves along elliptic orbits. In the case of the Coulomb field of a charged particle, the ion trajectory is closed and a test ion moves along the same trajectory after each rotation (Figure 7.2). If the Coulomb field of the particle is screened, the elliptic orbit of the ion rotates after each period [159, 160], as is shown in Figure 7.3. This ion motion continues until the next event of charge exchange, after which the ion can leave the ion field region, attach to the particle, or remain to be captured and move along another elliptic trajectory.

Formula (7.43) gives the number density of free ions located in the particle field for a rare gas in accordance with criterion (7.45). Along with these free ions, bound (or trapped) ions are located in the field of a charged particle that are formed in collisions with atoms as a result of the charge exchange process [161]. Such events are seldom, that is, the probability of trapped ion formation in the course of motion of a free ion near a charged particle is small. But because trapped ions located in bound orbits have a long lifetime, they can contribute to the number density of ions near a charged particle as follows from the numerical calculation of certain condi-

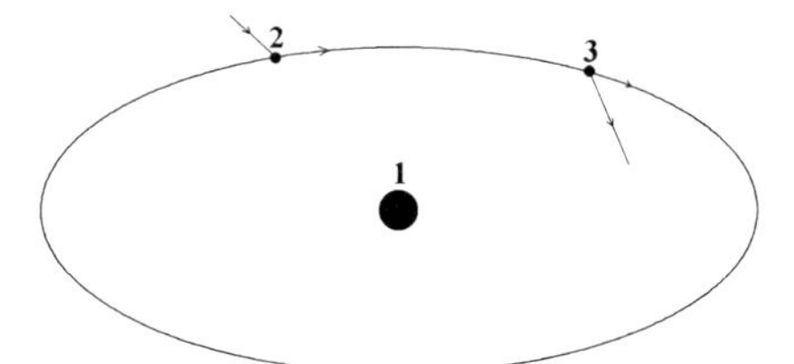

Fig. 7.2 Trajectories of a captured positive ion in the Coulomb field of a negatively charged particle. 1 – particle, 2 – point of ion capture as a result of a resonant charge exchange event, 3 – point of a subsequent resonant charge exchange event on a gas atom. Arrows indicate the motion direction.

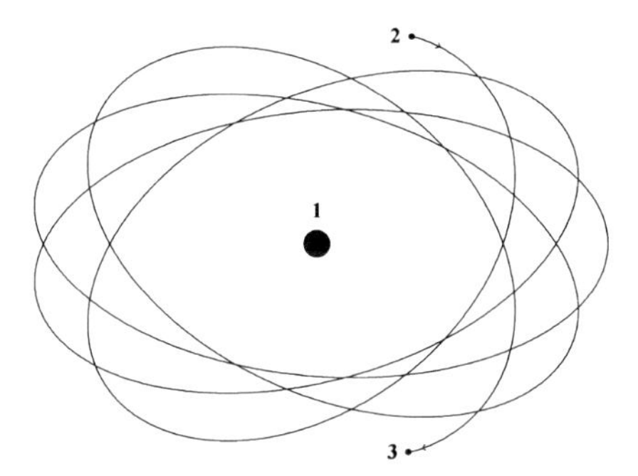

Fig. 7.3 Trajectories of a captured positive ion in the field of a negatively charged particle that differs from the Coulomb one, such that the elliptic orbit of the ion rotates. 1 – particle, 2 – point of ion capture as a result of a resonant charge exchange event, 3 – point of a subsequent resonant charge exchange event on a gas atom. Arrows indicate the motion direction.

tions of a dusty plasma [162–164]. Therefore, some models are elaborated [165–167] to describe the formation and evolution of trapped ions.

The number density of trapped ions follows from the balance equation

$$N_a \sigma_{res} N_i P_{tr} v_i = N_a \sigma_{res} N_{tr} v_{tr} , \qquad (7.46)$$

where N_i, N_{tr} are the number densities of free and trapped ions, respectively, v_i is the relative velocity of a free ion and atom partaking in the resonant charge exchange process, v_{tr} is the relative velocity of a trapped ion and atom, and P_{tr} is the probability of formation of a trapped ion a result of resonant charge exchange involving a free ion. We assume that the resonant charge exchange of a trapped ion leads to its capture by a particle. Because in a region of ion capture $P_{tr} \sim 1$, we obtain, taking $v_i \sim v_{tr}$, that the number density of trapped ions in the region of ion capture is on the order of the number density of free ions. Indeed, though the probability of formation of a trapped ion per single incident ion is small, the long lifetime of trapped ions according to criterion (7.45) compensates this smallness.

Problem 7.31

Determine the probability that an atomic ion will be captured in the field of a charged particle after the resonant charge exchange event in collisions with atoms of a gas.

When charge transfer results from ion–atom collision, the ion formed obtains the velocity of the incident atom because the cross section of resonant charge exchange exceeds significantly a thermal cross section of elastic ion–atom scattering. Let us assume that this process proceeds at a distance R from a charged particle, and denote by θ the angle between the velocity of the ion formed and the line joining the ion and the charged particle, as is shown in Figure 7.4, and denote by ε the kinetic energy of the ion formed. We have two conditions of ion capture by a charged particle, so the first one requires a negative ion energy, that is,

$$U(R) \geq \varepsilon \, .$$

Let us introduce a distance R_o such that $-U(R_o) = \varepsilon$. As is seen, the formation of a trapped ion is possible when the resonant charge exchange event takes place at distances from the charged particle $R \leq R_o$.

The second condition for the formation of a trapped ion requires that the distance of closest approach $r_{\min}$ of the ion and particle exceeds the particle radius r_o. We use the conservation of the total ion energy E captured at distance R if this energy is taken at distance r from the particle

$$E = U(R) + \varepsilon = \frac{m_i v_r^2}{2} + U(r) + \varepsilon \frac{R^2 \cos^2\theta}{r^2} \, ,$$

where m_i is the ion mass and v_r is the component of the ion velocity toward the particle. The ion momentum with respect to the particle center L that is conserved in the course of ion motion is

$$L = m_i v(\infty)\rho = m_i v(R) R \cos\theta \, ,$$

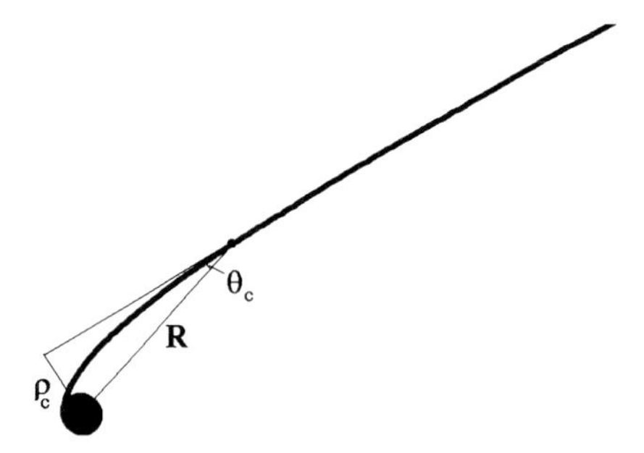

Fig. 7.4 Geometry of the formation of a captured ion in a charge exchange event. R is the distance from the particle center for the point of ion formation, ε is the energy of the ion formed (or the energy of an atom that loses an electron), and θ is the angle between the trajectory of the ion formed and a line that joins this point and the particle.

and the impact parameter ρ for ion motion far from the particle is $\rho = R\sin\theta$. From this we have the relation between the angle θ and the distance of closest approach $r_{\min}$ (Figure 7.4):

$$\sin\theta = \frac{r_{\min}}{R}\sqrt{1 + \frac{U(R) - U(r_{\min})}{\varepsilon}} .$$

This gives the probability of formation of a trapped ion if the charge exchange process proceeds at distance R from the charged particle,

$$P_{\mathrm{tr}}(R) = \cos\theta_{\mathrm{c}} , \quad \sin\theta_{\mathrm{c}} = \frac{r_{\mathrm{o}}}{R}\sqrt{1 + \frac{U(R) - U(r_{\mathrm{o}})}{U(R_{\mathrm{o}})}} ,$$

and the formation of a trapped ion is possible in a range of distances $R_* < R < R_{\mathrm{o}}$, where R_* is given by the solution of the following equation:

$$\frac{R_*^2}{r_{\mathrm{o}}^2} - 1 = \frac{U(R_*) - U(r_{\mathrm{o}})}{U(R_{\mathrm{o}})} .$$

For simplicity, we consider below the case where screening the particle charge by ions is not important, so the interaction potential of an ion and a charged particle is

$$U(r) = -\frac{Ze^2}{r} ,$$

where Z is a negative particle charge. Then according to the above relations we have

$$P_{\mathrm{tr}}(R) = \sqrt{1 - \frac{r_{\mathrm{o}}R_{\mathrm{o}}}{R^2}\left(1 - \frac{r_{\mathrm{o}}}{R}\right)} .$$

In the range of distances $R_* = \sqrt{r_{\mathrm{o}}R_{\mathrm{o}}} < R$, where capture is possible on an orbit without touching the particle, the capture probability is

$$P_{\mathrm{tr}}(R) = \sqrt{1 - \frac{r_{\mathrm{o}}R_{\mathrm{o}}}{R^2}} , \quad R_{\mathrm{o}} \gg R \geq R_* = \sqrt{r_{\mathrm{o}}R_{\mathrm{o}}} . \tag{7.47}$$

Problem 7.32

Determine the average probability of ion capture in the field of a charged particle if the resonant charge exchange event leads to the formation of an ion with a negative energy.

We use in formulas the above notations, so that r_{o} is the charged particle radius, Z is its charge, ε is the kinetic energy of the ion formed as a result of resonant charge exchange, $R_{\mathrm{o}} = Ze^2/\varepsilon$, and $r_{\mathrm{o}} \ll R_{\mathrm{o}}$. By definition, we have for the average probability of formation of a trapped ion

$$\overline{P_{\mathrm{tr}}} = \frac{3}{4\pi R_{\mathrm{o}}^3}\int\limits_{r_{\mathrm{o}}}^{R_{\mathrm{o}}} P_{\mathrm{tr}}(R)\cdot 4\pi R^2 dR .$$

Using formula (7.47), we obtain

$$\overline{P_{tr}} = \frac{3}{R_o^3}\int_{R_*}^{R_o} \sqrt{R^2 - R_*^2}\, R\, dR = \left(1 - \frac{r_o}{R_o}\right)^{3/2} .$$

Since $r_o \ll R_o$, we obtain that the average probability is close to one for the formation of a trapped ion in the region where a the ion becomes bound.

Problem 7.33

Show that the number density of trapped positive ions can exceed that of free ions.

Let us rewrite the balance equation (7.46), ignoring the dependence of the cross section of resonant charge exchange on the collision velocity. We then take into account that the probability of resonant charge exchange is identical for free and trapped ions. Then we take into account that the probability that a trapped ion will be captured in an elliptic orbit is close to one for this range of distances. To take this into account, we rewrite the balance equation (7.46) in the form

$$N_a \sigma_{res} N_i P_{tr} v_i = N_a \sigma_{res} N_{tr} v_{tr} (1 - P_{tr}) . \qquad (7.48)$$

If we take for estimation the Coulomb interaction between a trapped ion and a particle, one can use formula (7.47) for the probability that an ion will be located in an elliptic orbit. Then we have

$$1 - \overline{P_{tr}} \approx \frac{3 r_o}{2 R_o} ,$$

and the number density of trapped ions N_{tr} is connected to the number density N_i of free ions in this region by the relation

$$N_{tr} \sim \frac{R_o}{r_o} N_i ,$$

that is, the number density of trapped ions N_{tr} exceeds significantly the number density N_i of free ions.

7.4 Kinetics of Cluster Charging in Ionized Gas

Problem 7.34

Derive the kinetic equation for the charge distribution function of clusters.

In addition to the above analysis of particle charging in a dense gas, we find the charge distribution function of small particles. Let us define the distribution function f_Z of clusters on charges Z such, that f_Z is the probability for a cluster to have a charge Ze. The normalization condition gives

$$\sum_Z f_Z = 1 , \tag{7.49}$$

and the kinetic equation for the distribution function f_Z has the form

$$f_Z(J_{Z,Z+1} + J_{Z,Z-1}) = f_{Z-1} J_{Z-1,Z} + f_{Z+1} J_{Z+1,Z} , \tag{7.50}$$

where $J_{Z,Z+1}$ is the rate (the probability per unit time) of change of the cluster charge from Z to $Z+1$ as a result of collision with plasma ions. In equation (7.50) we account for the fact that the cluster charge changes by one as a result of contact with an electron or ion.

On the basis of formulas (7.2) and (7.4) we have the following expressions for the rates of charging:

$$J_{Z,Z+1} = \frac{4\pi D_+ N_o r_o Z x}{\exp(Zx) - 1} , \quad J_{Z,Z-1} = \frac{4\pi D_- N_o r_o Z x}{1 - \exp(-Zx)} , \tag{7.51}$$

where $x = e^2/r_o T$, N_o is the number density of ions of each sign far from the cluster, and the other notations are explained above. This leads to the following kinetic equation for the charge distribution function of clusters:

$$\frac{\left[D_+ + D_- \exp(Zx)\right] Zx}{\exp(Zx) - 1} f_Z = \frac{D_+(Z-1)x}{\exp[(Z-1)x] - 1} f_{Z-1} + \frac{D_-(Z+1)x}{\exp[(Z+1)x] - 1} f_{Z+1} . \tag{7.52}$$

Problem 7.35

Determine the charge distribution function of small spherical particles if they are located in a quasineutral plasma and their radius satisfies the relations $\lambda \ll r_o \ll e^2/T$.

In the case under consideration, $x = e^2/r_o T \gg 1$, the average particle charge is small, and particles are mostly neutral. Then, in particular, from the set of equations (7.50) it follows for the probability f_2, if we assume $f_3 \ll f_2$, that

$$\frac{f_2}{f_1} = \frac{D_+(e^{2x} - 1)}{2(D_+ + D_- e^{2x})(e^x - 1)} = \frac{D_+}{2D_-} e^{-x}$$

and f_Z with $Z \geq 2$ are exponentially small. The same relation holds for $Z \leq -2$. Thus, we can restrict our attention to neutral and singly charged ions, which comprise the main contribution to the total charge of particles. Under this assumption, we obtain from the set of equations (7.52)

$$\frac{f_1}{f_0} = \frac{D_+(e^x - 1)}{(D_+ + D_- e^x)x} = \frac{D_+}{D_- x} , \quad \frac{f_{-1}}{f_0} = \frac{D_-(e^x - 1)}{(D_- + D_+ e^x)x} = \frac{D_-}{D_+ x} ,$$

and in this limiting case $f_0 \approx 1$. The average charge of particles is

$$\overline{Z} = f_1 - f_{-1} = \frac{D_+^2 - D_-^2}{D_+ D_- x},$$

in accordance with formula (7.38). In particular, if $D_+ - D_- = \Delta D \ll D_+$, this equation gives $\overline{Z} = 2\Delta D/(Dx)$.

Problem 7.36

Determine the charge distribution function of clusters if they are located in a quasineutral plasma under the conditions $r_o \gg \lambda$, $r_o \gg e^2/T$. Consider a range of large charges $Z \gg 1$ and take the cluster charge Z as a continuous parameter.

Let us consider f_Z to be a continuous function of a particle's charge. Using the variable $z = Zx$, we have the normalization condition in the form

$$\int_{-\infty}^{\infty} f(z)dz = x .$$

Then the kinetic equation (7.52) takes the form

$$-f(z)[D_+ F(z) + D_- G(z)] + f(z-x) D_+ F(z-x) + f(z+x) D_- G(z+x) = 0,$$

where

$$F(z) = \frac{z}{e^z - 1}, \quad G(z) = F(-z) = e^z F(z) = \frac{z}{1 - e^{-z}} .$$

Let us expand the equation over a small parameter x. We get

$$-D_+ x \frac{d}{dz}[f(z)F(z)] + D_- x \frac{d}{dz}[f(z)F(z)e^z] + D_+ \frac{x^2}{2}\frac{d^2}{dz^2}[f(z)F(z)] + D_- \frac{x^2}{2}\frac{d^2}{dz^2}[f(z)F(z)e^z] = 0 .$$

Restricting ourselves to linear terms, we obtain the equation

$$\frac{d}{dz}\left[\left(\frac{D_+}{D_-} - e^z\right) f(z)F(z)\right] = 0 ,$$

the solution of which has the form

$$f = \frac{C(e^z - 1)}{\left|\frac{D_+}{D_-} - e^z\right|} .$$

We see that the charge distribution function of clusters becomes infinite at $z = \ln(D_+/D_-)$, which corresponds to the average cluster charge according to formula (7.7).

Accounting for the terms of the next order of expansion over a small parameter x, we reduce the kinetic equation to the form

$$\frac{d}{dz}\left[\left(\frac{D_+}{D_-}-e^z\right)f(z)F(z)\right]-\frac{x}{2}\frac{d^2}{dz^2}\left[\left(\frac{D_+}{D_-}+e^z\right)f(z)F(z)\right]=0. \quad (7.53)$$

First we consider the case $D_+ = D_-$, where the average cluster charge is zero and the distribution function is an even function of z: $f(z) = f(-z)$. Then the kinetic equation has the form

$$\frac{d}{dz}[z f(z)]+\frac{x}{2}\frac{d^2}{dz^2}\left[\frac{e^z+1}{e^z-1}z f(z)\right]=0\,.$$

Integration of this equation gives

$$z f(z)+\frac{x}{2}\frac{d}{dz}\left[\frac{e^z+1}{e^z-1}z f(z)\right]=C_1\,.$$

From the symmetry of the distribution function $f(z) = f(-z)$ we have $C_1 = -C_1 = 0$. Integrating this equation, we obtain

$$f(z)=\frac{C}{2}\frac{e^z-1}{e^z+1}\exp\left(-\frac{2}{x}\left[2\ln\left(\frac{1+e^z}{2}\right)-z\right]\right),$$

where the integration constant C is determined on the basis of the normalization condition of the distribution function. One can see that for $x \ll 1$ the main contribution to the normalization of the distribution function gives $z \sim \sqrt{x} \ll 1$. Expanding this expression for small z and using the normalization condition, we transform the distribution function to the form

$$f(z)=\sqrt{\frac{1}{2\pi x}}\exp\left(-\frac{z^2}{2x}\right).$$

From this it follows that the probability of a particle charge ze in the limiting case $x = e^2/(r_o T) \ll 1$ and $zx \ll 1$ is given by

$$f(z)=\sqrt{\frac{x}{2\pi}}\exp\left(-\frac{Zx^2}{2}\right). \quad (7.54)$$

Let us now consider the kinetic equation in the general case $D_+ \neq D_-$. Let us introduce the parameter $z_o = (\ln D_+/D_-)$, that is the average value of z according to formula (7.7) and write equation (7.53) in the range $|z - z_o| \ll 1$. In these variables the kinetic equation (7.53) obtains the form

$$\frac{d}{dz}\left[(z-z_o)f(z)\right]+x\frac{d^2 f}{dz^2}=0\,.$$

The normalized solution of this equation is

$$f(z)=\sqrt{\frac{1}{2\pi x}}\cdot\exp\left[-\frac{(z-z_o)^2}{2x}\right],\quad |z-z_o|\ll 1\,,\ x\ll 1\,. \quad (7.55)$$

Note that the normalization of the distribution function for $x \ll 1$ is determined by just this range of the variable $|z - z_o| \ll 1$.

Problem 7.37

Under a conditions of the kinetic equation (7.53), analyze the evolution of charging of small particles in a quasineutral plasma and estimate a typical time for this process assuming the diffusion coefficients of positive and negative ions to be close.

Based on the kinetic equation (7.53), we have this equation in the following form for the time-dependent case:

$$\frac{\partial f}{\partial t} = 4\pi D N_o r x \left[\frac{\partial}{\partial z}(z f) + x \frac{\partial^2 f}{\partial z^2} \right] ,$$

where the diffusion coefficient is $D = D_+ = D_-$ and N_o is the number density of ions of each charge sign far from the particles. Multiplying this equation by z^2 and integrating the result over dz, we obtain

$$\frac{d\overline{Z^2}}{dt} = -\frac{2}{\tau}(\overline{Z^2} - x) ,$$

where the typical time of establishment of the equilibrium charge distribution of particles is

$$\tau = \frac{T}{4\pi N_o D e^2} . \tag{7.56}$$

The solution of the above equation is

$$\overline{Z^2} = x \left[1 - \exp\left(-\frac{2t}{\tau} \right) \right] .$$

This solution describes the character of the establishment of equilibrium for the charge distribution function of particles.

8
Ionization Equilibrium of Clusters in a Gas

Clusters occupy a position intermediate between the positions of small atomic particles, atoms and molecules, and macroscopic atomic systems, bulk solids, and liquids. Therefore, the properties of these physical objects also may relate to clusters. Namely, like small atomic objects, hot clusters may be ionized, and their equilibrium charge is determined by the Saha equation, though the ionization potential of clusters is closer to that of macroscopic atomic systems rather than that of atoms. Similar to hot bulk metals, hot clusters may emit electrons, that is, the thermoemission process is typical for hot clusters. A charge equilibrium of dielectric clusters proceeds similarly to that in bulk dielectrics – through bound states of ions with active centers on the cluster surface. In addition, ionization processes involving clusters are effective in collisions with photons of short wavelengths or fast electrons. These processes are like those that include the participation of atoms or molecules, but because a cluster consists of many atoms, these cluster processes are more effective.

8.1
Ionization Equilibrium for Large Metal Clusters

Problem 8.1

Derive the relation between the numbers of metal clusters whose charges differ by one in a hot vapor by analogy with the Saha relationship.

Usually, the ionization potential of clusters is lower than that for atoms and molecules. Hence, ionization of metal clusters proceeds at relatively low electron temperatures or at typical electron energies at which ionization of atoms or molecules does not occur. Therefore, clusters located in a plasma can contribute to the formation of free electrons. Next, metal clusters give and accept electrons from the surrounding plasma, and they can be positively or negatively charged. At high electron temperatures metal clusters have a positive charge, whereas at low temperatures they are negatively charged.

We assume metal clusters to be isolated in a hot gas or plasma and consider the ionization equilibrium for an individual cluster. Note that the ionization po-

Cluster Processes in Gases and Plasmas. Boris M. Smirnov

ISBN: 978-3-527-40943-3

tential of clusters lies between the ionization potential of metal atoms I and the work function of metals W (the binding energy of electrons with a metal surface), and because $I > W$, the ionization potential of clusters is below the ionization potential of atoms. But in contrast to atoms, which are usually singly ionized in a plasma, clusters may have an arbitrary charge. Below we consider the ionization equilibrium between charged clusters that results from collisions with electrons:

$$M_n^{+Z+1} + e \leftrightarrow M_n^{+Z} \,, \tag{8.1}$$

where M is an atom and M_n^Z is a cluster consisting of n atoms and having a charge Z. This equilibrium is typical for metal clusters, and hence below we consider the ionization equilibrium of clusters with an electron component of a surrounding plasma. This equilibrium is analogous to the Saha ionization equilibrium in an atomic gas. In this consideration we assume for simplicity that the temperatures of free and bound electrons are identical. This means an equilibrium for an electron component both in a plasma and in clusters, which allows us to introduce the electron temperature and the equilibrium between these two components, so this temperature is the same for free and bound electrons.

Let us denote by $P_Z(n)$ the probability that a cluster consisting of n atoms will have a charge Z. Assuming ionization equilibrium for the electron subsystem inside a cluster, we use the Saha relation for this quantity in the form

$$\frac{P_Z(n) N_e}{P_{Z+1}(n)} = 2\left(\frac{m_e T_e}{2\pi\hbar^2}\right)^{3/2} \exp\left[-\frac{I_Z(n)}{T_e}\right] . \tag{8.2}$$

Here T_e is the electron temperature, m_e is the electron mass, N_e is the electron number density, and $I_Z(n)$ is the ionization potential of the cluster consisting of n atoms and having a charge Z.

Note that the ionization equilibrium (8.1) in an ionized gas results from stepwise ionization of charged clusters in collisions with electrons and three body recombination of charged clusters and electrons. But because the number density of the internal electrons of a metal cluster greatly exceeds that of the plasma electrons, a third particle in the recombination process may be a bound electron. Under equilibrium between internal and plasma electrons, the Saha formula (8.2) is valid, but electron release is connected to internal electrons.

Problem 8.2

Within the framework of the ionization equilibrium (8.1) determine the charge distribution function of large metal clusters in a hot vapor in the limit $e^2/r_o < T_e$ and $r_o \ll \lambda$.

The equilibrium (8.1) between free electrons and bound electrons in a metal cluster corresponds to a large mean free path of electrons in a gas compared with the cluster radius. Hence, the interaction energy of the electron and a charged cluster, when an electron is removed from the charged cluster, must be included in the ionization potential of this cluster. The ionization potential $I_Z(n)$ of a large cluster $n \gg 1$ differs from that of a neutral cluster $I_0(n)$ in the energy that is consumed as a result of the electron's removal from the cluster's surface to infinity. Hence we have

$$I_Z(n) - I_0(n) = \frac{Ze^2}{r_o} \,. \tag{8.3}$$

From this it follows that

$$I_Z(n) = \frac{Ze^2}{r_o} + I_0(n) \,, \tag{8.4}$$

where $I_0(n)$ is the ionization potential of a neutral particle consisting of n particles.

Using this relation in formula (8.2), one can reduce it to the form of the Gauss formula in the case $Z \gg 1$. We have

$$\frac{P_Z}{P_{Z-1}} = A \exp\left(-\frac{Ze^2}{r_o T_e}\right) ,$$

where

$$A = \frac{2}{N_e}\left(\frac{m_e T_e}{2\pi\hbar^2}\right)^{3/2} \exp\left(-\frac{I_0}{T_e}\right) . \tag{8.5}$$

From this we have for the ionization equilibrium under consideration

$$\frac{P_Z}{P_0} = A^Z \exp\left(-\frac{Z^2 e^2}{2 r_o T_e}\right) . \tag{8.6}$$

For large Z it is convenient to expand the function $\ln P_Z(n)$ in formula (8.6), which allows us to reduce this expression to the Gauss formula

$$P_Z(n) = P_{\overline{Z}}(n) \exp\left[-\frac{(Z-\overline{Z})^2}{2\Delta^2}\right] , \tag{8.7}$$

where the average cluster charge is given by

$$\overline{Z} = \frac{r_o T_e}{e^2}\left\{\ln\left[\frac{2}{N_e}\left(\frac{m_e T_e}{2\pi\hbar^2}\right)^{3/2}\right] - \frac{I_0(n)}{T_e}\right\} , \quad \Delta^2 = \frac{r_o T_e}{e^2} . \tag{8.8}$$

This equation holds true if $\Delta \gg 1$, which is valid for large cluster charges.

Problem 8.3

Within the framework of a simple model, express the cluster ionization potential through the atom ionization potential I and the work function W of bulk metal.

Let us take the ionization potential $I_0(n)$ of a neutral metal cluster as a monotonic function of the number n of cluster atoms. Assuming the variation of the cluster ionization potential to be proportional to the cluster radius, we obtain

$$I_0(n) = W + \frac{\text{const}}{n^{1/3}} ,$$

where the constant follows from the relation $I_0(1) = I$. Accounting for the dependence (8.4) of the cluster ionization potential on the cluster charge, we find for the cluster ionization potential within the framework of this simple model

$$I_Z(n) = \frac{Ze^2}{r_o} + W + \frac{I - W}{n^{1/3}} . \tag{8.9}$$

Problem 8.4

Obtain the expression for the average cluster charge in a plasma on the basis of formula (8.9) for the cluster ionization potential.

If we define the average cluster charge on the basis of the relation $P_Z(n) = P_{Z+1}(n)$, we obtain from the ionization equilibrium (8.2)

$$Z = \frac{T_e r_o}{e^2} \left\{ \ln \left[\frac{2}{N_e} \left(\frac{m_e T_e}{2\pi\hbar^2} \right)^{3/2} \right] - \frac{I_0(n)}{T_e} \right\} .$$

From this, on the basis of formula (8.9), we obtain for the average cluster charge

$$Z = \frac{T_e r_o}{e^2} \left\{ \ln \left[\frac{2}{N_e} \left(\frac{m_e T_e}{2\pi\hbar^2} \right)^{3/2} \right] - \frac{W}{T_e} - \frac{I - W}{T_e n^{1/3}} \right\} . \tag{8.10}$$

According to this equation, the basic dependence of the cluster charge on the cluster size has the form $Z \sim n^{1/3}$, and the proportionality coefficient depends on the cluster temperature.

Problem 8.5

Determine the number density of free electrons of an ionized gas if it is determined by an equilibrium with metal clusters consisting of n atoms.

The ionization equilibrium under consideration results from the emission of electrons by metal clusters and the absorption of electrons as a result of recombination. We take the limit of large clusters when the parameter $e^2/(r_o T_e)$ is small. Then formula (8.8) gives, in the limit $Z \to 0$, $n \to \infty$ for the number density of free electrons due to their equilibrium with metal clusters,

$$N_e = 2 \left(\frac{m_e T_e}{2\pi\hbar^2} \right)^{3/2} \exp \left[-\frac{I_0(n)}{T_e} \right] . \tag{8.11}$$

Problem 8.6

Determine the conditions for which large metal clusters or particles are charged positively or negatively in a plasma.

An equilibrium charge of a metal cluster results from the balance of the fluxes of electrons emitted from the cluster surface and electrons absorbed by the cluster surface. An equilibrium number density of electrons corresponds to the case where the cluster charge is zero at a given electron temperature. If the number density of electrons exceeds this value, the cluster is charged negatively, and in the opposite case it has a positive charge. Let us denote by N_+ the number density of positively charged clusters with a single positive charge, and by N_- the number density of clusters with a single negative charge. Evidently, the sign of the mean cluster charge coincides with the sign of the value $\ln(N_+/N_-)$. In other words, if $N_+/N_- > 1$, the mean cluster charge is positive, and for $N_+/N_- < 1$ it is negative. Let us use the Saha relation for the ratio of the number densities of positively charged N_+ and neutral N_o clusters:

$$\frac{N_+ N_e}{N_o} = 2\left(\frac{m_e T_e}{2\pi\hbar^2}\right)^{3/2} \exp\left[-\frac{I_0(n)}{T_e}\right], \tag{8.12}$$

where N_e is the electron number density and $I_0(n)$ is the ionization potential of a neutral cluster consisting of n atoms. In the same way one can express the ratio of the number densities of neutral and negatively charged clusters through the cluster electron affinity.

Let us connect the cluster ionization potential with the atom ionization potential $I_0(1)$ and the work function of the corresponding surface W as was done in the previous problem. Let us also connect the cluster electron affinity and the electron affinity EA of the atom. Then we get

$$\frac{N_+}{N_-} = \zeta^2 \exp\left(-\frac{\Delta}{T_e n^{1/3}}\right),$$

$$\zeta = \frac{2}{N_e}\left(\frac{m_e T_e}{2\pi\hbar^2}\right)^{3/2} \exp\left(-\frac{W}{T_e}\right), \quad \Delta = I_0(1) + EA - 2W. \tag{8.13}$$

Let us consider as an example the case of copper, when $I_0(1) = 7.73$ eV, $EA = 1.23$ eV, and $W = 4.4$ eV. Then $\Delta = 0.16$ eV, and the ratio $\Delta/(T_e n^{1/3})$ is small for large clusters and the electron temperatures under consideration. Hence, the sign of the mean cluster charge is determined by the sign of the function $\ln \zeta(T_e)$. This is a monotonic function of the electron temperature, so at high electron temperatures clusters are positively charged, and at low electron temperatures they have a mean negative charge. The electron temperature of the transition T_* is determined by the equation $\zeta(T_*) = 1$. In particular, in the case of large copper clusters it is $T_* = 2380$ K for $N_e = 1 \cdot 10^{11}\,\text{cm}^{-3}$, $T_* = 2640$ K for $N_e = 1 \cdot 10^{12}\text{cm}^{-3}$, and $T_* = 2970$ K for $N_e = 1 \cdot 10^{13}\,\text{cm}^{-3}$. As is seen, the temperature T_* of the change of the cluster charge sign increases with an increase in the number density of plasma electrons.

Thus, a metal cluster is positively charged at high temperatures and negatively charged at low temperatures. As a demonstration of this fact, Figure 8.1 gives the average charge of tungsten clusters in a plasma as a function of the cluster temperature, and also the cluster temperature T_* at which this cluster becomes neutral on average, that is, $\zeta(T_*) = 1$.

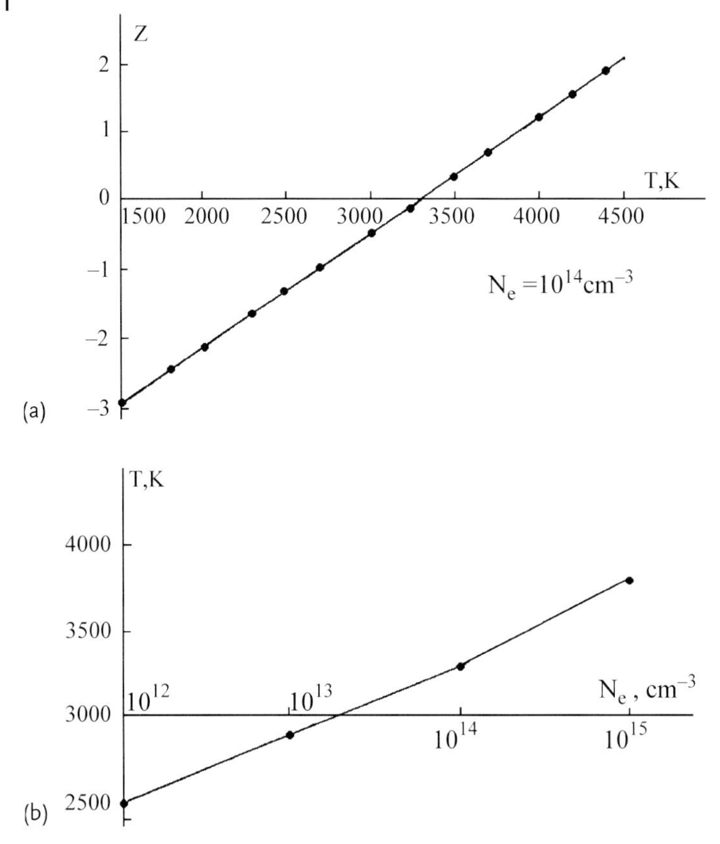

Fig. 8.1 The dependence on the electron temperature of the mean charge of a tungsten cluster consisting of $n = 1000$ atoms in a plasma with a number density of electrons of $N_e = 10^{14}\,\mathrm{cm}^{-3}$ (a) and the dependence of the electron temperature at which this charge is zero on the electron number density (b).

Note that for a large metal cluster const in first formula of Problem 8.3 in units e^2/r_W is equal to 1/2 [168, 169] or 3/8 [170–172] for different versions of the theory. In the same manner, we have for this const in the expression for the electron affinity of a large cluster the value $-1/2$ or $-5/8$ for the corresponding versions of the theory. According to deeper versions of the theory [173–175] and experimental results [176–179], these constants depend on the cluster material. In particular, Figure 8.2 gives the size dependence for the ionization potential and electron affinity of aluminum clusters.

Problem 8.7

Determine the electron temperature T_* at which the average charge of metal clusters is zero, that is, the number densities of positively and negatively charged clusters are equal.

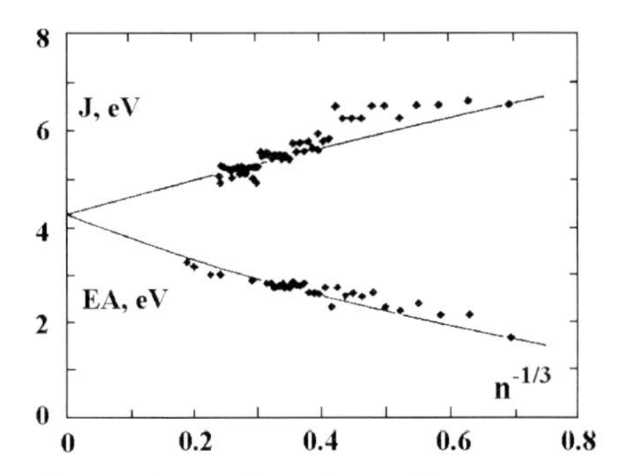

Fig. 8.2 The size dependence of the ionization potential (1) and the electron affinity (2) according to experimental (symbols) and theoretical (solid curve) data [179] for aluminium clusters.

A metal cluster is positively charged at high temperatures and negatively charged at low temperatures. As a demonstration of this fact, Figure 8.1 gives the average charge of tungsten clusters in a plasma as a function of the cluster temperature, and also the cluster temperature T_* at which this cluster becomes neutral on average $N_+ = N_-$, that according to formula (8.13) corresponds to the relation,

$$\ln \zeta(T_*) = \frac{\Delta}{2T_e n^{1/3}} .$$

Table 8.1 contains the parameters of formula (8.13) for some clusters and a plasma.

From formula (8.10) it follows that the basic dependence of the cluster charge on the cluster size is $Z \sim n^{1/3}$, and the proportionality coefficient depends on the cluster temperature. In the limit of large clusters, this formula can be represented in the form

$$\overline{Z} = z n^{1/3}(T_e - T_*) , \quad z = \frac{W r_W}{e^2 T_*} , \tag{8.14}$$

where r_W is the Wigner–Seitz radius and T_* is the electron temperature at which the cluster charge is zero. Note that z depends on the electron number density through the value T_*.

The number density of electrons N_e in formulas (8.10) and (8.14) uses a free parameter that does not depend on cluster parameters. Now we consider a case where this value is determined by the ionization of clusters. This is realized at a high density of clusters. We consider the case where $T > T_*$, and the number density of electrons results from their thermoemission from the cluster surface. Assuming the plasma to be quasineutral, we have $N_e = \overline{Z} N_{cl}$, where $\overline{Z}$ is the average cluster charge and N_{cl} is the number density of clusters. The balance of rates per unit volume for the processes of thermoemission of electrons and attachment of electrons to clusters has the form

$$\nu_{em} N_{cl} n^{2/3} = N_e N_{cl} k_e n^{2/3} ,$$

Table 8.1 Parameters of charged metal clusters. I is the atom's ionization potential, W is the metal work function, EA is the atom's electron affinity, $\Delta = I + EA - 2W$, T_* is the temperature when the number densities of positively and negatively charged clusters are equalized at the average cluster size $n = 1000$ and electron number density $N_e = 10^{13}\,\mathrm{cm}^{-3}$, and Z is the average cluster charge at a temperature of 1000 K for argon as a buffer gas if the cluster charging is determined by processes of attachment of electrons and positive ions to the cluster. The rate constant of electron–cluster collision is given at $T = 1000$ K [36].

Element	I, eV	W, eV	EA, eV	Δ, eV	T_*, 10^3 K	$Z/n^{1/3}$	k_e, $10^{-8}\mathrm{cm}^3/\mathrm{s}$
Ti	6.82	3.92	0.08	–0.96	2.51	0.084	1.7
V	6.74	4.12	0.52	–0.98	2.63	0.078	1.5
Fe	7.90	4.31	0.15	–0.57	2.75	0.074	1.3
Co	7.86	4.41	0.66	–0.30	2.82	0.073	1.3
Ni	7.64	4.50	1.16	–0.20	2.87	0.072	1.3
Zr	6.84	3.9	0.43	–0.53	2.51	0.093	2.1
Nb	6.88	3.99	0.89	–0.21	2.57	0.085	1.7
Mo	7.10	4.3	0.75	–0.75	2.74	0.081	1.6
Rh	7.46	4.75	1.14	–0.90	3.00	0.078	1.5
Pd	8.34	4.8	0.56	–0.70	3.04	0.080	1.5
Ta	7.89	4.12	0.32	–0.03	2.65	0.085	1.7
W	7.98	4.54	0.82	–0.28	2.90	0.081	1.6
Re	7.88	5.0	0.2	–1.92	3.12	0.080	1.5
Os	8.73	4.7	1.1	0.43	3.01	0.078	1.5
Ir	9.05	4.7	1.56	1.21	3.03	0.080	1.5
Pt	8.96	5.32	2.13	0.45	3.38	0.081	1.6
Au	9.23	4.30	2.31	2.94	2.85	0.083	1.9

where ν_{em} is the rate of the thermoemission process and $k_e = (8T_e/\pi m_e)^{1/2} * \pi r_W^2$ is the rate constant of electron–cluster collision in accordance with formula (2.7). From this we have for the average cluster charge

$$\overline{Z} = \frac{\nu_{em}(T)}{k_e N_{cl}} .$$

As follows from this formula, clusters tend to be neutral at low temperatures and large densities of clusters.

In the limit $Ze^2/r_o \ll T_e$ the above formula takes the form

$$\frac{N_{-Z} N_e}{N_{-(Z+1)}} = 2\left(\frac{m_e T_e}{2\pi\hbar^2}\right)^{3/2} \exp\left(-\frac{EA}{T_e}\right) \exp\left(-\frac{Ze^2}{r_o T_e}\right) .$$

Note that Z in this formula has a positive value. A low rate constant of electron attachment to a strongly negatively charged cluster does not allow a metal cluster to have a very high negative charge in a rare ionized gas.

Problem 8.8

Evaluate the average negative charge of a large metal cluster when the equilibrium process for the cluster charge proceeds near the cluster surface.

We now find the charge distribution for negatively charged metal clusters if the temperatures of internal and free electrons are identical. The cluster radius is small compared with the mean free path of electrons, and the Coulomb interaction of the electron with the cluster charge on its surface greatly exceeds an electron thermal energy. Assume the same character of electron thermoemission for charged and neutral metal clusters.

As above, the balance of rates for the formation and destruction of negatively charged metal clusters has the form

$$N_{-Z} N_e \pi r_o^2 \cdot \overline{\left(v - \frac{2Ze^2}{r_o m_e v}\right)} = N_{-(Z+1)} \cdot 4\pi r_o^2 \frac{e m_e T_e^2}{2\pi^2 \hbar^2} \exp\left(-\frac{EA}{T_e}\right) ,$$

where the bar indicates the average over electron velocities. Here we assume that the electron release proceeds near the cluster's surface, and hence it is determined by the electron affinity EA of the neutral cluster. The cluster field accelerates the released electron and removes it from the surface. Therefore, the difference in the electron thermoemission process for clusters of different charges takes place far from the cluster surface.

Let us find the charge of a negatively charged metal cluster on the basis of the above equation. We define the average cluster charge as $-(Z + 1/2)$ if $N_{-Z} = N_{-(Z+1)}$. Then the above equation gives

$$\overline{Z} = \frac{1}{2} + \frac{r_o T_e}{e^2} \ln \frac{1}{\zeta} , \tag{8.15}$$

for the mean negative charge $-\overline{Z}$ of a metal cluster, where the value ζ is defined in formula (8.13). As is seen, a high negative charge of a metal cluster can occur at high temperatures, large cluster sizes, or small number densities of electrons when $\zeta \ll 1$. Because we neglect here the formation of positive ions, this equation differs from formula (8.14) for $\zeta \sim 1$.

Problem 8.9

Analyze the character of recombination of free electrons with a metal cluster.

Above we considered different conditions of charging of a cluster. If the cluster radius r_o is large compared with the mean free path λ of ions and electrons in a buffer gas where the cluster is located, the charging process proceeds in the diffusion regime of collision between charged atomic particles (electrons and ions) and clusters. Since this takes place at large distances from the cluster surface compared with the mean free path λ of atomic particles, the character of the interaction of electrons and ions with the cluster surface is not essential for the charging process. Hence, the cluster charge is determined by plasma parameters only, and the character of the cluster charging is identical for both metallic and dielectric clusters.

In the kinetic regime of collision between charged atomic particles and clusters ($\lambda \gg r_0$) the character of cluster charging is determined by processes involving electrons, ions, and the cluster surface. Then emission of electrons from the cluster surface is important, and therefore the charging processes for dielectric and metallic particles are different. In the case of a metallic particle, the charging process is determined by the balance of fluxes of emission and attachment for electrons to the cluster surface. At high temperatures clusters acquire a positive charge, which promotes the approach of positive ions to the cluster surface. Hence, positive ions of a surrounding plasma do not influence the charging process. This condition is valid if the cluster charge is positive or not too negative, so the electron flux to the cluster surface from the plasma greatly exceeds the ion flux. For this reason one can neglect the influence of positive ion currents toward the cluster surface on the charging process of metal clusters. On the other hand, attachment of positive ions to the cluster surface can be important for the charging of dielectric clusters.

8.2 Electron Thermoemission of Metal Clusters

Problem 8.10

Derive the Richardson–Dushman formula for the thermoemission current density from a hot metallic surface.

Let us consider the ionization equilibrium between a plasma and a metal surface, considering the latter as a large metal cluster. The electron current from this cluster, the thermoemission current, is equal to the current of electrons absorbed by a cluster, that is,

$$i = e \cdot N_e \sqrt{\frac{T_e}{2\pi m_e}} ,$$

where we assume the probability of electron attachment to the cluster surface as a result of their contact to be one. Using formula (8.11) for the number density of electrons due to equilibrium and replacing in this formula the ionization cluster potential $I_0(n)$ by the metal work function W, we obtain the Richardson–Dushman formula for the thermoemission current density:

$$i = eN_e \sqrt{\frac{T_e}{2\pi m_e}} = \frac{e m_e T_e^2}{2\pi^2 \hbar^2} \exp\left(-\frac{W}{T_e}\right) . \tag{8.16}$$

Table 8.2 Parameters of the electron current for thermoemission of metals [145, 181–183]; i_b is the electron current density at the boiling point, T_b is the metal boiling point.

Material	A_R, A/(cm^{2}K^2)	W, eV	T_b, K	i_b, A/cm^2
Ba	60	2.49	1910	600
Cs	160	1.81	958	0.045
Cu	60	4.4	2868	4.6
Mo	51	4.3	5070	$8.8 \cdot 10^4$
Nb	57	4.0	5170	$2.1 \cdot 10^5$
Pd	60	4.8	3830	240
Re	720	5.0	5870	$2.3 \cdot 10^6$
Ta	55	4.1	5670	$3.3 \cdot 10^5$
Th	70	3.3	4470	$2.2 \cdot 10^5$
Ti	60	3.9	3280	750
W	75	4.5	5740	$2.8 \cdot 10^5$
Y	100	3.3	3478	$3.5 \cdot 10^3$
Zr	330	3.9	4650	$2.4 \cdot 10^5$

It is convenient to rewrite formula (8.16) for the thermoemission current density in the form

$$i = A_R T_e^2 \exp\left(-\frac{W}{T_e}\right), \quad A_R = \frac{e m_e}{2\pi^2 \hbar^2}, \tag{8.17}$$

and the Richardson parameter A_R according to formula (8.12) is equal to 120 A/(cm^{2}K^2). Table 8.2 contains the values of this parameter for real metals. Figures 8.3 and 8.4 give examples where the thermoemission process is important for the balance of parameters of metal clusters.

We note two mechanisms of electron emission from the surface of metal clusters. In the analysis of the thermoemission of electrons, we model electrons as being located in a potential well with the depth of the work function W, and electrons are almost free in this well. Along with thermoemission, resonant emission of electrons is possible [180], which take into account that the electron spectrum is not continuous, and the energy consumed in the release of certain electrons exceeds the binding energy. In reality, both mechanisms are possible, but at low temperatures thermoemission of electrons is more preferable.

Problem 8.11

Niobium and tungsten clusters of size $n = 10^3$ are heated by a laser pulse and then are cooled owing to the emission of radiation and thermoemission of electrons. Compare the contributions of these mechanisms to the rates of cluster cooling.

Note that the emission of electrons by neutral metal clusters consists of two parts [180]; the first one is thermoemission, and the second one has a resonant

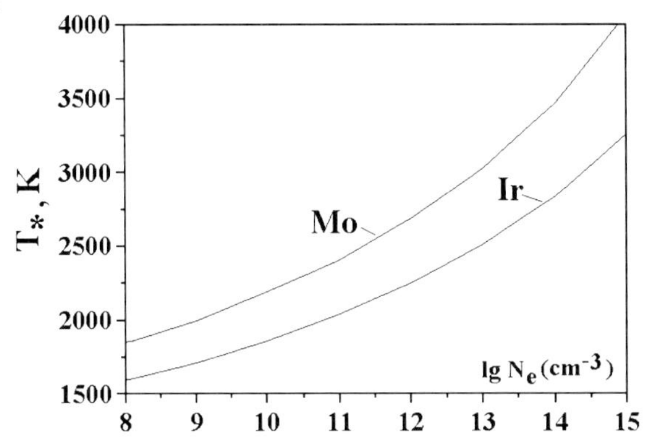

Fig. 8.3 The dependence of the effective electron temperature on the number density of electrons N_e (cm^{-3}) in a plasma for molybdenum and iridium clusters consisting of 1000 atoms. At this number density the mean charge of clusters is zero and the flux of electrons attached to a neutral cluster is equal to the flux of electrons emitted as a result of thermoemission [186].

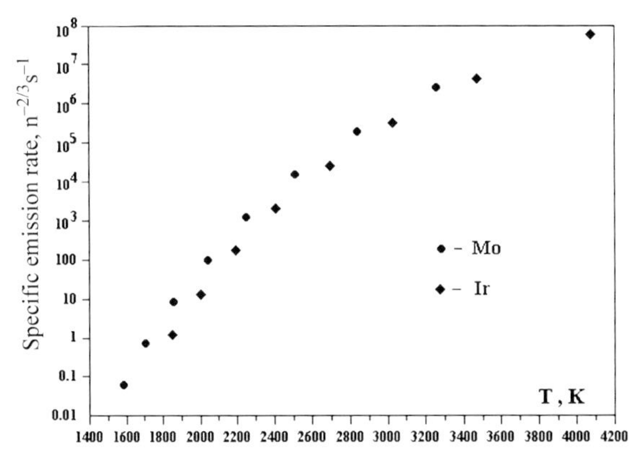

Fig. 8.4 The specific rate of electron thermoemission from the surface of large molybdenum (closed circle) and iridium (closed rhombus) clusters as a function of the cluster temperature [186].

character and is caused by the structure of electron levels above the boundary of the continuous spectrum. The resonance electron emission is important for cluster excitation by fast electrons and ions. In considering thermal cluster excitation and assuming that bound electrons are found in equilibrium in the course of excitation by a laser beam, we ignore the resonance part of electron emission.

The cooling power P_{emis} of a neutral cluster due to thermoemission of electrons is based on the Richardson–Dushman formula (8.17) and in neglecting the contribution of the surface energy to the total cluster energy is

$$P_{\text{emis}} = 4\pi r_o^2 W i = 4\pi r_W^2 n^{2/3} A_R T^2 W \exp\left(-\frac{W}{T}\right) , \qquad (8.18)$$

Table 8.3 The cooling power (W) of a neutral cluster due to the thermoemission of electrons P_{emis} at the temperatures indicated and due to radiation emission P_{rad}, given in parentheses. The notation is such that 1.8^{-10} means $1.8 \cdot 10^{-10}$.

	$T = 3000$ K	$T = 3500$ K	$T = 4000$ K
Nb	1.8^{-10} (2.3^{-12})	2.2^{-9} (4.2^{-12})	1.5^{-8} (7.3^{-12})
W	2.7^{-11} (2.7^{-12})	4.4^{-10} (5.1^{-12})	3.7^{-9} (8.7^{-12})

where W is the work function, which is 4.0 and 4.5 eV for niobium and tungsten, respectively [34, 183], r_W is the Wigner–Seitz radius, n is the number of cluster atoms, i is the thermoemission electron current density in accordance with formula (8.16), and A_R is the Richardson–Dushman constant, which is 57 and 75 A/cm^2K^2 for niobium and tungsten, respectively (Table 8.2). The Wigner–Seitz radii are 1.68 and 1.60 Å for niobium and tungsten, respectively (Table 2.1). The values of the power P_{emis} of cluster cooling due to the thermoemission of electrons are given in Table 8.3.

We compare this with the cooling power P_{rad} due to cluster radiation, which is

$$P_{rad} = \sigma_{abs} \kappa T^4 ,$$

where κ is the Stefan–Boltzmann constant, σ_{abs} is the absorption cross section, which is assumed to be independent of the photon frequency, and according to the data in Tables 5.2 and 5.3 the specific cross sections for niobium and tungsten clusters are $\sigma_{abs}/n = (5 \pm 1) \cdot 10^{-18}\,\text{cm}^2$ and $\sigma_{abs}/n = (6 \pm 1) \cdot 10^{-18}\,\text{cm}^2$, respectively. Values of the quantity P_{rad} at appropriate temperatures are given in Table 8.3 in parentheses.

As is seen, in the range of temperatures under consideration the cooling power due to electron thermoemission for a neutral power exceeds significantly that due to radiation. But this does not mean that cluster cooling is determined by electron thermoemission. Indeed, as electrons are released, the cluster is charged, and then the rate of thermoemission decreases owing to interaction of a released electron and the cluster charge. When a cluster has charge Z, the cooling power P_Z due to electron thermoemission is connected to P_{emis} of a neutral cluster and is given by

$$P_Z = P_{emis} \exp\left(-\frac{Ze^2}{r_o T}\right) ,$$

where P_{emis} is given by formula (8.18).

In particular, from this we have for the cluster charge Z_* when the cooling power due to radiation and thermoemission of electrons are equalized ($P_Z = P_{rad}$)

$$Z_* = \frac{r_o T}{e^2} \ln \frac{P_{emis}}{P_{rad}} .$$

For the niobium cluster at $T = 4000$ K, $Z_* = 3$. In other cases it is less.

Problem 8.12

On the basis of properties of a metal plasma near a metal surface, determine the thermoemission current from this surface.

Above we evaluated the emission current from a metal surface on the basis of the principle of detailed balance by using parameters of the surrounding plasma. Now we obtain this value from parameters of a metal plasma. We assume a metal plasma to be similar to a degenerate electron gas, and the momentum distribution function of electrons is given by the Fermi–Dirac formula:

$$f(\mathbf{p})d\mathbf{p} = \frac{2d\mathbf{p}}{(2\pi\hbar)^3}\left[1+\exp\left(\frac{\varepsilon-\mu}{T}\right)\right]^{-1} .$$

Here $\mathbf{p} = m_e\mathbf{v}$ is the electron momentum, $\mathbf{v}$ is the electron velocity, $\varepsilon = p^2/(2m_e)$ is the electron kinetic energy, and $\mu = \varepsilon_F$ is the chemical potential or the Fermi energy for this distribution. The electron is released if the electron kinetic energy in the direction toward the metal surface exceeds the value $\varepsilon_F + W$, where W is the metal's work function. Hence, the flux of released electrons from the metal surface is equal to $\int v_x f(\mathbf{p})d\mathbf{p}$, where the integral is taken over velocities of electrons $m_e v_x^2/2 \geq \varepsilon_F + W$ and v_x is the component of the velocity toward the surface.

Note that because of $W \gg T$, one can neglect unity compared with the exponent in the Fermi–Dirac equation, so this takes the form of the Boltzmann formula. In this case the number density of electrons in the momentum interval from $\mathbf{p}$ to $\mathbf{p}+d\mathbf{p}$ in the energy range $\varepsilon-\mu \gg T$ is given by

$$f(\mathbf{p})d\mathbf{p} = \frac{2d\mathbf{p}}{(2\pi\hbar)^3}\exp\left(-\frac{\varepsilon-\mu}{T}\right) .$$

Using frame of the cylindrical coordinates $d\mathbf{p} = 2\pi p_\rho dp_\rho dp_x$ and $\varepsilon = p_x^2/(2m_e)+p_\rho^2/(2m_e)$, we obtain for the flux of released electrons

$$j = 2\pi m_e T\int\frac{m_e v_x dv_x}{4\pi^3}\exp\left(-\frac{m_e v_x^2}{2T}+\frac{\mu}{T}\right) = \frac{m_e T^2}{2\pi^2\hbar^3}\exp\left(-\frac{W}{T}\right) .$$

Accounting for the electron current density $i = ej$, we find that this expression coincides with the Richardson–Dushman formula (8.16). Thus, we obtain the same result for the electron thermoemission current density both from the equilibrium of a metal surface with a surrounding plasma and from the evaporation of electrons of a metal plasma that is modeled by a degenerate electron gas.

Problem 8.13

Ascertain the role of internal and plasma electrons for the electron release process in establishing the cluster equilibrium charge.

Release of electrons from the surface of a metal particle or cluster is determined by two processes – thermoemission of electrons and ionization by plasma electrons. Below we find the contribution of these processes in the release of bound electrons of a metal particle or cluster. Usually, the temperature of internal electrons coincides with the cluster temperature, so the thermoemission current of electrons from the surface of a metal cluster is determined by the cluster temperature, while the current of released electrons resulting from ionization by the plasma electrons is determined by the electron temperature of the plasma electrons. The equilibrium cluster charge follows from the equality of the total current of released electrons and that of attached electrons. Below we estimate the role of plasma electrons in the ionization of a metal cluster.

We take a simple model for the interaction of electrons with a metallic surface, according to which the effective surface potential has the form of rectangular walls in a spatial region and the depth of this potential well equals W, the metal's work function. For collisions of plasma electrons with internal ones, we use the Thomson model of pair collisions. Within the framework of this model, the internal electron energy is zero, and interaction with surrounding electrons and ions is absent in the course of collision of the plasma and internal electrons. The energy ε of a plasma electron satisfies the relation $\varepsilon \sim T_e \ll W$, where T_e is the temperature of the plasma electrons. Penetrating in the metal region, this electron obtains the energy $\varepsilon + W$. The cross section for transfer of energy from a plasma electron to an internal electron in the interval from $\Delta\varepsilon$ to $\Delta\varepsilon + d\Delta\varepsilon$ is given by the Rutherford formula:

$$d\sigma = \frac{\pi e^4}{\varepsilon + W} \frac{d\Delta\varepsilon}{\Delta\varepsilon^2} .$$

The capture of an incident electron by the metal surface takes place if the energy transfer to an internal electron exceeds ε but does not exceed W, so both colliding electrons are trapped in the metal potential well. Hence, the rate constant of this process is

$$k_{\text{cap}} = \int f(\varepsilon) d\varepsilon \int_{\varepsilon}^{W} d\sigma .$$

Here $f(\varepsilon)$ is the Maxwell distribution function of plasma electrons, which is normalized to unity. Introducing $x = \varepsilon / T_e$, we have $f(\varepsilon) d\varepsilon = 2\pi^{-1/2} x^{1/2} \exp(-x) dx$. From this we have

$$k_{\text{cap}} = \frac{\pi e^4}{W T_e} \sqrt{\frac{8 T_e}{\pi m_e}} ,$$

for the effective rate constant of capture of a plasma electron resulting from its collision with an internal electron of the metal surface if we take into account $T_e \ll W$.

Problem 8.14

Compare the rates of electron release from a metal surface due to internal metal electrons and plasma electrons.

The rate constant of the ionization process for the release of electrons from a metal surface under the action of plasma electrons is given by the relation

$$k_{\text{ion}} = \int_W^\infty f(\varepsilon)d\varepsilon \int_W^\varepsilon d\sigma = \frac{\pi e^4}{W^2}\sqrt{\frac{8T_e}{\pi m_e}} \exp\left(-\frac{W}{T_e}\right) = k_{\text{cap}}\frac{T_e}{W}\exp\left(-\frac{W}{T_e}\right).$$

We take into account within the framework of the Thomson model that the release of an internal electron takes place when the energy obtained by this electron exceeds W, but does not exceed the energy of the incident electron ε, so both colliding electrons abandon the metal potential well. Next, the cluster temperature is relatively small, that is, a thermal energy of internal electrons does not influence this process.

From this it follows that the ionization rate constant is small compared with that of electron capture, because $W \gg T_e$. We obtain the cluster equilibrium charge from the equality of the rate of thermoemission of internal electrons and the rate of capture of plasma electrons. Because the rate of capture of plasma electrons is large compared with the rate of release of internal electrons in collisions with plasma electrons, the ionization process due to plasma electrons is weak compared with the thermoemission process. Hence, the ionization process with participation of plasma electrons does not influence the charge equilibrium for metal clusters, and their equilibrium charge is determined by the cluster temperature.

Problem 8.15

In considering the charge equilibrium for small clusters, we assume the processes of electron evaporation and electron attachment to be identical for neutral and charged clusters. Find the correction for the average electron charge due to the fact that the electron attachment cross sections to charged and neutral metallic particles are different.

Within the framework of the equilibrium (8.1), we have the following equality between rates of formation and destruction of clusters of a given charge:

$$N_{Z+1}N_e k_{\text{at}} = N_Z \cdot 4\pi r_o^2 i\,.$$

Here N_Z is the number density of metal clusters having charge Z, r_o is the cluster radius, k_{at} is the rate constant of electron attachment to a cluster with charge $Z+1$, and i is the thermoemission current density. By definition, we have

$$\frac{N_Z}{N_{Z+1}} = \frac{P_Z}{P_{Z+1}}\,,$$

where P_Z is the probability that the cluster has charge Z.

We first consider the charge equilibrium for a neutral cluster when the rate constant of electron attachment to the cluster surface is given by formula (2.7)

$$k_{at} = \sqrt{\frac{8T_e}{\pi m_e}} \cdot \pi r_o^2 \,.$$

Using formula (8.16) for the thermoemission current density and replacing in this formula the metal's work function W by the cluster ionization potential I, we obtain from the above balance equation

$$\frac{N_{Z+1}N_e}{N_Z} = 2\left(\frac{m_e T_e}{2\pi\hbar^2}\right)^{3/2} \exp\left(-\frac{I}{T_e}\right) .$$

This is the Saha formula (8.2) for the charge distribution of clusters.

Now let us take into account the cluster charge in the electron attachment process. According to formula (4.1), the rate constant of this process is

$$k_{at} = \pi r_o^2 \left\langle v + \frac{2Ze^2}{r_o m_e v}\right\rangle .$$

Here we take into account the Coulomb interaction between an electron and a charged cluster; v is the electron velocity, and the average is taken over the electron velocities. Averaging over the Maxwell velocity distribution of electrons, we obtain

$$\frac{N_{Z+1}N_e}{N_Z} = \left(1 + \frac{Ze^2}{r_o T_e}\right)^{-1} \cdot 2\left(\frac{m_e T_e}{2\pi\hbar^2}\right)^{3/2} \exp\left(-\frac{I}{T_e}\right) ,$$

from the equilibrium for the charging processes. As is seen, this formula coincides with the Saha formula (8.2) in the limit $Ze^2/r_o \ll T_e$. Violation of this criterion leads to a decrease in the mean cluster charge because the cross section of electron attachment to a positively charged cluster exceeds that for a neutral cluster. Note that we assume the identical character of the thermoemission of charged and neutral metal clusters.

Problem 8.16

Find the charge distribution for negatively charged metal clusters if the temperatures of internal and free electrons are identical. The cluster radius is small compared with the mean free path of electrons, and the Coulomb interaction of an electron with the cluster charge on the cluster surface exceeds significantly an electron thermal energy. Assume the identical character of thermoemission for charged and neutral metal clusters.

We base our analysis on the balance equation for rates of formation and destruction of negatively charged metal clusters that has the form

$$N_{-Z}N_e\pi r_o^2\left\langle v - \frac{2Ze^2}{r_o m_e v}\right\rangle = N_{-(Z+1)} \cdot 4\pi r_o^2 \frac{e m_e T_e^2}{2\pi^2\hbar^2} \exp\left(-\frac{EA}{T_e}\right) .$$

Here we assume that the electron release proceeds near the cluster surface, and hence it is determined by the electron affinity EA of the neutral cluster. The cluster field accelerates a released electron and removes it from the surface. Therefore, the difference in the thermoemission rates for clusters of different charges is determined by processes far from the clusters' surfaces.

In the limit $Ze^2/r_o \gg T_e$ the above equation takes the form

$$\frac{N_{-Z} N_e}{N_{-(Z+1)}} = 2\left(\frac{m_e T_e}{2\pi\hbar^2}\right)^{3/2} \exp\left(-\frac{EA}{T_e}\right) \exp\left(-\frac{Ze^2}{r_o T_e}\right) .$$

Note that Z in this equation has a positive value. A low rate constant for electron attachment to a strongly negatively charged cluster does not allow a metal cluster to have a very high negative charge in a rare ionized gas.

Let us find the charge of a negatively charged metal cluster on the basis of the above equation. We consider the average cluster charge to be $-(Z + 1/2)$ if $N_{-Z} = N_{-(Z+1)}$. Then the above equation gives

$$\overline{Z} = \frac{1}{2} + \frac{r_o T_e}{e^2} \ln \frac{1}{\zeta} , \tag{8.19}$$

for the average negative charge $-\overline{Z}$ of a metal cluster, and this formula coincides with formula (8.15). In accordance with the above equations, we have here

$$\zeta = \frac{2}{N_e}\left(\frac{m_e T_e}{2\pi\hbar^2}\right)^{3/2} \exp\left(-\frac{EA}{T_e}\right) .$$

As is seen, a high negative charge of a metal cluster occurs at low temperatures or small number densities of electrons when $\zeta \ll 1$. Note that this formula is not correct for $\zeta \sim 1$ because in its derivation we ignored the formation of positive ions.

Problem 8.17

Determine the maximum charge of a metal cluster ionized by ultraviolet laser radiation, when the photon energy $\hbar\omega$ exceeds the metal work function W, in the kinetic regime of interaction between electrons and a charged cluster.

The maximum energy of a released electron is equal to $\hbar\omega - W$. In the kinetic regime of electron–cluster collisions, when the mean free path of electrons λ exceeds the cluster radius r_o, an electron must overcome the attraction energy of the cluster Ze^2/r_o to remove far from the cluster. From this, the cluster charge Z at which a released electron can remove far from the cluster is

$$Z = \frac{r_o(\hbar\omega - W)}{e^2} . \tag{8.20}$$

8.3
Ionization Equilibrium for Large Dielectric Clusters

Problem 8.18

Analyze the character of interaction for an isolated dielectric particle with an ionized gas.

The character of the ionization equilibrium for a dielectric cluster located in an ionized gas is determined by the properties of the cluster surface. Each dielectric particle or cluster has on its surface traps for electrons that we call active centers. Electrons are captured by these centers, and this leads to the formation of negative ions located at certain points on the particle surface. The ionization equilibrium of a dielectric particle in a plasma corresponds to the equilibrium of these bound negative ions and free plasma electrons. Though electrons can transfer between neighboring active centers, this process proceeds slowly. The ionization equilibrium in this case results from the detachment of bound negative ions on the particle's surface by electron impact and the capture of electrons by active centers on the particle's surface.

Thus, there are different processes involving clusters and small particles located in a plasma. At high temperatures ionization of metal particles results from the collision of internal electrons, which leads to the formation of fast electrons that are capable of being released from the cluster's surface. This process corresponds to the thermoemission of electrons for a bulk surface. Along with this, the attachment of plasma electrons is important for the charge equilibrium. As a result, metal particles become positively charged at high temperatures owing to the thermoemission process. At low temperatures they are negatively charged. Usually dielectric particles are charged negatively, because their ionization potential greatly exceeds the electron affinity.

At low temperatures the charge of dielectric clusters in an ionized gas results from attachment of electrons and ions to them. This process leads to the formation of chemical bonds of charged atomic particles on the surface of a small particle and can proceed through intermediate stages. For example, attachment of electrons to dielectric particles can proceed through the formation of free negative ions owing to dissociative attachment. These negative ions attach later to the surface owing to collisions with gas atoms or molecules. If electrons and positive or negative ions attach to the surface of a small particle, they recombine on the surface. The character of recombination depends on the type of the particle's material. In the case of a metal particle, electrons can move through the entire particle volume, and attachment of a positive or negative ion to a metal particle decreases or increases by one the total number of particle electrons. This means charge exchange of a positive ion on a small metal particle, so one electron transfers from the small particle to the positive ion. On the other hand, a weakly bound electron of a negative ion transfers to the small metal particle as a result of their contact. Thus, in the case of a small metal particle, recombination of positive and negative ions on the particle's surface proceeds through interaction with a collective of electrons of this particle.

In the case of the small dielectric particle, attachment of positive and negative ions to its surface leads to the formation of bound states of ions at some points on the particle's surface, i. e., at active centers. Hence, bound states of positive and negative ions can exist simultaneously on the particle's surface. Diffusion of these bound ions over the particle surface leads to their recombination. Thus, though the recombination of positive and negative charges on the surface of a metal particle differs from that of a dielectric particle, these processes are identical from the standpoint of fluxes of charged particles on the particle's surface because in the end opposite charges recombine. Hence, below we will not distinguish recombination of charges on a particle's surface for metal and small dielectric particles.

Problem 8.19

Determine the maximum negative charge of an isolated dielectric particle.

The character of charging of dielectric particles differs from that of metal ones. In the case of a metal cluster, valence electrons are distributed over the entire cluster surface. Hence, interaction of electrons is important, so the ionization potential of neutral metal clusters as well as the electron affinity of these clusters tend to the surface work function in the limit of large cluster sizes. In the case of a dielectric particle, there are active knots or centers on its surface that are traps for electrons. Negative atomic or molecular ions may be captured by these centers or be charged by them as a result of the charge exchange process. Other active centers of the particle surface can capture positive atomic or molecular ions. These centers can also transfer an electron to the ions, so the particle obtains a positive charge.

We consider the diffusion regime of interaction of electrons and ions with clusters when the mean free path of ions and electrons in a buffer gas is small compared with the cluster size. In comparison with the case where the cluster ionization equilibrium is determined by transport processes of electrons and ions near the cluster surface (e.g., formulas (7.3) and (7.4)), in the case under consideration the ionization equilibrium for a dielectric cluster results from the processes [184]

$$e + A_n^{-Z} \to \left(A_n^{-(Z+1)}\right)^{**} ,$$
$$\left(A_n^{-(Z+1)}\right)^{**} + A \to A_n^{-(Z+1)} + A \,, \quad B^+ + A_n^{-(Z+1)} \to B + A_n^{-Z} \,, \tag{8.21}$$

so an autodetachment state $\left(A_n^{-(Z+1)}\right)^{**}$ is quenched by collisions with surrounding atoms. Because the rate constant of pair attachment of an electron to a dielectric particle greatly exceeds the ionization rate constant of the particle by electron impact, these particles are charged negatively.

In contrast to metal particles, for active centers the binding energies do not depend on particle size, because the action of each center is concentrated in a small spatial region. Evidently, the number of such centers is proportional to the area of the particle's surface, and for micron-sized particles this value is large compared with that of the charges. Hence, above we considered and below we consider the regime of charging of a small dielectric particle far from the saturation of active centers. Then positive and negative charges can exist simultaneously on the particle's surface. They transfer over the surface and can recombine there.

Usually, the binding energy of electrons in negative active centers is in the range $EA = 2{-}4\,\mathrm{eV}$, and the ionization potential for positive active centers is $I_o \approx 10\,\mathrm{eV}$. Hence, attachment of electrons is more profitable in glow discharge, and a small dielectric particle has a negative charge in glow discharge.

The particle with charge Z obeys the electric potential $\varphi = Ze/r_o$, where r_o is the particle radius. If $e\varphi < EA$, the electron state is stable, whereas in the case $e\varphi > EA$, there is a barrier for the electron transition, so that an unstable state with electrons attached to active centers can have a long lifetime. This system is formed as a result of attachment of new electrons to the particle. An isolated charged particle emits electrons until it reaches the limiting charge [184]:

$$Z_* = r_o \cdot EA/e^2 \,. \tag{8.22}$$

In particular, for a dielectric particle with a radius of $r_o = 1\,\mu\mathrm{m}$ and $EA = 3\,\mathrm{eV}$ this charge is $Z_* = 2 \cdot 10^3$, and the cluster electric potential is equal to 3 V.

Problem 8.20

Determine the conditions of the charging regime for a small dielectric particle when this process is limited by fluxes of electrons and positive ions on the particle's surface.

Evidently, this regime exists if electrons or ions colliding with the particle's surface are found in the field of action of active centers. Then they attach to the surface or recombine with a center with charge of the opposite sign. In this case the number of active centers must be large compared with the particle's charge, that is, the particle is sufficiently large. This is valid for micron-sized particles. In particular, the numerical example of the previous problem corresponds to distances between neighboring charged centers of about $0.3\,\mu\mathrm{m}$, that is, larger by one or two orders of magnitude than the typical distance between neighboring active centers.

One more condition is required in the case where the particle's negative charge exceeds the value Z_*. Then the bound states of captured electrons become autodetachment states and can decay as a result of tunnel electron transitions. In this case the rate of capture of electrons must exceed the rate of decay of autodetachment states, that is, the lifetime of a captured electron with respect to its tunnel transition must be large compared with a typical time of electron capture per electron charge of the particle.

Let us estimate the dependence on parameters of the problem for the lifetime of a negatively charged center that is located on the surface of a dielectric particle. Its decay results from the electron tunnel transition, and the probability per unit time for the escape of an electron through the potential barrier has the following exponential dependence:

$$\frac{1}{\tau} \sim \exp(-2S)\,, \quad S = \int_{r_o}^{R_c} dR \sqrt{\frac{2m_e}{\hbar^2}} \left[EA - U(r_o) + U(R)\right].$$

Here EA is the electron binding energy (the electron affinity of an active center), $U(R) = Ze^2/R$ is the interaction potential of an electron with the Coulomb field of a particle if its distance from the particle's center is R, and R_c is the turning point, that is,

$$R_c = \frac{r_o}{1 - EA/\varepsilon_o}\,,$$

where $\varepsilon_o = Ze^2/r_o$. Thus, we have [184]

$$S = \frac{\pi}{2} r_o \cdot \sqrt{\frac{2m_e}{\hbar^2} \cdot \frac{\varepsilon_o}{1 - EA/\varepsilon_o}}\,.$$

Assuming the parameter ε_o to be of the order of a typical atomic value, we obtain $S \sim r_o/a_o$, where a_o is the Bohr radius. Being guided by small micron-sized particles, we obtain a very high lifetime of surface negative ions with respect to their barrier decay. Hence, the lifetime of a negatively charged particle with respect to the tunnel transition of a captured electron is very large for micron-sized particles located in a plasma, and a particle's charge can exceed the limit value Z_* given by formula (8.22).

Problem 8.21

Determine the charge of a small dielectric particle that results from the ionization equilibrium for this particle with a surrounding electron gas.

This equilibrium is similar to the Langmuir isotherm [185] for the equilibrium of a surface with a gas when gas atoms or molecules are absorbed by active centers of the surface. In this case the equilibrium is described by the scheme

$$A + e \longleftrightarrow A^-\,, \tag{8.23}$$

where A denotes an active center of the surface and A^- is a bound negative ion. Denoting the total number of active centers on the particle's surface by p and the electron binding energy by EA, we have from this equilibrium according to the Saha formula

$$\frac{(p - Z)N_e}{Z} = g \cdot \left(\frac{m_e T_e}{2\pi\hbar^2}\right)^{3/2} \exp\left(-\frac{EA}{T_e}\right).$$

Here N_e is the number density of free electrons, Z is the particle charge or number of electrons bound with active centers, and $g \sim 1$ is the combination of statistical weights of an electron, active center, and bound negative ion. Below, for simplicity, we take $g = 1$. From this we have for the particle charge in the limit $Z \ll p$ that

$$Z = N_e p \left(\frac{2\pi\hbar^2}{m_e T_e}\right)^{3/2} \exp\left(\frac{EA}{T_e}\right) . \tag{8.24}$$

Because the number of active centers on the surface is proportional to the area ($p \sim n^{2/3}$), we have

$$Z \sim n^{2/3} .$$

Since the particle radius is $r_o \sim n^{1/3}$, the electric potential of a large dielectric particle is

$$\varphi \sim \frac{Ze}{r_o} \sim n^{1/3} .$$

Therefore, the interaction of surface electrons with the electric potential of a negatively charged dielectric particle may be responsible for the negative charge of the large particle.

Problem 8.22

Compare the charges of strongly negatively charged metal and dielectric particles.

All the electrons of a metal particle participate in the thermoemission process, whereas in the case of a dielectric particle, electrons that are captured by active centers and create a negative particle charge contribute to the current of released electrons. Hence, the negative charge of a dielectric particle exceeds significantly that of a metal particle if this charge is determined by processes at the particle's surface. In the case of a dielectric particle, a typical negative charge is on the order of the limiting charge Z_* below which a stable negative ion exists. This charge is, according to formula (8.22),

$$Z_* = \frac{r_o \cdot EA}{e^2} .$$

The formula in the previous problem for a negatively charged metallic particle can be represented in the form

$$\overline{Z} = Z_* + \frac{1}{2} + \frac{rT_e}{e^2} \ln \frac{1}{\chi} , \quad \chi = \frac{2}{N_e} \left(\frac{m_e T_e}{2\pi\hbar^2}\right)^{3/2} .$$

Since

$$\frac{1}{\chi} \ll N_e a_o^3 \ll 1 ,$$

where a_o is the Bohr radius, we have

$$Z_* - \overline{Z} \gg 1 , \tag{8.25}$$

for a strongly negatively charged metallic particle located in an ionized gas.

Thus, the character of cluster charging and its charge equilibrium with a surrounding plasma is determined by the processes of thermoemission of electrons from the cluster surface and collision processes that establish the cluster charge. These processes and possible regimes of cluster charging were given above.

Problem 8.23

Analyze the character of detachment of surface negative ions of a dielectric particle that is located in a plasma.

We consider the interaction of an individual dielectric particle with a surrounding plasma when collisions of plasma electrons and ions with the particle's surface determine its charge. Evidently, this case corresponds to the situation in which the mean free path of electrons and ions in a gas is large compared with the particle size. Under condition (8.25), the detachment of surface negative ions results from collisions with free electrons of a surrounding plasma. Another mechanism of this process is determined by the charge exchange process with the participation of positive ions of the plasma. In this process the bound electron of the negative ion transfers in the field of the positive ion.

Let us compare the rates of detachment of surface negative ions due to these processes. The rate of detachment of a bound negative ion in collisions with electrons is estimated as

$$\nu_e \sim Z v_e \sigma_o \exp\left(-\frac{\varepsilon}{T_e}\right) N_e ,$$

where T_e is the electron temperature, σ_o is of the order of the cross section of the negative ion, ε is the electron binding energy in the surface negative ion, and $v_e \sim \sqrt{T_e/m_e}$ is a typical electron velocity, where m_e is the electron mass. There are two versions for the recombination mechanism of free positive ions with bound negative ions. In the first one, a positive ion captures a weakly bound electron of a surface negative ion as a result of the charge exchange process. The rate constant of this recombination process is of the order of magnitude

$$\nu_i \sim Z v_i \sigma_{ex} N_i ,$$

where σ_{ex} is the cross section of the charge exchange process for an incident positive ion and a bound negative ion, N_i is the number density of positive ions, and $v_i \sim \sqrt{T/m_i}$ is a typical ion velocity, where T is the ion temperature and m_i is the ion mass, Assuming $\sigma_{ex} \sim \sigma_o$ and accounting for the plasma's quasineutrality $N_i = N_e$, we obtain for $T_e > T$ that the criterion $\nu_e \gg \nu_i$ has the form

$$T_e \gg \frac{2\varepsilon}{\ln\left(\frac{T_e}{T}\frac{m_i}{m_e}\right)} .$$

From this it follows that formula (8.24) for the charge of a dielectric particle is valid at high electron temperatures.

In the case of the second version for the recombination mechanism of positive and bound negative ions, each contact of a positive ion with the particle's surface leads to its attachment. Then recombination takes place as a result of drift of a bound electron along the particle's surface. In this case the rate of the recombination process for positive and bound negative ions is

$$\nu_{\mathrm{i}} \sim \pi r_{\mathrm{o}}^{2} v_{\mathrm{i}} N_{\mathrm{i}} ,$$

where r_o is the particle's radius. In this case detachment of bound negative ions as a result of capture of positive ions by the particle's surface is more preferable.

9
Kinetics of Cluster Growth

Cluster growth in gases and plasmas has the same mechanisms as nucleation processes in other systems. In analyzing clusters as consisting of atoms M (we are guided by metal atoms) of some element located in a buffer gas, we assume that the condensed phase (solid or liquid) of this element is thermodynamically profitable under the conditions chosen, so that the cluster growth process characterizes the transition from the gaseous to the condensed phase. There are three possible mechanisms of cluster growth that are taken from nucleation in other physical systems and represented in Figure 9.1 [186].

Let us denote by N the number density of free metal atoms and by N_b the number density of bonded metal atoms in clusters. If free atoms dominate,

$$N \gg N_b ,$$

the first mechanism (Figure 9.1a) for cluster growth is realized as a result of atom attachment to clusters according to the following scheme:

$$M + M_n \to M_{n+1} . \tag{9.1}$$

In the opposite case, where the number density of free atoms becomes small,

$$N \ll N_b ,$$

cluster growth results from the joining of clusters, that is, the coagulation process proceeds as

$$M_j + M_n \to M_{j+n} , \tag{9.2}$$

which is represented in Figure 9.1b. This is an important process for the Earth's atmosphere [187].

But the coagulation process is restricted by cluster motion in a buffer gas, that is, by a slow process. There is another mechanism of cluster growth, the coalescence process, with the transport of atoms. Indeed, an equilibrium is established between a gas of clusters and an atomic vapor, and at low temperatures the number density of free atoms is relatively small ($N \ll N_b$). This equilibrium is supported by

Cluster Processes in Gases and Plasmas. Boris M. Smirnov

ISBN: 978-3-527-40943-3

(a) Attachment of atoms

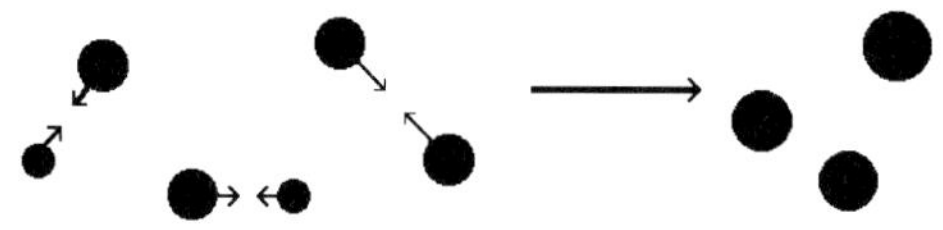

(b) Coagulation

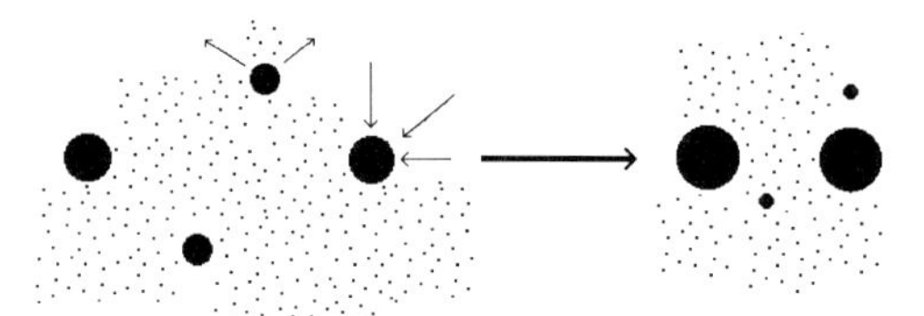

(c) Coalescence

Fig. 9.1 Mechanisms of cluster growth. (a) Attachment of free atoms to clusters; (b) coagulation resulting in the joining of clusters upon contact; (c) coalescence that proceeds owing to cluster equilibrium with a parent vapor; as a result of atom attachment to clusters and cluster evaporation, small clusters decompose and large clusters grow. As a result, the average cluster size increases.

cluster evaporation and attachment of atoms to clusters, and the total rates of these processes are equal under equilibrium conditions. But if the cluster size is below the critical radius, the rate of evaporation for this cluster is higher than the rate of atom attachment. In contrast, the evaporation rate of large clusters is below the rate of atom attachment to this cluster. Therefore, small clusters evaporate under these equilibrium conditions, and large clusters grow. This mechanism of cluster growth is also know as Ostwald ripening [5] and is depicted in Figure 9.1c. This process was studied first in solid solutions, where it leads to the growth of grains of a solute through the evaporation of atoms and their diffusion in a solution with a subsequent attachment to grains. Then the total process results in the growth of large grains and the evaporation of small grains, so that the average grain size grows over time. The rate of this process is restricted by atom diffusion in the solid. In contrast to these conditions, in the case of a cluster plasma the diffusion process is fast, and the rate of the coalescence process is expressed through the rate of cluster evaporation. Note that heat release results from nucleation processes, and because of the high rate of nucleation processes in a cluster plasma, heat release is important for the heat balance and in turn may influence the rate of the nucleation process.

In analyzing processes of cluster growth, we will be guided by metal clusters that are located in a buffer gas (inert or molecular gas). This buffer gas plays a stabilization role, but atoms of a buffer gas have no bonds in clusters.

9.1 Cluster Growth Involving Free Atoms

Problem 9.1

Derive the balance equation for the number density of clusters of different sizes that results from the equilibrium between clusters and their vapor.

Let us introduce the size distribution function f_n of clusters, that is, the number density of clusters consisting of n atoms. Taking into account that the cluster growth and evaporation processes have a stepwise character with attachment or evaporation of one atom only, we obtain the Smoluchowski kinetic equation [188] for the size distribution function of atoms:

$$\frac{\partial f_n}{\partial t} = N k_{n-1} f_{n-1} - N k_n f_n - \nu_n^{\mathrm{ev}} f_n + \nu_{n+1}^{\mathrm{ev}} f_{n+1} \,. \tag{9.3}$$

Here N is the number density of free attaching atoms, k_n is the rate constant for attachment of an atom to the cluster surface, and ν_n^{ev} is the rate of evaporation of a cluster consisting of n atoms.

Note that for a solid cluster the rate constant k_n and the rate ν_n^{ev} are nonmonotonic functions of the number n of cluster atoms. Within the framework of the liquid drop model for a cluster these values are given by formulas (2.7) and (4.18), which are valid for large liquid clusters and solid clusters on average. Assuming the monotonic character of these dependencies, one can reduce the kinetic equation for the size distribution function of large clusters to the form of the continuity equation

$$\frac{\partial f_n}{\partial t} = -\frac{\partial j_n}{\partial n} \,.$$

Accounting for formulas (2.7) and (4.18) for the rates of the processes of cluster growth and evaporation leads to the following form of this kinetic equation:

$$\frac{\partial f_n}{\partial t} = -\frac{\partial j_n}{\partial n} \,, \quad j_n = \pi r_o^2 \sqrt{\frac{8T}{\pi m}} \xi \cdot \left[N - N_{\mathrm{sat}}(T) \exp\left(-\frac{\varepsilon_n - \varepsilon_o}{T}\right)\right] . \tag{9.4}$$

One can see the similarity between this kinetic equation and the balance equation (4.19).

Note that this equation corresponds to the kinetic regime of the cluster growth process. Transition to a general case on the basis of formula (4.26) gives the following expression for a flux in a space of the cluster size:

$$j_n = \frac{\pi r_o^2}{1+\alpha} \cdot \sqrt{\frac{8T}{\pi m}} \xi \cdot \left[N - N_{\mathrm{sat}}(T) \exp\left(-\frac{\varepsilon_n - \varepsilon_o}{T}\right)\right] , \tag{9.5}$$

where the parameter α is determined by formula (4.24).

Problem 9.2

Derive the expression for the collision integral of clusters in a space of the cluster size.

For large cluster sizes we represent the collision integral of clusters in the form

$$I_{\text{col}}(f_n) = -\frac{\partial j_n}{\partial n},$$

and according to the kinetic equation (9.3) we have in the limit $n \gg 1$

$$j_n = k_o(T)\xi n^{2/3}\left[N f_n - N_{\text{sat}}(T) f_{n+1}\exp\left(\frac{2A}{3Tn^{1/3}}\right)\right].$$

Thus, the collision integral has the form of a flux in a space of the cluster size. Taking $f_{n+1} = f_n + \partial f_n/\partial n$, one can represent the collision integral in the form of the sum of two fluxes: the first is the hydrodynamic flux, expressed through the first derivative of the distribution function over n, and the second is the diffusion flux, which includes the second derivative of the distribution function over n. The diffusion flux is small compared with the hydrodynamic one, but it is responsible for the width of the size distribution function of clusters. Neglecting the diffusion flux for large n, we get

$$I_{\text{col}}(f_n) = -\frac{\partial}{\partial n}\left\{k_o(T)\xi n^{2/3} f_n\left[N - N_{\text{sat}}(T)\exp\left(\frac{\Delta\varepsilon}{Tn^{1/3}}\right)\right]\right\}, \tag{9.6}$$

and the expression inside the square brackets is zero at the critical cluster size.

This formula corresponds to the following expression in equation (9.4):

$$j_n = A_n f_n, \quad A_n = N k_n - \nu_n^{\text{ev}},$$

which follows from the right-hand side of formula (9.3) with expansion over large n if we ignore the difference between terms corresponding to n and $n+1$. Therefore, this expression for the collision integral is valid for a system that is sufficiently far from equilibrium. If the state of a gas of clusters is close to equilibrium, we use the flux in the cluster size space given by the Fokker–Planck equation [189, 190], which holds true for large n and for a monotonic size dependence of the parameters of this cluster system. Then it is necessary to include the diffusion term in the flux expression, so the above expression is transformed into the form

$$j_n = A_n f_n - B_n\frac{\partial f_n}{\partial n}.$$

The relation between A_n and B_n follows from the condition that under thermodynamic equilibrium $j_n = 0$ [191]. Under thermodynamic equilibrium $f_n \backsim \exp(-G_n/T)$, where G_n is the free enthalpy of a cluster consisting of n atoms that is given by formula (2.61) within the framework of the liquid drop model for clusters.

Thus, we have the following relation [75]:

$$A_n = \frac{B_n\mu_n}{T}, \quad \mu_n = \frac{\partial G_n}{\partial n}, \tag{9.7}$$

where μ_n is the cluster chemical potential given by formula (2.69). One can see that expression (9.6) for the collision integral is valid far from the critical point.

Problem 9.3

Analyze the behavior of the size distribution functions of clusters depending on the size dependence for the binding energy ε_n of cluster atoms under equilibrium between clusters and their parent atomic vapor where clusters are located.

According to the kinetic equation (9.3), an equilibrium between clusters whose sizes differ by one atom due to processes of atom attachment to the cluster surface and cluster evaporation leads to the following relation for the size distribution functions:

$$f_n N k_n = f_{n+1} \nu_{n+1}^{ev} . \tag{9.8}$$

Using formulas (2.7) and (4.18) for the rates of these processes, we obtain for the ratio of the distribution functions in the limit $n \gg 1$

$$\frac{f_n}{f_{n-1}} = S \exp\left(\frac{\varepsilon_n - \varepsilon_o}{T}\right) ,$$

where ε_n and ε_o are the atom binding energies for this cluster and for a bulk system and S is the supersaturation degree defined according to formula (2.58), $S = N_{sat}(T)/N$. Within the framework of the liquid drop model for large clusters, the atom binding energy in a cluster ε_n is given by formula (2.10):

$$\varepsilon_n = \varepsilon_o - \frac{\Delta\varepsilon}{n^{1/3}} , \quad \Delta\varepsilon = \frac{2}{3}A , \quad n \gg 1 .$$

In this case we have for the ratio of the equilibrium size distribution functions for clusters whose atom numbers differ by one

$$\frac{f_n}{f_{n-1}} = S \exp\left(-\frac{\Delta\varepsilon}{n^{1/3}T}\right) . \tag{9.9}$$

This shows that for a supersaturated vapor $S > 1$, a monotonous size dependence of the atom binding energy ε_n leads to a decrease in the size distribution function with a size increase at low n, and growth of the size distribution function with n increase at large n, and the minimum of the distribution function corresponds to the critical cluster size n_{cr} in accordance with formula (2.63). This behavior of the size distribution function is represented in Figure 9.2 by a solid curve, and the critical size n_{cr} is given by

$$n_{cr} = \left(\frac{\Delta\varepsilon}{T \ln S}\right)^3 . \tag{9.10}$$

Correspondingly, the atom binding energy ε_{cr} for the critical cluster size is

$$\varepsilon_{cr} = \varepsilon_o - T \ln S , \tag{9.11}$$

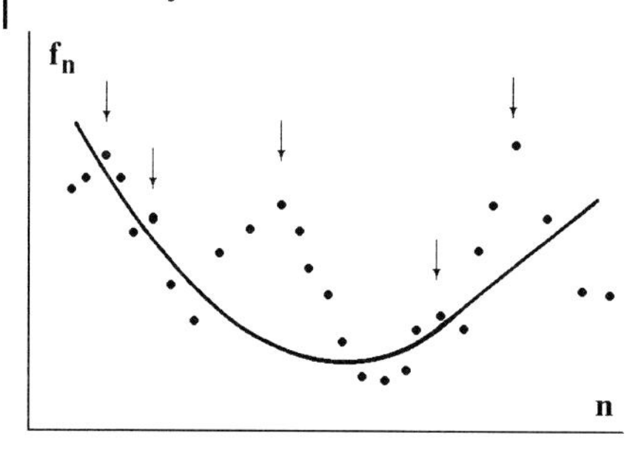

Fig. 9.2 Equilibrium size distribution function for a solid (closed circles) and liquid (solid curve) cluster. Arrows show the cluster magic numbers.

so that if the atom binding energy is below $\varepsilon_{\rm cr}$, the size distribution function drops with cluster growth and increases for the inverse relation between these parameters.

Another behavior of the size distribution function of clusters takes place for solid clusters where the atom binding energy ε_n is a nonmonotonic size function. For example, if m is a magic number of atoms and $\varepsilon_{\rm m} > \varepsilon_{\rm cr}$, we have $f_m > f_{m-1}$. If for a cluster whose size differs by one atom we have $\varepsilon_{m+1} < \varepsilon_{\rm cr}$, this gives $f_{m+1} < f_m$. Then the distribution function has a local maximum at a magic number of cluster atoms. Such behavior of the size distribution function of solid clusters is represented in Figure 9.2 by closed circles.

Problem 9.4

Using an equilibrium for clusters whose sizes differ by one atom, find the relation (4.16) between the rates of atom attachment and evaporation within the thermodynamic terms.

As follows from the kinetic equation (9.3), an equilibrium leads to relation (9.8) and under the thermodynamic equilibrium the size distribution function is $f_n \backsim \exp(-G_n/T)$. This gives the following relation between the rates:

$$N k_n = \nu_n^{\rm ev} \exp\left(\frac{\mu_n}{T}\right) , \tag{9.12}$$

where $\mu_n = \partial G_n/\partial n$ is the cluster chemical potential. This coincides with relation (4.18).

Correspondingly, from this we have for the parameters of the Fokker–Planck equation

$$\begin{aligned} A_n &= N k_n - \nu_n^{\rm ev} = N k_n \left[1 - \exp\left(-\frac{\mu_n}{T}\right)\right] , \\ B_n &= N k_n \frac{T}{\mu_n} \left[1 - \exp\left(-\frac{\mu_n}{T}\right)\right] . \end{aligned} \tag{9.13}$$

In particular, at the critical point $n = n_{\rm cr}$ it follows from these relations that

$$A(n_{\rm cr}) = 0 , \quad B(n_{\rm cr}) = N k(n_{\rm cr}) . \tag{9.14}$$

Problem 9.5

Consider the behavior of the size distribution function of clusters growing in a supersaturated parent atomic vapor.

The evolution of a supersaturated vapor will lead to equilibrium when the excess of a vapor is transformed into a condensed phase. When this process proceeds in a gas, this corresponds to cluster growth up to equilibrium. One can see that thermodynamic equilibrium is not attained for large clusters, and therefore the size distribution function for large clusters is determined by the kinetics of cluster growth. In contrast, this equilibrium takes place for clusters of nearby sizes because it is attained by the attachment or evaporation of one atom. Hence, the thermodynamic equilibrium for the size distribution of clusters is violated for large clusters and is supported between clusters of nearby sizes.

In considering solid clusters from this standpoint, we find that the size distribution function has local maxima at magic numbers of cluster atoms. This behavior of the size distribution function is the basis of an experimental method for determining the cluster magic numbers on the basis of the cluster mass spectrum [23, 26]. As a demonstration of this method, we give in Figure 9.3 the mass spectrum of inert gas clusters. To obtain this spectrum, a beam of neutral clusters is irradiated by a beam of ultraviolet photons, and the charged clusters that are formed are selected. In another case, the beam of neutral clusters is crossed by a weak electron beam. In both cases the assumption is used that cluster ionization does not significantly change the cluster size.

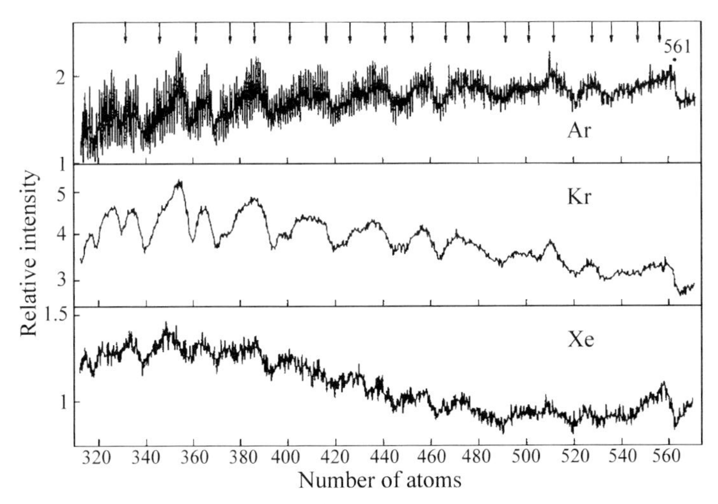

Fig. 9.3 Mass spectra of inert gas clusters resulting from free jet expansion of inert gases through a nozzle [30]. Arrows indicate magic numbers for the icosahedral cluster structure.

Problem 9.6

Formulate the method to determine the average melting point for clusters with a given atomic shell on the basis of the measured mass spectrum of clusters in a hot gas [192, 193].

As is seen, the mass spectrum of liquid clusters is monotonic, whereas the mass spectrum of solid clusters exhibits local maxima corresponding to magic numbers of clusters. This may be used to determine the cluster melting point [192, 193] when clusters pass through a thermostat set at a given temperature, and then the mass spectrum of clusters is recorded by the standard method using cluster ionization by a crossed electron beam. This thermostat is a tube with metallic walls that are supported at a certain temperature. The tube is filled with helium, so that the

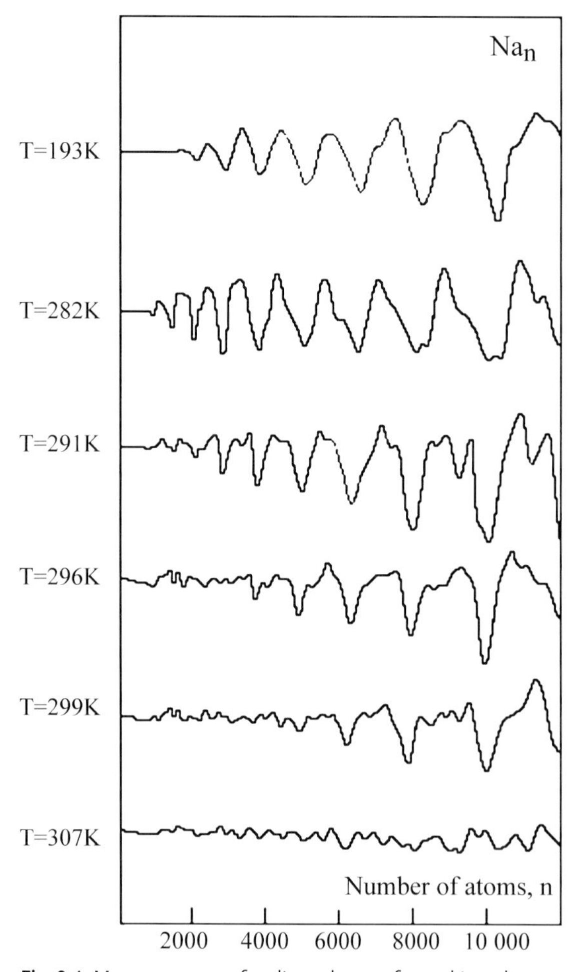

Fig. 9.4 Mass spectrum of sodium clusters formed in a thermostat at the temperature indicated [192, 193].

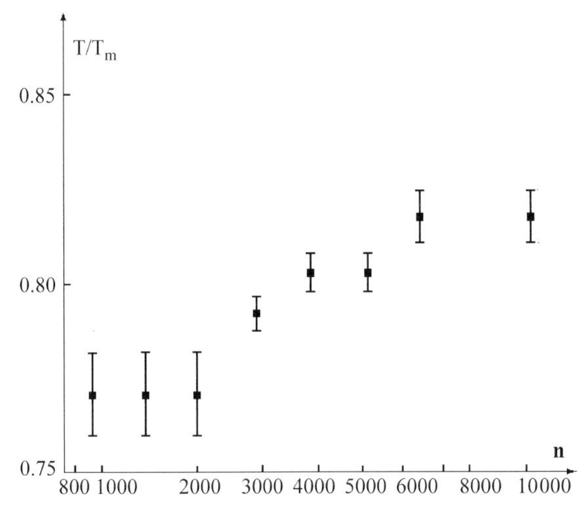

Fig. 9.5 The melting point of sodium clusters with respect to the melting point of bulk sodium follows from their mass spectra, shown in Figure 9.4 [192, 193].

collision of helium atoms with walls and clusters establishes the wall temperature for clusters.

Under these conditions, the mass spectrum of clusters exhibits their phase state. This is demonstrated in Figure 9.4. Because the melting point of clusters depends on their size, and small clusters melt at lower temperatures, a nonmonotonic spectrum with a temperature increase disappears first for small clusters. As a result, one can find the cluster melting point depending on the cluster size, as is given in Figure 9.5. This dependence does not admit a classical approximation for the size dependence of the melting point of a macroscopic particle.

Problem 9.7

Formulate the classical theory of condensation of an atomic vapor [194–197] when the size distribution function of clusters is close to the thermodynamic one in the basic range of sizes including the critical size.

The classical theory of gas condensation is based on the one-step character of attachment and evaporation processes, and the equilibrium size distribution function of clusters is valid up to large sizes. The condensation process violates this equilibrium, but this violation is weak in the basic size range, so that the size distribution function of clusters is given by

$$f_n = C \exp\left(-\frac{G_n}{T}\right) ,$$

where G_n is the enthalpy of cluster formation of n atoms and C is the normalized coefficient that follows from the normalization relation

$$\int f_n dn = N . \tag{9.15}$$

Though we obviously do not use the liquid drop model for clusters, it is implied because we assume the parameters of a cluster of n atoms to be independent of the history of its formation. For large clusters this equilibrium size distribution is violated, and we take for definiteness $f_{m+1} = 0$, where $m \gg n_{cr}$. Correspondingly, we take $f_n = 0$ for $n > m + 1$.

Let us introduce the flux J of nucleating atoms, and, accounting for the one-step character of cluster growth and evaporation, we find the following set of equations for the condensation process, where we start from $n = l$:

$$\begin{aligned}
f_l N k_l - f_{l+1}\nu_{l+1} &= J\,, \\
f_{l+1} N k_{l+1} - f_{l+2}\nu_{l+2} &= J\,, \\
f_{l+2} N k_{l+2} - f_{l+3}\nu_{l+3} &= J\,, \\
&\ldots\ldots\ldots\ldots \\
f_{m-1} N k_{m-1} - f_m \nu_m &= J\,, \\
f_m N k_m &= J\,, \\
f_m &= \frac{J}{N k_m}\,, \\
f_{m-1} &= f_m \exp\left(\frac{G_m - G_{m-1}}{T}\right) + \frac{J}{N k_{m-1}}\,, \\
&\ldots\ldots\ldots\ldots \\
f_{l+1} &= f_{l+2} \exp\left(\frac{G_{l+2} - G_{l+1}}{T}\right) + \frac{J}{N k_{l+1}}\,, \\
f_l &= f_{l+1} \exp\left(\frac{G_{l+1} - G_l}{T}\right) + \frac{J}{N k_l} \\
&\ldots\ldots\ldots\ldots
\end{aligned}$$

Here N is the number density of free atoms, k_n is the rate constant of atom attachment to a cluster consisting of n atoms, and ν_n is the rate of cluster evaporation. Under equilibrium conditions we have $k_n f_n = \nu_{n+1} f_{n+1}$, which also leads to the principle of detailed balance for the rates of these processes:

$$N k_n \exp\left(-\frac{G_n}{T}\right) = \nu_{n+1} \exp\left(-\frac{G_{n+1}}{T}\right)\,.$$

One can find from this set of equations the flux J of nucleating atoms in the limit where this value is small and the nucleation process does not violate the equilibrium size distribution in the basic range of sizes. Solving subsequently each equation of this set, we obtain for the flux of attaching atoms

$$J = \frac{f_l N}{\sum\limits_{n=l}^{m} \frac{1}{k_n} \exp\left(\frac{G_n - G_l}{T}\right)} = \frac{C N}{\sum\limits_{n=l}^{m} \frac{1}{k_n} \exp\left(\frac{G_n}{T}\right)}\,.$$

Replacing the summation by integration, we obtain from this formula

$$J = \frac{CN}{\int\limits_0^\infty \frac{dn}{k_n} \exp\left(\frac{G_n}{T}\right)}, \tag{9.16}$$

where we account for the convergence of this integral at sizes near the critical size, and therefore the result depends weakly on a lower limit l and upper limit m, which allows us to replace them by 0 and ∞.

Since the main contribution to integral (9.16) gives a size range near the critical size, one can represent this expression in the form

$$J = Nk(n_{cr})C\Gamma \exp\left(-\frac{G(n_{cr})}{T}\right) = Nk(n_{cr})f(n_{cr})Z\,, \quad \text{where}$$
$$Z = \left[-\frac{1}{2\pi T}\frac{\partial^2 G_n}{\partial n^2}\Big|_{n=n_{cr}}\right]^{1/2}. \tag{9.17}$$

Here $f(n_{cr})$ is the equilibrium size distribution function of clusters, values of the cluster free enthalpy and its derivations near the critical point are given by formula (2.62) for the liquid drop model, and the Zeldovich parameter Z is given by [194]

$$Z = r_W\sqrt{8\pi\gamma r_{cr}} = \sqrt{\frac{2Ar_W}{n_{cr}}}. \tag{9.18}$$

Introducing the temperature T_* at which the vapor pressure is equal to the saturated vapor pressure and using expression (2.67) for the cluster free enthalpy within the terms of this temperature, it is convenient to rewrite the above formula in the form

$$J = \left(\frac{8T}{\pi m}\right)^{1/2} \pi r_W^2 N^2 \frac{2\gamma^{1/2} r_W^3}{3r_{cr}^2} \exp\left[-\frac{1}{2T}\left(\frac{8\pi\gamma r_W^2}{3\varepsilon_o}\right)^2 \cdot \left(\frac{T_*}{T-T_*}\right)^2\right]. \tag{9.19}$$

Note that in the case under consideration, the size distribution function for small clusters, including the critical one, coincides with that under thermodynamic equilibrium. Large sizes at which this is violated do not contribute to the rate of cluster growth.

Problem 9.8

Find the rate of condensation of an atomic vapor under the conditions of the previous problem that are close to the thermodynamic equilibrium on the basis of the Fokker–Planck equation [197].

We represented above the classical theory of condensation of an atomic vapor [194–197] as is usually given in books on nucleation (e.g., [75, 198]). This method exhibits the character of this process. Then we used the fact that cluster parameters are monotonic size functions, and a typical cluster size is large: $n \gg 1$. The analysis convinces us that a basic time of cluster nucleation is found in a size range near the critical size. We now determine the nucleation rate J on the basis of the Fokker–Planck equation, which describes the evolution of the cluster size distribution function and gives the following expression for the nucleation rate:

$$J = j_n = A_n f_n - B_n \frac{\partial f_n}{\partial n} \, .$$

Considering this relation as an equation for the size distribution function and accounting for

$$A_n = \frac{\mu_n}{T} B_n = \frac{1}{T} \frac{\partial G_n}{\partial n} \, ,$$

one can reduce this equation to the form

$$J = -B_n \exp\left(\frac{G_n}{T}\right) \frac{d}{dn} \left[f_n \exp\left(-\frac{G_n}{T}\right) \right] .$$

We have the solution of this equation:

$$f_n = -J \exp\left(-\frac{G_n}{T}\right) \left[\int \frac{dn}{B_n} \exp\left(\frac{G_n}{T}\right) + \text{const} \right] ,$$

and taking the boundary condition $f_n \to 0$, for $n \to \infty$, we obtain from this

$$f_n = J \exp\left(-\frac{G_n}{T}\right) \int_n^{\infty} \frac{dn}{B_n} \exp\left(\frac{G_n}{T}\right) .$$

Taking a range of small sizes where the size distribution function is

$$f_n = C \exp\left(-\frac{G_n}{T}\right) ,$$

and the normalization coefficient as determined by condition (9.15), we find from this relation

$$J = C \left[N \int_0^{\infty} \frac{dn}{B_n} \exp\left(\frac{G_n}{T}\right) \right]^{-1} .$$

Substituting into this equation the value of B_n near the critical point $B_n = N k_n$, we find for the flux of nucleating atoms

$$J = \frac{C N}{\int_0^{\infty} \frac{dn}{k_n} \exp\left(\frac{G_n}{T}\right)} \, .$$

As is seen, this formula coincides with formula (9.16), that is, we obtain the same result under identical conditions through different methods. Note that cluster growth in this case is determined by a size range close to the critical size where the rates of atom attachment and cluster evaporation are nearby, and a small difference in these values determines the rate of nucleation and corresponds to a weak violation of the thermodynamic equilibrium for the system consisting of free atoms and clusters of different sizes.

Problem 9.9

The growth of metal clusters in a plasma where an atomic metal vapor is an admixture to a buffer gas results from the formation of diatomic metal molecules in three body collisions, and later these diatomic molecules are the nuclei of condensation. Find the size distribution function of clusters at times until the number density of free metal atoms varies weakly.

Under given conditions, the growth of metal clusters proceeds according to the scheme [35, 199]

$$2\mathrm{M} + \mathrm{A} \rightarrow \mathrm{M}_2 + \mathrm{A}\,, \quad \mathrm{M}_n + \mathrm{M} \longleftrightarrow \mathrm{M}_{n+1}\,. \tag{9.20}$$

Here M and A are a metal atom and an atom of a buffer gas, respectively, and n is the number of cluster atoms. The rate constant of the pairwise attachment process, the second process in scheme (9.20), is given by formula (2.7) $k_n = k_o n^{3/2}$, where the cluster is modeled on the liquid drop model. Because the first stage of this process has a three body character, the formation of a diatomic metal molecule is a slow process. Subsequently forming diatomic molecules become the nuclei of condensation and are quickly converted into large clusters as a result of attachment of free metal atoms in pairwise collision processes. From this it follows that a typical cluster size is large at any stage of the cluster growth process.

Introducing the parameter of cluster growth

$$G = \frac{k_o}{N_a K} \gg 1\,, \tag{9.21}$$

we have that this parameter is usually large. Here N_a is the number density of buffer gas atoms, K is the rate constant for formation of a diatomic molecule in three body collisions, and its typical value in this case is $K \sim 10^{-32}\,\mathrm{cm}^6/\mathrm{s}$, whereas a typical rate constant of the attachment process is $k_o \sim 3 \cdot 10^{-11}\,\mathrm{cm}^3/\mathrm{s}$ for metal atoms in a buffer gas. Hence, cluster growth ($G \gg 1$) takes place if $N_a \gg 3 \cdot 10^{21}\,\mathrm{cm}^{-3}$, that is, it is realized in a nondense buffer gas.

The scheme of processes (9.21) leads to the following set of balance equations for the number density N_m of free metal atoms, the number density N_{cl} of clusters, the number density of bound atoms N_b in clusters, and cluster size n (the number of cluster atoms):

$$\begin{aligned} \frac{dN_b}{dt} &= -\frac{dN_m}{dt} = \int N_m k_o n^{2/3} f_n\, dn + K N_m^2 N_a\,, \\ \frac{dN_{cl}}{dt} &= K N_m^2 N_a\,, \end{aligned} \tag{9.22}$$

where f_n is the cluster size distribution function, which satisfies the normalization conditions

$$N_{cl} = \int f_n dn\,, \quad N_b = \int n f_n dn\,.$$

It is necessary to add to this set the equation of cluster growth

$$\frac{dn}{dt} = k_o n^{2/3} N_m\,,$$

and the solution of this equation for the case under the assumption $N_m = \text{const}$ is

$$n = \left(\frac{N k_o t}{3}\right)^3, \tag{9.23}$$

where t is the growth time if a diatomic molecule, a condensation nucleus for this cluster, is formed at $t = 0$. One can obtain from this $n \leq n_{max}$, where n_{max} is the size of a cluster whose nucleus of condensation is formed at $t = 0$.

Introducing the size distribution of clusters f_n, we have that $f_n dn$, the number density of clusters with a size range between n and $n + dn$, is proportional to the time range dt when diatomic molecules, the condensation nuclei for these clusters, are formed. This relation $f_n dn \backsim dt$ leads to the following size distribution function of clusters:

$$f_n = \frac{C}{n^{2/3}}\,, \quad n < n_{max}\,, \tag{9.24}$$

where C is the normalization constant, n_{max} is the maximum cluster size at this time, and $f_n = 0$, if $n > n_{max}$.

This formula may be obtained directly from the kinetic equation for the distribution function f_n if we neglect evaporation processes

$$\frac{\partial f_n}{\partial t} = -\frac{\partial}{\partial n}\left(N k_o n^{2/3} f_n\right)\,,$$

and hence this equation confirms formula (9.24).

Problem 9.10

For nucleation of a metal vapor in a buffer gas, find the connection between the average and the maximum cluster size at a given time until the number density of free metal atoms does not vary.

We have the following relations for the number density of clusters N_{cl} and the total number density of bound atoms N_b:

$$N_{cl} = \int f_n dn\,, \quad N_b = \int n f_n dn\,,$$

and using expression (9.24) for the size distribution function of clusters, we get for these parameters

$$N_{cl} = 3Cn_{max}^{1/3}, \quad N_b = \frac{3}{4}Cn_{max}^{4/3}.$$

On the basis of the average cluster size $\overline{n} = N_b/N_{cl}$, we obtain from this

$$\overline{n} = \frac{n_{max}}{4}. \qquad (9.25)$$

Problem 9.11

On the basis of the model in which the number density of free metal atoms is a constant in the course of cluster growth, determine the time of cluster growth and the maximum cluster size when at the beginning nucleating metal atoms are located in a buffer gas.

Cluster growth proceeds according to scheme (9.20), and because the parameter (9.21) is large, the set of balance equations (9.22) leads to the following set of balance equations for the number density N_m of free metal atoms, the number density N_{cl} of clusters, the number density of bound atoms N_b in clusters, and cluster size n (the number of cluster atoms):

$$\frac{dN_m}{dt} = -\int N_m k_o n^{2/3} f_n dn, \quad \frac{dN_{cl}}{dt} = KN_m^2 N_a. \qquad (9.26)$$

Here we account for the fact that the ratio of the second and third terms of the right-hand side of the first balance equation (9.22) is on the order of $Gn^{2/3}$, and since a typical number n of cluster atoms and the cluster parameter G are large, this ratio is small.

The set (9.22) of balance equations describes the character of cluster growth for the conditions under consideration. In addition, we assume here that a typical cluster size n is large compared with the critical cluster size, which allows us to neglect the processes of cluster evaporation in the set of balance equations (9.22). We add to this set of equations the conditions $N_m = \text{const}$ and $N_b(\tau) = N_m$, where τ is the time of the nucleation process.

We find below the parameters of this process. From the normalization condition

$$N_b = \int n f_n dn$$

we obtain for a normalization constant in formula (9.24)

$$C = \frac{4N_b}{3n_{max}^{4/3}},$$

where n_{max} is taken at the end of the process. Next, solving balance equations (9.26) for the end of the cluster growth process and using the size distribution function

(9.24) of clusters, we obtain

$$C = \frac{1}{N_m k_o \tau n_{max}}, \quad N_{cl} = K N_m^2 N_a \tau = \frac{N_m^2 G \tau}{k_o}.$$

Adding to this formula (9.25) for the average cluster size $\overline{n} = N_b/N_{cl}$, we find for the maximum cluster size n_{max} at the end of this process and the duration τ of the cluster growth process [35, 208]

$$n_{max} = 1.2 G^{3/4}, \quad \tau = \frac{3.2 G^{1/4}}{N_m k_o}, \tag{9.27}$$

Because $G \gg 1$, the cluster size is large, $n \gg 1$, and the reduced nucleation time τ is large compared with a time of order of $(k_o N_m)^{-1}$ – that is a time of attachment of one atom.

9.2 Kinetics of Cluster Coagulation

Problem 9.12

Clusters are located in a buffer gas and grow as a result of pairwise collisions of clusters. Assuming that the rate constant of this process depends only on cluster sizes, derive the kinetic equation for the cluster growth.

Under the given conditions, coagulation of clusters proceeds according to scheme (9.2),

$$M_{n-m} + M_m \to M_n,$$

and a buffer gas does not partake in this process but plays a stabilization role. If liquid clusters partake in this process, the parameters of a given cluster do not depend on the history of its formation and are determined by its size. Therefore, the kinetic equation for the evolution of the size distribution function of clusters f_n as a result of cluster coagulation is described by the Smoluchowski equation [188]:

$$\frac{\partial f_n}{\partial t} = -f_n \int k(n, m) f_m dm + \frac{1}{2} \int k(n-m, m) f_{n-m} f_m dm. \tag{9.28}$$

Here $k(n-m, m)$ is the rate constant of the process (9.2), the factor 1/2 accounts for the fact that collisions of clusters consisting of $n-m$ and m atoms are present in the equation twice, and the distribution function is normalized as $\int f_n dn = N_{cl}$, where N_{cl} is the number density of clusters.

Problem 9.13

Show that the total number density N_b of bound atoms is conserved in the course of the coagulation process.

The total number density of bound atoms is $N_b = \int_0^\infty n f_n dn$. We obtain the equation for this value, multiplying equation (9.28) by n and integrating over dn, which is

$$\frac{dN_b}{dt} = -\int nk(n,m) f_n dn f_m dm + \frac{1}{2}\int nk(n-m,m) f_{n-m} f_m dn dm ,$$

and in the second integral $n > m$. Replacing $n - m$ by n on the right-hand side of this relation, we obtain the mutually cancelled terms, and $dN_b/dt = 0$, that is, within the framework of the Smoluchowski equation, the total number density N_b of bound atoms is conserved in this process.

Problem 9.14

Find the time dependence of the average cluster size in the course of cluster coagulation if the rate constant of the joining of two clusters is independent of the cluster size and the typical cluster size is large compared with the initial cluster size.

Taking the rate constant of two cluster joining k_{as} to be independent of cluster size, we reduce equation (9.28) to

$$\frac{\partial f_n}{\partial t} = -k_{as} f_n \int_0^\infty f_m dm + \frac{1}{2} k_{as} \int_0^n f_{n-m} f_m dm .$$

Multiplying this equation by n and integrating over dn, we obtain

$$\frac{d}{dt}\int_0^\infty n f_n dn = -k_{as}\int_0^\infty n f_n dn \int_0^\infty f_m dm + \frac{1}{2} k_{as} \int_0^\infty n dn \int_0^n f_{n-m} f_m dm$$
$$= 0 .$$

This means that the total number density of bound atoms $N_o = \int_0^\infty n f_n dn$ is conserved during cluster growth. If we multiply the kinetic equation by n^2 and integrate over dn, we get

$$\frac{d}{dt}\int_0^\infty n^2 f_n dn = N_b \frac{d\overline{n}}{dt}$$
$$= -k_{as}\int_0^\infty n^2 f_n dn \int_0^\infty f_m dm + \frac{1}{2} k_{as} \int_0^\infty n^2 dn \int_0^n f_{n-m} f_m dm = \frac{1}{2} k_{as} N_b ,$$

where we define the average cluster size $\overline{n}$ as

$$\overline{n} = \frac{\int_0^\infty n^2 f_n dn}{\int_0^\infty n f_n dn} .$$

This gives for the evolution of the average cluster size

$$\overline{n} = \frac{1}{2} k_{as} N_b t \, , \tag{9.29}$$

if at the beginning the average size is relatively small.

Problem 9.15

Assuming a rate constant for joining of two clusters to be independent of the cluster size, find the size distribution function of clusters in the course of their coagulation when a typical cluster size is large compared with the initial cluster size.

Let us introduce the concentration of clusters of a given size $c_n = f_n/N_b$, where N_b is the total number density of bound atoms in clusters. The normalization condition for the cluster concentration is $\sum_n n c_n = 1$, and the kinetic equation (9.28) in terms of cluster concentrations has the form

$$\frac{\partial c_n}{\partial \tau} = -c_n \int_0^\infty c_m dm + \frac{1}{2} \int_0^n c_{n-m} c_m dm \, , \tag{9.30}$$

where the reduced time is $\tau = N_b k_{as} t$. The solution of this equation is

$$c_n = \frac{4}{\overline{n}^2} \exp\left(-\frac{2n}{\overline{n}}\right) . \tag{9.31}$$

This expression satisfies the normalization condition $\int_0^\infty n c_n dn = 1$ and the average cluster size $\overline{n}$ corresponds to formula (9.29). Indeed, substituting expression (9.31) into the kinetic equation, we confirm formula (9.29), $\overline{n} = \tau$.

Problem 9.16

Within the framework of the liquid drop model for the association of two clusters, determine the time dependence for the average cluster size if it is small compared with the mean free path of gas atoms.

The rate constant of association of two clusters $k(n, m)$ consisting of n and m atoms within the framework of the liquid drop model for clusters is given by formula (4.3):

$$k(n, m) = k_o (n^{1/3} + m^{1/3})^2 \sqrt{\frac{n+m}{nm}} \, ,$$

where k_o is given by formula (2.7). Multiplication of the Smoluchowski equation (9.28) by n^2 and integration over dn and using the normalization condition $\int n f_n dn = N_b$ (N_b is the number density of bound atoms in clusters) leads to the following equation:

$$\frac{d\overline{n}}{dt} = k_o N_b I \,\overline{n}^{1/6} ,$$

$$I = \frac{1}{2} \int_0^\infty \int_0^\infty (x^{1/3} + y^{1/3})^2 \sqrt{\frac{x+y}{xy}} \exp(-2x - 2y) x\, dx\, y\, dy = 5.5 ,$$

where we use the notation $x = n, y = m$ and formula (9.31) for the size distribution function of clusters. This gives for the mean cluster size [200, 201]

$$\overline{n} = 6.3(N_b k_o t)^{1.2} . \tag{9.32}$$

Because of the assumption used, $n \gg 1$, these formulas are valid under the condition

$$k_o N_b t \gg 1 .$$

Problem 9.17

Metal clusters are formed in a buffer gas from an atomic vapor according to scheme (9.20), and the subsequent cluster growth is determined by the coagulation process (9.2). Show that the attachment and coagulation stages of cluster growth are separated if a significant variation in the cluster size due to coagulation proceeds after transformation of an atomic metal vapor into clusters.

These stages of the cluster growth process are separated if a typical time τ_{at} for transformation of an atomic vapor into clusters is less than a typical time τ_{coag} of the coagulation process. A typical time for atom attachment to clusters is given by formula (9.27)

$$\tau_{at} \sim \frac{G^{1/4}}{N_m k_o} ,$$

where N_m is the number density of free metal atoms at the beginning. According to formula (9.27) a typical cluster size after transformation of an atomic metal vapor into clusters is estimated as

$$\overline{n} \sim G^{3/4} .$$

A typical time for a significant increase in this size as a result of the coagulation process according to formula (9.32) is given by

$$\tau_{coag} \sim \frac{\overline{n}^{5/6}}{N_b k_o} \sim \frac{G^{5/8}}{N_b k_o} ,$$

where N_b is the total number density of bound atoms. We have $N_m = N_b$ because all the free metal atoms form clusters. Hence, the ratio of these typical times is

$$\frac{\tau_{at}}{\tau_{coag}} \sim G^{-3/8} ,$$

and because $G \gg 1$, this ratio is small, which allows us to separate these stages of the cluster growth process.

9.3
Cluster Growth During Gas Expansion in a Vacuum

Problem 9.18

During expansion of a buffer gas with an admixture of a nucleating (metal) vapor, a small portion of the atoms of the vapor are converted into clusters owing to three body formation of diatomic molecules. Approximating the time dependence of the number density of atoms as $n \sim \exp(-t/\tau_{ex})$ (τ_{ex} is a typical expansion time), find the criterion when a small portion of nucleating atoms will be converted into clusters. Assume the average size of the clusters at the end of the process to be large, and determine the size distribution function of clusters within the framework of the liquid drop model for clusters.

Diatomic molecules, which are the nuclei of condensation, are formed as a result of the three body process (9.20)

$$2M + A \to M_2 + A ,$$

and subsequently are converted into clusters. Correspondingly, the number density of diatomic molecules M_2 is determined by the balance equation

$$\frac{dN_2}{dt} = K N_b N_m^2 ,$$

where K is the rate constant of this three body process, N_a is the number density of atoms of the buffer gas, and N_m is the number density of metal atoms. We assume that starting from diatomic molecules, the cluster growth process proceeds as a result of pairwise processes [the second process of scheme (9.20)]. In addition, the expansion process leads to a strong drop in temperature, and at typical temperatures of cluster growth one can neglect the evaporation of clusters. Taking the time dependence for the number density of buffer gas atoms and metal atoms as $N_a \sim \exp(-t/\tau_{ex})$, $N_m \sim \exp(-t/\tau_{ex})$, where τ_{ex} is a typical expansion time, we find for the total number density of diatomic metal molecules and clusters $[N_2]$ at time t is

$$[N_2] = \frac{1}{3}\left[1 - \exp\left(-\frac{3t}{\tau_{ex}}\right)\right] K N_a N_m^2 \tau_{ex} ,$$

where N_a, N_m are the initial number densities of the corresponding atoms. The total number N_b of bound atoms is $N_2\overline{n}$, where $\overline{n}$ is the average number of cluster atoms.

Within the framework of the liquid drop model for a cluster, the balance equation for the cluster size has the form

$$\frac{dn}{dt} = k_o n^{2/3} N_m ,$$

in accordance with formula (2.7), where the rate constant k_o is defined. Taking $N_m \sim \exp(-t/\tau_{ex})$ and solving this equation, we get for the maximum number of cluster atoms

$$n_{max} = \left(\frac{k_o N_m \tau_{ex}}{3}\right)^3 .$$

Here the atomic number density N_m corresponds to the initial time, and a diatomic molecule as a nucleus of condensation for this cluster is formed at the initial time $t = 0$. If the diatomic molecule is formed at time t and is subsequently grown in a cluster, then the number of atoms in this cluster is

$$n(t) = n_{max} \exp(-t/\tau_{ex}) .$$

We have from this the criterion that the number density of bound atoms is small compared with that of free metal atoms:

$$N_b \sim N_2 n_{max} \ll N_m .$$

This criterion may be rewritten in the form

$$N_a N_m^4 K k_o^3 \tau_{ex}^4 \ll 1 . \tag{9.33}$$

Problem 9.19

During expansion of a buffer gas with an admixture of a nucleating metal vapor, when a small portion of the nucleating atoms are converted into clusters, and approximating the time dependence for the number density of atoms as $N_m \sim \exp(-t/\tau_{ex})$ (τ_{ex} is a typical expansion time), find the size distribution of clusters within the framework of the liquid drop model for clusters.

We have a single-valued connection between a time t of formation of a diatomic molecule, which is subsequently a nucleus of condensation, and the cluster size at the end of the process. Correspondingly, the number density of clusters with number of atoms between n and $n + dn$ is

$$d f_n = K N_a N_m^2 \exp\left(-\frac{3t}{\tau_{ex}}\right) dt ,$$

where N_a is the number density of buffer gas atoms and N_m relates to the initial time, and the relation between the cluster size n at the end of the process and a time t of formation of a diatomic molecule has the form

$$n = n_{max} \exp\left(-\frac{3t}{\tau_{ex}}\right) ,$$

and

$$dn = \frac{3 n_{max}}{\tau_{ex}} \exp\left(-\frac{3t}{\tau_{ex}}\right) dt .$$

Taking the ratio of the equation for $d f_n$ to the last equation, we obtain

$$\frac{d f_n}{dn} = \frac{K N_a N_m^2 \tau_{ex}}{3 n_{max}} . \tag{9.34}$$

From this we have for the total number of bound atoms N_b and the average number of cluster atoms $\overline{n}$

$$N_b = \int n d f_n = \frac{1}{6} K N_a N_m^2 \tau_{ex} n_{max} = \frac{1}{162} K k_o^3 N_a N_m^5 \tau_{ex}^4 ;$$
$$\overline{n} = \frac{\int n d f_n}{\int d f_n} = \frac{n_{max}}{2} . \tag{9.35}$$

Note these expressions are based on the assumption that only a small portion of a metal vapor is converted into clusters.

Problem 9.20

A buffer gas expands through a nozzle with a radius r and contains an admixture of a metal vapor that is converted into clusters in a region with a low temperature. Assuming that the expanding gas has the form of a conic stream with conic angle α and nucleation starts near the nozzle, find the size distribution function of clusters far from the nozzle where the nucleation process finishes.

Under the given conditions, the buffer gas flux flows through a nozzle with a velocity that is close to the sound speed u_s. Assuming this velocity is conserved after the nozzle, we obtain for the number density of atoms for a time t after passage the nozzle

$$N_m = \frac{N_o}{1 + \left(\frac{t}{t_o}\right)^2} , \quad t_o = \frac{r}{u_s \tan \alpha} .$$

In the course of the adiabatic expansion, the temperature of an expanding gas decreases, so that the temperature T varies as $T \backsim N^{2/3}$. Finally, the temperature becomes so low that the nucleation process starts. Under the given conditions, this takes place near the nozzle.

Repeating the method of derivation of formula (9.35) under the given conditions, we find now

$$d f_n = \frac{K N_a N_m^2}{[1 + (t/t_o)^2]^3} dt = \frac{F d\xi}{(1 + \xi^2)^3} , \quad F = K N_a N_m^2 t_o , \quad \xi = t/t_o$$

where the values of the number densities N_a and N_m relate to the nozzle plane, t denotes a time of formation of a diatomic molecule M_2 that is a nucleus of subsequent nucleation. This time is connected to the size of a formed cluster n at the end of the process by the relation

$$3n^{1/3} = \int_t^{\infty} \frac{k_o N_m dt}{1 + (t/t_o)^2} .$$

This gives for the maximum cluster size

$$n_{\max} = \left(\frac{\pi}{6} k_o N_m t_o\right)^3 ,$$

and

$$n = n_{\max}\left(\left[\frac{2}{\pi}\,\text{arccotan}(t/t_o)\right]\right)^3 .$$

We find from the above relations the total number of clusters (and diatomic molecules)

$$N_{cl} = \int_0^\infty \frac{K N_a N_m^2 t_o}{(1+\xi^2)^3} d\xi = \frac{3\pi}{16} K N_a N_m^2 t_o , \tag{9.36}$$

and the reduced average cluster size is

$$\frac{\overline{n}}{n_{\max}} = \frac{16}{3\pi}\int_0^\infty \frac{2}{\pi}\frac{(\text{arccotan}\,\xi)^3}{(1+\xi^2)^3} d\xi = 0.52 . \tag{9.37}$$

The total number of bound atoms is $N_b = N_{cl}\overline{n}$. One can see that a typical time t_o in expressions (9.36) and (9.37) plays the role of a typical expansion time, as does the parameter τ_{ex} in formulas (9.33)–(9.35).

Problem 9.21

Consider the limiting regime of expansion of a buffer gas with an admixture of a metal vapor if the rate of nucleation exceeds the expansion rate.

If the nucleation rate exceeds the expansion rate, the evolution of a uniform mixture of a buffer gas and a nucleating metal vapor will proceed as in a motionless gas, and the yield parameters of the nucleation process are determined approximately by formulas (9.27). Using these formulas, we give the criterion of their validity for an expanding gas:

$$\tau_{nuc} \sim \frac{1}{k_o N_m} \ll \tau_{ex} .$$

9.4 Cluster Growth through Coalescence

Problem 9.22

Metal clusters are located in a hot buffer gas where the equilibrium is supported between clusters and their atomic vapor. Show that the critical cluster size varies more slowly than the size distribution function of clusters far from the critical size.

When the equilibrium is established between clusters and their atomic vapor, the rates of cluster evaporation and atom attachment to atoms become equal. But they do not coincide for clusters of different sizes. If the cluster size is equal to the critical size, these rates coincide, and the number density of clusters of such a size does not vary until the gas temperature and the number density of free atoms do not vary. But for cluster sizes other than the critical one this equality is not fulfilled. Therefore, small clusters evaporate and decrease in size, and large clusters grow owing to attachment of atoms to them. This process leads to an increase in the average cluster size.

Let us consider the character of the variation of the average cluster size in detail, assuming this variation to be relatively small. One can see that the critical size does not vary in the first approximation because of the equilibrium of such clusters and atomic vapor. Let us divide clusters into two groups, whose sizes are below and above the critical size, and there are no transitions between clusters of these groups. We assume for simplicity that the critical size coincides with the maximum size of the size distribution function. The difference between the total rates ΔJ of cluster evaporation and growth due to atom attachment for small clusters is compensated by the same difference $-\Delta J$ for clusters of the second group. As a result, the equilibrium between clusters and atomic vapor conserves the number density of free metal atoms and, correspondingly, the critical cluster size (4.19) in the first approximation for expansion over a small parameter $\Delta J/J$, where J is the total rate of atom attachment to clusters or the total rate of cluster evaporation.

In the second approximation we obtain that the number density of clusters of the second group drops, which decreases the rate of formation of free atoms, and the number density of free metal atoms N_{m} varies according to the balance equation

$$\frac{dN_{\mathrm{m}}}{dt} \backsim -\frac{(\Delta J)^2}{J} .$$

A decrease in the number density of free metal atoms leads to an increase in both the average cluster size and the critical cluster size according to formula (4.19), which corresponds to the second approximation in expansion over a small parameter $\Delta J/J$.

Problem 9.23

Metal clusters are in equilibrium with their atomic vapor and the size distribution function has the Gauss form whose width Δ is small compared with the critical cluster size n_{o}. Find the relative rate difference for small clusters that is responsible for cluster growth as a result of coalescence.

The size distribution function f_n of clusters has the form

$$f_n = \frac{N_{\mathrm{cl}}}{\sqrt{2\pi\Delta}} \exp\left[-\frac{(n-n_{\mathrm{o}})^2}{2\Delta^2}\right] , \quad \Delta \ll n_{\mathrm{o}} , \tag{9.38}$$

where N_{cl} is the number density of clusters. The total rate of atom attachment to clusters according to formula (2.7) is

$$J = \int_0^\infty N_m k_o n^{2/3} f_n dn = N_{cl} N_m k_o n_o^{2/3} ,$$

and it is equal to the rate of cluster evaporation. We divide clusters into two groups by size, and for small clusters $n \leq n_o$ let us determine the excess of the evaporation rate over the atom attachment rate. We have

$$\Delta J = \int_0^{n_o} N_m k_o n^{2/3} f_n dn$$
$$= \int_0^{n_o} N_m k_o \frac{2(n_o - n)}{3n_o^{1/3}} f_n dn = N_{cl} N_m k_o \cdot \frac{2\sqrt{2\Delta}}{3\sqrt{\pi} n_o^{1/3}} ,$$

and the relative variation in the rate difference for small (and for large) clusters is

$$\frac{\Delta J}{J} = \frac{\sqrt{2}}{3\sqrt{\pi}} \cdot \frac{\Delta}{n_o} . \tag{9.39}$$

Problem 9.24

On the basis of formula (9.39), estimate a typical time of cluster growth due to the coalescence processes when metal clusters are found in equilibrium with their atomic vapor and grow in a hot buffer gas.

In the case of a narrow size distribution function according to formula (9.39), a decrease in the number density of free metal atoms is determined from the balance equation

$$-\frac{dN_m}{dt} = \Delta J = N_{cl} N_m k_o \cdot \frac{\sqrt{2\Delta}}{3\sqrt{\pi} n_o^{1/3}} ,$$

and the number density of free metal atoms varies as

$$N_m = N_m(0) \exp(-t/\tau_{coal}) ,$$

where a typical coalescence time is

$$\tau_{coal} = \frac{3\sqrt{\pi}}{\sqrt{2}} \cdot \frac{n_o^{4/3}}{N_b k_o \Delta} ,$$

and $N_b = n_o N_{cl}$ is the number density of bound atoms in clusters. If the width of the size distribution function is comparable with the critical cluster size n_{cr}, this equation leads to the following estimate for a typical coalescence time:

$$\tau_{coal} \sim \frac{n_o^{1/3}}{N_b k_o} .$$

Problem 9.25

Analyze the possibility of the automodel solution for the size distribution function of clusters in the case of coalescence of metal clusters when they are found in equilibrium with their atomic vapor in a hot buffer gas.

The automodel character of evolution of the size distribution function assumes a size distribution function of clusters in the form

$$f_n(t) = C(t) f\left[\frac{n}{g(t)}\right] . \tag{9.40}$$

This distribution does not depend on the initial size distribution of clusters and is established through a certain time in the course of cluster evolution. The dependence $C(t)$ is the normalized function, and $g(t)$ is characterized by processes that determine the evolution of the size distribution function of clusters. In particular, in the case of nucleation of a vapor of free atoms and coagulation we have the automodel size distribution functions (9.24) and (9.31). In the case of coalescence, the size distribution function of clusters differs slightly for the coalescence process when grains grow in a solid solution owing to the diffusion of their atoms [197, 202–205] and when liquid drops grow in a supersaturated vapor [206, 207].

A peculiarity of the coalescence process involving metal clusters in a buffer gas is that proceeds in a cluster plasma and consists in competition between the processes of atom attachment to clusters and cluster evaporation. The collision integral of the kinetic equation in this case is given by expression (9.6), and the kinetic equation has the form

$$\frac{\partial f_n}{\partial t} = -\frac{\partial j_n}{\partial n} , \quad j_n = k_o n^{2/3} f_n \left[N_m - N_{sat}(T) \exp\left(\frac{\Delta\varepsilon}{T n^{1/3}}\right)\right] .$$

One can add this equation by equation for the number density N_m of free metal atoms

$$-\frac{d N_m}{dt} = \frac{d N_b}{dt} = \sum_n n \frac{\partial f_n}{\partial t} = \int_0^{n_{cr}} k_o n^{2/3} f_n \left[N_m - N_{sat} \exp\left(\frac{\Delta\varepsilon}{T n^{1/3}}\right)\right] dn .$$

The critical cluster size n_{cr} is important for this formula and is given by formula (4.19),

$$N_m = N_{sat}(T) \exp\left(\frac{\Delta\varepsilon}{T n_{cr}^{1/3}}\right) .$$

The critical size divides clusters into two groups: smaller clusters evaporate and larger clusters grow. Therefore, transitions between these two cluster groups are absent, and this provides an additional relation to analyze the size distribution function of clusters.

Problem 9.26

The form of the size distribution function of clusters f_n, the evolution of which results from the coalescence process, is conserved in time and in its simplest form is

$$f_n = C n^{-2/3} \exp(-n/n_o) . \tag{9.41}$$

Analyze the character of variation of the parameters of this function in time. Assume the number density of free atoms N_m to be small compared with the number density of bound atoms N_b ($N_m \ll N_b$).

The automodel size distribution function results from the solution of the Smoluchowski equation (9.3). Simplifying this problem, we try to "guess" the size distribution function, and the parameters of this function follow from the balance of evaporation and attachment processes. This example allows us to analyze the character of the evolution of the size distribution function. We have from the normalization condition for the constant of formula (9.41)

$$C = \frac{N_b}{n_o^{4/3} \Gamma(4/3)} = \frac{1.12 N_b}{n_o^{4/3}} .$$

On the basis of the collision integral (9.6) we have for the rate J of atom attachment to clusters

$$J_{at} = N_m k_o \int_0^\infty n^{2/3} f_n dn = N_m C k_o n_o .$$

We also have for the rate of atom evaporation J_{ev}

$$J_{ev} = N_{sat}(T) k_o \int_0^\infty n^{2/3} f_n dn \exp\left(\frac{\Delta\varepsilon}{T n^{1/3}}\right) = N_{sat}(T) C k_o n_o \varphi(a) ,$$

$$a = \frac{\Delta\varepsilon}{T n_o^{1/3}} , \quad \varphi(a) = \int_0^\infty dx \exp\left(-x + \frac{a}{x^{1/3}}\right) .$$

Because clusters are in equilibrium with free atoms, these rates are equal, which gives

$$N_m = N_{sat}(T) \varphi(a) .$$

On the other hand, introducing the critical cluster size n_{cr} on the basis of formula (4.19), we have

$$N_m = N_{sat}(T) \exp\left(\frac{\Delta\varepsilon}{T n_{cr}^{1/3}}\right) = N_{sat}(T) \exp\left(\frac{a}{x_{cr}^{1/3}}\right) ,$$

where $x_{cr} = n_{cr}/n_o$. From this we have

$$\varphi(a) \equiv \int_0^\infty dx \exp\left(-x + \frac{a}{(x_o + x)^{1/3}}\right) = \exp\left(\frac{a}{(x_o + x_{cr})^{1/3}}\right) . \tag{9.42}$$

Here we take into account that the integral diverges at small x because of a heightened evaporation rate at small cluster sizes. To avoid an integral divergence, the final rate of evaporation is governed by the parameter x_o.

We assume that in the course of the coalescence process the temperature T of a buffer gas is preserved, that is, $N_{sat} = \text{const}$, and the critical size n_{cr}, the parameter n_o of the size distribution function, and the number density of free atoms N_m vary in this process. As follows from equation (9.42), the critical cluster size n_{cr} varies proportionally to n_o, and the ratio of these parameters is

$$x_{cr} = \frac{n_{cr}}{n_o} = \left[\frac{a}{\ln \varphi(a)}\right]^3 . \tag{9.43}$$

It is convenient to reduce the cluster parameters to the average cluster size $\overline{n}$. The average cluster size $\overline{n}$ for the distribution function (9.24) is

$$\overline{n} = \frac{1}{N_b} \int n^2 f_n dn = \frac{4}{3} n_o ,$$

and the new reduced parameters are

$$y = \frac{n_{cr}}{\overline{n}} = \frac{3}{4} x_{cr} , \quad b = \frac{\Delta\varepsilon}{T \overline{n}^{1/3}} = (3/4)^{1/3} a = 0.909 a .$$

One can see that a typical cluster size grows in time. Under the given conditions this means that parameter a decreases in time. One can solve equation (9.43) in the limit of small a and x_o. Then we have

$$x_{cr} = [\Gamma(2/3)]^{-3} = 1.36 n_{cr} = 1.02 \overline{n} .$$

Thus, the critical size divides clusters into two groups, those with a size below the critical radius and with a size above it, and during the coalescence process transitions between clusters of these groups are absent, and the critical cluster radius is close to the average cluster size in the limit of long times.

Problem 9.27

Assuming the size distribution function of clusters f_n in the course of coalescence is given by formula (9.41), find the rate of variation of the critical cluster radius. The gas temperature is constant during this process, and the number density of free atoms N_m is small compared with the number density of bound atoms N_b ($N_m \ll N_b$).

We have according to formula (9.43) that under a constant temperature the parameter x_{cr} does not vary in the course of coalescence, and formula (4.19) gives the connection between the rates of variation of the critical radius n_{cr} and the number density of free atoms N_m:

$$\frac{dn_{cr}}{dt} = -3\frac{Tn_{cr}^{4/3}}{\Delta\varepsilon}\frac{dN_m}{N_m dt} .$$

Next, we have the equation for the evolution of the critical size:

$$f(n_{cr})\frac{dn_{cr}}{dt} = \int_0^{n_{cr}} k_o n^{2/3} f_n dn \left[N_m - N_{sat}\exp\left(\frac{\Delta\varepsilon}{Tn^{1/3}}\right)\right] .$$

Let us introduce the reduced time τ

$$\tau = \frac{J}{1.12 N_b x_{cr}^{4/3}} \int_{x_{cr}}^{\infty} dx \left[\exp\left(x_{cr} - x + \frac{a}{(x_o + x)^{1/3}} - \frac{a}{(x_{cr} + x_{cr})^{1/3}}\right) - 1\right],$$

where $J = J_{at} = CN_m k_o n_o$ is the total rate of atom attachment to clusters. The solution of the above equation has the form

$$n_{cr} = \frac{n_*}{\sqrt{1 + n_*^2 \tau}} ,$$

where n_* is the initial critical cluster size. We add to this the above equation, which follows from relation (4.19):

$$\frac{d\ln N_m}{dt} = -\frac{a}{3}\frac{d\ln n_{cr}}{dt} .$$

As is seen, in the limit of large times the number density of free atoms varies slowly.

Problem 9.28

Find the reduced critical radius of a cluster if the automodel size distribution function of clusters f_n is approximated by the following equation

$$f_n = Cn^{1/3}\exp(-n/n_o) , \tag{9.44}$$

and this distribution function results from the coalescence process when the number density of free atoms N_m is small compared with the number density of bound atoms N_b ($N_m \ll N_b$).

The automodel size distribution function of clusters is established through a long time and is determined by kinetic equation (9.3), which takes into account the processes of atom attachment to clusters and evaporation of clusters. We use a simpler method to guess the form of the automodel distribution function, and above we considered the case where it has the form of (9.41). Taking now another form (9.44) of the distribution function, one can analyze the sensitivity of the results to this form. We now repeat the operations of Problem 9.26.

We first express the average cluster size $\overline{n}$ through the parameter n_o of the distribution function (9.44):

$$\overline{n} = \frac{1}{N_b} \int n^2 f_n dn = \frac{7}{3} n_o .$$

Next we find the total rate J_{at} of atom attachment to clusters and atom evaporation J_{ev}, in this case

$$J_{at} = N_m k_o \int_0^\infty n^{2/3} f_n dn = N_m C k_o n_o^2 = J_{ev} = N_{sat}(T) C k_o n_o^2 \Phi(a) ,$$

$$a = \frac{\Delta\varepsilon}{T n_o^{1/3}} , \quad \Phi(a) = \int_0^\infty x dx \exp\left(-x + \frac{a}{x^{1/3}}\right) .$$

According to the definition of the critical radius, we have the equation for the reduced critical radius $x_{cr} = n_{cr}/n_o$:

$$\Phi(a) \equiv \int_0^\infty x dx \exp\left(-x + \frac{a}{(x_o + x)^{1/3}}\right) = \exp\left(\frac{a}{(x + x_{cr})^{1/3}}\right) . \tag{9.45}$$

In the limit of small a and x_o we have by analogy with the operations of Problem 9.26

$$n_{cr} = [\Gamma(2/3)]^{-3} n_o = 0.294 n_o .$$

9.5
Heat Regime of Cluster Growth

Problem 9.29

Determine the number of atoms released as a result of the joining of two large clusters assuming that the temperature does not vary as a result of the joining of the clusters and the release of several atoms.

The release of energy in the coagulation process (9.2) leads to an increase in the temperature of a formed cluster compared with that of the joining clusters and can cause the release of surface atoms. Therefore, the coagulation process proceeds according to the following scheme:

$$M_{n-m} + M_m \rightarrow M_{n-q} + qM , \tag{9.46}$$

where q is the number of evaporated atoms. Because the evaporation of diatomic molecules or fragments consisting of several atoms is characterized by a small probability compared with the evaporation of one atom, the liberation of several atoms as a result of cluster joining proceeds in series.

Take the binding energy of a large cluster consisting of n atoms in accordance with formula (2.9), $E = \varepsilon_o n - A n^{2/3}$. The process (9.46) leads to the energy release

$$\Delta E = A\left[m^{2/3} + (n-m)^{2/3} - n^{2/3}\right] .$$

In particular, for $m = n/2$, when this function of m at a given n has a maximum, the above equation gives

$$\Delta E_{\max} = 0.25 A n^{2/3} .$$

As is seen, this energy released is created by the cluster surface energy. Hence, it is small compared with the total binding energy of the cluster atoms, but it can exceed the binding energy of one surface atom.

Assume that at the end of these processes the temperature of a formed cluster is equal to the temperature of colliding clusters. This means that the energy released is used for the liberation of atoms, so the number of atoms released is given by

$$q = \Delta E\,(dE/dn)^{-1} ,$$

where dE/dn is the binding energy of the surface atoms. In the case of a large cluster, we take $dE/dn = \varepsilon_o - 2A/(3n^{1/3}) \approx \varepsilon_o$, so that the maximum number of atoms released as a result of the formation of a cluster containing n atoms is given by

$$q_{\max} = 0.25 A n^{2/3}/\varepsilon_o .$$

In particular, for a large face centered cubic cluster with a short range interaction of atoms we have $q_{\max} = 0.32 n^{2/3}$. For a large icosahedral cluster with a short range interaction of atoms this formula gives $q_{\max} = 0.28 n^{2/3}$, and for a large icosahedral cluster with the Lennard-Jones interaction of atoms it follows from this formula that $q_{\max} = 0.40 n^{2/3}$.

Thus, this effect of cluster heating resulting from the joining of clusters can be responsible for the formation of free atoms in an expanding nucleating vapor, but the number of atoms released is small compared with the number of atoms in a formed cluster.

Problem 9.30

A buffer gas containing liquid clusters expands in a vacuum. Taking into account the nucleation of small clusters (in comparison with the mean free path) according to the scheme (9.2), determine the mean size of small clusters at the end of the process. Assume the diffusion character of the motion of clusters in a buffer gas, and take a typical size of clusters to be larger than the mean free path of gas atoms.

Under these conditions, the normalization condition takes the form $\sum_n n f_n = N_o \exp(-t/\tau_{ex})$, where τ_{ex} is a typical expansion time, and N_o is the initial total number density of bound atoms. Correspondingly, the kinetic equation (9.30) is transformed into

$$\frac{\partial c_n}{\partial \tau} = -\frac{c_n}{\tau_{ex}} - N_o c_n \int_0^\infty k_{n,m} c_m dm + \frac{1}{2} N_o \int_0^n k_{n-m,m} c_{n-m} c_m dm \,,$$

According to this equation, in the course of expansion the rates of cluster collisions decrease, and at the end of the process the buffer gas becomes so rare that cluster collisions cease. From this one can estimate the mean cluster size at the end of the process of $\overline{n} \sim N_o k_{dif} \tau_{ex}$.

For a more accurate determination of the mean cluster size at the end of the expansion process let us multiply the above equation by n^2 and integrate the result over dn. Using the normalization condition $\sum_n n c_n = \exp(-t/\tau_{ex})$, we obtain from this

$$\frac{d\overline{n}}{dt} = N_o k_o \overline{n}^{1/6} I e^{-t/\tau_{ex}} \,,$$

where the integral I is

$$I = \frac{1}{2} \int_0^\infty \int_0^\infty (x^{1/3} + y^{1/3})^2 \sqrt{\frac{x+y}{xy}} \exp(-x-y) dx dy = 5.5 \,.$$

The solution of the above equation in the limit of large t leads to formula (9.32).

Problem 9.31

Determine the variation in temperature of an expanding buffer gas with an admixture of a nucleating vapor if this gas expands through a jet.

A jet expansion is an adiabatic process that is accompanied by transitions of energy between different degrees of freedom of an expanding gas. For the analysis of this process let us extract a gas volume V in which are located n_b atoms of a buffer gas and n_v atoms of a vapor, and these atoms are initially free. The variation in the total energy of this volume is

$$dE = dQ + p dV \,,$$

where Q is the thermal energy of particles in this volume and p is the gas pressure. Assuming the concentration of vapor atoms in a buffer gas to be small, we neglect the thermal energy of vapor atoms. Then the variation in the thermal energy is

$$dQ = \frac{3}{2} n_b dT - \sum_k E_k dn_k \,,$$

where dT is the temperature variation, n_k is the number of clusters consisting of k atoms and located in a given volume, and E_k is the total binding energy of the atoms in a cluster containing k atoms. Introducing the number density of a buffer gas N_b and using its definition $N_b = n_b/V$, we obtain $p dV = -n_b T d N_b/N_b$, where we use the equation of the gaseous state $p = N_b T$. On the basis of the adiabatic character of the expansion process $dE = 0$, we obtain, accounting for the above relations,

$$dE = dQ + p dV = \frac{3}{2} n_b dT - n_b T \frac{dN_b}{N_b} - \sum_k E_k dn_k = 0 . \tag{9.47}$$

Ignoring the nucleation process in equation (9.47), we get the adiabatic law of expansion of a monatomic gas

$$N_b \sim T^{3/2} .$$

Let us assume that the flow of a buffer gas conserves cylindrical symmetry during a free jet expansion. Denote the beam radius by R and take into account that the atom flux $J = \pi R^2 N_b u$ is conserved, where the drift velocity of atoms u is assumed to be independent of the temperature. Then we have $N_b/N_o = R_o^2/R^2$, where N_o, R_o are the initial values of the beam parameters. Substituting this into equation (9.47), we get

$$\frac{3}{2} n_b dT - 2 n_b T dR/R - \sum_k E_k dn_k = 0 .$$

Now let us assume that initially all vapor atoms are free and that condensation takes place in a narrow temperature range near T_*. The solution of this equation under these conditions has the form

$$T = T_o \left(\frac{R_o}{R}\right)^{4/3} \exp\left(\frac{2\varepsilon c}{3 T_*}\right) , \tag{9.48}$$

where $c = n_v/n_b$ is the concentration of vapor atoms and ε is the cluster binding energy per atom.

From this it follows that the nucleation process does not influence the character of gas expansion if the criterion

$$c \ll T_*/\varepsilon$$

is fulfilled. In the opposite case, the thermal effect of the nucleation process stops this process, so only a portion of the free atoms can form clusters. For this reason, a buffer gas is required for the total condensation of an expanding vapor.

Problem 9.32

Determine the maximum concentration of atoms in formed clusters as a result of free expansion of a pure vapor. Assume that the formation of clusters proceeds in a narrow range of temperatures near T_*.

For this goal we use equation (9.47), which takes the following form under the conditions considered:

$$\frac{3}{2}n\,dT - nT\frac{dN}{N} - \sum_k E_k\,dn_k = 0\,,$$

where n is the number of atoms in the expanding volume and N is the number density of atoms, and T is the temperature. Let the initial values of these parameters be N_o and T_o, and at the beginning of the nucleation process these values are N_* and T_*. During the first stage of expansion, when cluster growth is absent, the vapor state is governed by the equation

$$\frac{3}{2}dT - T\frac{dN}{N} = 0\,,$$

so $N \sim T^{3/2}$, and at the beginning of cluster growth we have

$$N_* = N_o\left(\frac{T_*}{T_o}\right)^{3/2}\,.$$

At the next stage of the process $dT = 0$, and we obtain

$$dN/N + \sum_k E_k\,dn_k/n = 0\,.$$

The nucleation process starts at temperature T_* when the formation of diatomic molecules, the nuclei of condensation, is possible. Hence, this temperature can be estimated from the relation $N_2(T_*) \sim N$, where $N_2(T_*)$ is the number density of molecules at thermodynamic equilibrium. Let us assume that the clusters formed are large, so the binding energy per atom is close to that of the bulk, which we denote by ε_o. Then the above equation of the heat balance takes the form

$$\frac{3}{2}n\,dT - nT\,dN/N - \varepsilon_o n\,dc = 0\,,$$

where c is the concentration of bound atoms, that is, the ratio of the number of bound atoms to the total number of atoms in a given volume.

Let us consider that the nucleation process is fulfilled under the criterion

$$KN^2\tau_{ex} \ll 1\,.$$

Here K is the rate constant of the three body process $3A \to A_2 + A$, and τ_{ex} is the expansion time. If the nucleation process is limited by heat release, it proceeds at constant temperature and leads to the maximum concentration of bound atoms at the end of the process, which according to the solution of this equation is given by

$$c_{max} = \frac{T_*}{\varepsilon_o}\ln(N_*/N_f)\,,$$

where N_f is the final number density of free atoms when the cluster growth process is complete. This corresponds to the relation

$$k_o N_f \tau_{ex} n_{max}^{2/3} \sim 1 ,$$

where n_{max} is a typical number of cluster atoms at the end of the process, which is given by

$$n_{max} \sim (k_o N_* \tau_{ex})^3 \sim (k_o N_o \tau_{ex})^3 (T_*/T_o)^{9/2} .$$

Thus, the maximum concentration of bound atoms at the end of the expansion process is given by

$$c_{max} = \frac{T_*}{\varepsilon_o} \ln n_{max} ,$$

with an accuracy up to a numerical factor under logarithm. In reality $T_*/\varepsilon_o \sim 0.1$, so the real maximum degree of nucleation in a pure vapor is on the order of 10%.

Problem 9.33

Compare the rates of cluster growth on the basis of processes (9.1) and (9.2).

If cluster growth is determined by the attachment of free atoms to nuclei of condensation, the average cluster size is

$$\overline{n} = \frac{2}{3}(k_o N \tau_{ex}/3)^3 , \tag{9.49}$$

where N is the initial number density of nucleating atoms, and this result is valid under the condition $N \gg N_o$, where N_o is the total number density of bound atoms in clusters. In the other limiting case, where cluster growth is determined by the cluster coagulation (9.2), the average cluster size is given by formula (9.32):

$$\overline{n} = 6.3(k_o N_o \tau_{ex})^{1.2} ,$$

where N_o is the initial total number density of bound atoms. This formula is valid if the total concentration of bound atoms does not vary in the course of the expansion of gases, that is, $N_b \gg N_m$. Though these equations correspond to different regimes of expansion, one can see from a comparison of the corresponding values $\overline{n}$ that formula (9.49) is valid if the expansion time τ_{ex} is small compared with a typical condensation time, that is, a time during which all the vapor can be transformed into clusters. Formula (9.32) corresponds to the opposite relation between these times. Thus, in the first case we have that during a typical time of cluster growth $N \gg N_o$, and cluster growth proceeds mostly in accordance with the scheme (9.1). In the opposite case, if the typical condensation time is smaller than the expansion time, we have $N \ll N_o$ during a typical time of cluster growth, which proceeds according to the process (9.2).

9.6
Cluster Growth in a Hot Gas with Metal-Containing Molecules

Problem 9.34

Metal-containing molecules MX_k (M is a metal atom, X is a halogen atom) are injected into a hot plasma. Find the criteria under which the decomposition of molecules into atoms leads to the formation of metal clusters.

The chemical equilibrium for metal-containing molecules MX_k in a buffer gas is described by the scheme

$$MX_k \longleftrightarrow M + kX\,, \quad M + M_{n-1} \longleftrightarrow M_n\,, \tag{9.50}$$

where the formation of metal clusters is a result of decay of metal-containing molecules. Near the walls, halogen molecules are found in the form X_2. At higher temperatures they are decomposed into atoms, and at relatively low temperatures a gaseous compound MX_k is formed. The equilibrium between components is described by the scheme

$$M + \frac{k}{2}X_2 \longleftrightarrow MX_k\,. \tag{9.51}$$

From this it follows that for the equilibrium constant at a temperature T we have

$$\frac{[M][X_2]^{k/2}}{[MX_k]} \sim N_o^k \exp\left(-\frac{E}{T}\right),$$

where $[Y]$ denotes the number density of atomic particles Y, E is the enthalpy per molecule for the formation of the compound MX_k, and the value N_o is on the order of a typical atomic magnitude. In the same fashion, we have from the equilibrium for clusters

$$\frac{[M][M_{n-1}]}{[M_n]} \sim N_{sat}(T)\,,$$

where $N_{sat}(T) \sim N_o \exp(-\varepsilon_M/T)$ is the saturated vapor number density of metal atoms at a given temperature and ε_M is the atom binding energy for bulk metal. Using these relations, we have the criterion that the compound MX_k is favorable at low temperatures:

$$E > \varepsilon_W\,.$$

Let us introduce the binding energy of MX_k per halogen atom:

$$\varepsilon_X = \frac{E}{k} + \frac{D}{2}\,,$$

where D is the dissociation energy of a halogen molecule. If we assume that the dissociation energy of the halogen molecules provides a small contribution to the enthalpy of formation of the molecule MX_k, one can rewrite the above criterion in the form

$$k\varepsilon_X > \varepsilon_M .$$

From the chemical equilibrium for molecules MX_k it follows that this compound is decomposed into atoms at temperatures on the order of

$$T_1 = \frac{\varepsilon_X}{\ln(N_o/[X])} , \tag{9.52}$$

where [X] is the total number density of free and bound halogen atoms. In turn, metal clusters are transformed into atomic vapor at temperatures on the order of

$$T_2 = \frac{\varepsilon_W}{\ln(N_o/[M])} , \tag{9.53}$$

where [M] is the total number density of free and bound metal atoms. Evidently, clusters remain stable in the temperature range [258, 288]

$$T_1 < T < T_2 . \tag{9.54}$$

Because of the halogen excess $[X] \gg [M]$, the possibility of the existence of clusters and gas molecules in the system under consideration corresponds to the criterion

$$\varepsilon_X < \varepsilon_M < k\varepsilon_X . \tag{9.55}$$

One can see that these criteria are compatible if the gaseous compound of the heat-resistant metal contains several halogen atoms. The data related to this analysis are given in Table 9.1. This table shows the values of ε_X and ε_M that are obtained on the basis of the Gibbs thermodynamic potential for the compounds under consideration, and the values that follow from the enthalpy values for these compounds are given in parentheses. The temperatures T_1 and T_2 are found from formulas (9.52) and (9.53), and the temperature T_3 coincides with T_2, but this value results from the saturated vapor pressure of the metal as a function of temperature. The data in Table 9.1 relate to the total number density of halogen atoms $1 * 10^{16}$ cm^{-3} and the total number density of metal atoms, which that is four to six times less depending on the type of metal-containing molecules. If criteria (9.54) are fulfilled, clusters of this metal exist in the temperature range $T_1 < T < T_2$, and such cases are marked in the last column of Table 9.1 by the + sign.

Problem 9.35

Assuming the binding energies of bonds $MX_k - X$ for different radicals to be identical, analyze the kinetics of radical formation when metal-containing molecules MX_6 are injected into a hot dense plasma and metal clusters are produced after decomposition of these molecules.

Table 9.1 Parameters of some compounds of heat-resistant metals.

Compound	ε_X	ε_M	T_1, 10^3 K	T_2, 10^3 K	T_3, 10^3 K	Cluster existence
$HfCl_4$	(4.2)	6.0 (6.4)	(2.4)	3.0(3.2)	3.15	+
HfF_4	(6.3)	6.0(6.4)	(3.5)	3.0(3.2)	3.15	–
$ThCl_4$	(4.7)	5.8(6.2)	(2.6)	3.0(3.1)	3.12	+
ThF_4	6.2(6.5)	5.8(6.2)	3.5(3.6)	3.0(3.1)	3.12	–
$TiBr_4$	3.0(3.2)	4.4(4.9)	1.7(1.8)	2.2(2.5)	2.28	+
$TiCl_4$	3.5(3.8)	4.4(4.9)	2.0(2.1)	2.2(2.5)	2.28	+
UCl_4	3.8(4.1)	5.1(5.5)	2.1(2.3)	2.6(2.8)	2.73	+
UF_4	5.7(5.9)	5.1(5.5)	3.2(3.3)	2.6(2.8)	2.73	–
$ZrCl_4$	4.3(4.8)	5.9(6.3)	2.4(2.7)	3.0(3.2)	3.08	+
ZrF_4	6.5(6.9)	5.9(6.3)	3.6(3.9)	3.0(3.2)	3.08	–
IrF_6	2.2(2.5)	6.4(6.9)	1.2(1.4)	3.2(3.5)	3.15	+
MoF_6	3.9(4.2)	6.3(6.8)	2.2(2.4)	3.2(3.4)	3.19	+
UCl_6	3.0(3.3)	5.1(5.5)	1.7(1.8)	2.6(2.8)	2.73	+
UF_6	4.7(5.0)	5.1(5.5)	2.6(2.8)	2.6(2.8)	2.73	–
WCl_6	(3.0)	8.4(8.8)	(1.7)	4.2(4.4)	4.04	+
WF_6	4.5(4.9)	8.4(8.8)	2.5(2.7)	4.2(4.4)	4.04	+

In the course of decay, a metal-containing molecule MX_k is transformed into atoms and radicals, and then metal atoms are converted into metal clusters, whereas halogen atoms and molecules leave a region where metal-containing molecules MX_6 are located at the beginning (we assume the dimensions of this region to be small in comparison with the plasma dimension). Assuming identical binding energies ε_X for each bond $MX_k - X$ ($k = 1 \div 5$), we use the following rates of decay of molecules and radicals in collisions with buffer gas atoms:

$$\nu_d = N_a k_{gas} \exp\left(-\frac{\varepsilon_X}{T}\right) . \tag{9.56}$$

Here N_a is the number density of buffer gas atoms, T is the temperature of buffer gas atoms, k_{gas} is the gas-kinetic rate constant for collisions of molecules and radicals with atoms of a buffer gas, and we assume the rate ν_d to be independent of the number of halogen atoms k in molecules or radicals. On the basis of this model, we analyze below the kinetics of destruction of metal-containing molecules in a hot buffer gas.

Under the above conditions, the set of balance equations for the radical concentrations c_k, if a radical contains k atoms, has the form

$$\frac{dc_0}{dt} = \nu_d c_1 , \quad \frac{dc_k}{dt} = -\nu_d (c_k - c_{k+1}) , \quad k = 1 \div 5 ; \quad \frac{dc_6}{dt} = -\nu_d c_6 . \tag{9.57}$$

At the beginning the system consists of molecules that correspond to the following initial condition:

$$c_6(0) = c_M ; \quad c_k = 0 , \quad k \neq 6 ,$$

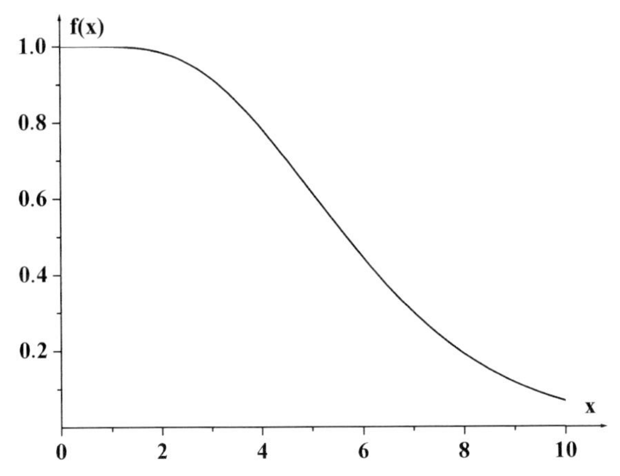

Fig. 9.6 The function $f(x)$ in accordance with formula (9.58).

where c_M is the total concentration of free and bound metal atoms, and we ignore here transport processes.

This set of balance equations with given initial conditions may be solved analytically, and the solution is

$$c_k = c_M \frac{x^{6-k}}{(6-k)!} e^{-x}\,, \quad c_0 = c_M[1 - f(x)]\,; \quad f(x) = e^{-x} \sum_{k=0}^{5} \frac{x^k}{k!}\,, \tag{9.58}$$

where $k = 1 \div 6$, $x = \int_0^t \nu_d dt$, and $f(x)$ is the portion of molecules and radicals, that is, $1 - f(x)$ is the portion of free metal atoms. Figure 9.6 gives the dependence $f(x)$.

Problem 9.36

Metal-containing molecules MX_6 are injected into a hot buffer gas plasma. Find the cooling of a buffer gas as a result of the transformation of metal-containing molecules in metal clusters.

Assuming a prompt cooling of a buffer gas after the decomposition of molecules and radicals, we obtain from the set of balance equations (9.57) and their solution (9.58) the following heat balance equation in the course of molecule destruction:

$$\frac{dT}{dt} = \delta T \sum_{k=0}^{5} \left(\frac{dc_{k+1}}{dt} - \frac{dc_k}{dt} \right) = \delta T \left(\frac{dc_6}{dt} - \frac{dc_0}{dt} \right) = -\nu_d c_M \delta T f(x)\,,$$

where δT is the cooling temperature for a buffer gas per metal-containing molecule. We also include in this parameter the heat released as a result of the conversion of metal atoms into clusters.

The solution of this set of equations gives the temperature of a buffer gas after the destruction of molecules and the transformation of metal atoms into metal

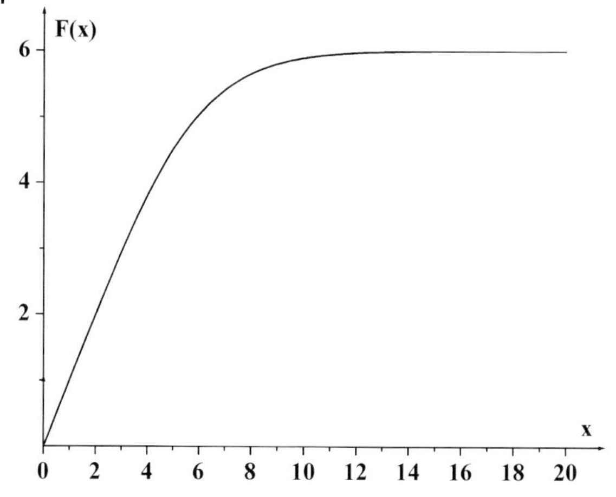

Fig. 9.7 The function $F(x)$ defined by formula (9.60).

clusters:

$$T = T_o - \frac{\delta T}{6} \sum_{k=0}^{5} (6-k) c_k = T_o - \frac{1}{6} c_M \delta T F(x) , \tag{9.59}$$

where

$$F(x) = 6 - e^{-x} \left(\frac{x^5}{120} + \frac{x^4}{12} + \frac{x^3}{2} + 2x^2 + 5x + 6 \right) . \tag{9.60}$$

Here the function $F(x)$, given in Figure 9.7, is the number of broken bonds per molecule, $T_o = T(0)$ is the initial temperature, and we take into account that the cooling of a buffer gas as a result of broken halogen–metal bonds is compensated partially by the joining of metal atoms in clusters. The total temperature change when all halogen bonds are broken and metal atoms join with large clusters is

$$T_o - T(\infty) = c_M \delta T ,$$

and this process corresponds to the cooling of a buffer gas. This formula is based on the assumption that a buffer gas has enough high heat capacity, which is valid at low concentrations of metal-containing molecules.

Problem 9.37

Metal-containing molecules MX_6 injected into a flow of hot dense plasma are located initially in a cylindrical region with a radius of ρ_o. Analyze the thermal regime of transformation of metal-containing molecules into clusters if heat release is compensated by the thermal conductivity of a buffer gas.

Since the binding energy of metal-containing molecules ε_X is large in comparison with a thermal energy $\sim T$, the process of molecule destruction acts strongly on the heat balance of a buffer gas with a metal admixture despite there being small concentration of metal-containing molecules in the buffer gas. On the basis of formula (9.59), the heat balance equation, accounting for the thermal conductivity of the buffer gas, is

$$\frac{\partial T}{\partial t} = \chi \Delta T - \nu_{d} c_{M} \delta T f(x) \,,$$

where χ is the thermal diffusivity coefficient of a buffer gas, and we ignore the diffusion of molecules and radicals outside this region.

In considering this formula in the case where metal-containing molecules occupy a cylindrical region in a plasma of radius ρ_o, we reduce this formula to a quasistationary regime. Let the buffer gas temperature be T_o in the region $\rho \geq \rho_o$, and T_* at the center. Then this equation is reduced to

$$\frac{T_o - T_*}{\tau_{eq}} = \nu_{d} c_{M} \delta T f(x) \,, \quad \tau_{eq} = \frac{0.17 \rho_o^2}{\chi} \,, \tag{9.61}$$

where τ_{eq} is a typical time for equilibrium establishment.

Another typical time of this process is the total time of destruction of metal-containing molecules, which is

$$\tau_o = \frac{6}{\nu_d(T_*)} = \frac{6}{N_a k_{gas} \exp\left(-\frac{\varepsilon_X}{T_o}\right)} \,. \tag{9.62}$$

Since our analysis is based on the quasistationary regime of destruction of metal-containing molecules and heat transport, the following criterion must be fulfilled:

$$\tau_{eq} \ll \tau_o \,.$$

Table 9.2 demonstrates of the above results and gives both the parameters of compounds of some metal-containing molecules (ρ is the density at room temperature, and T_m, T_b are the melting and boiling temperatures, respectively) and the parameters of the kinetics of destruction of these molecules in hot argon. We represent there an initial stage of molecule destruction $x = 0$ and take the optimal buffer gas temperature far from the destruction region such that $N_m/N_{sat}(T_*) \approx 10$ for MoF_6 and WF_6, and $N_m/N_{sat}(T_*) \approx 1000$ for IrF_6 and WCl_6. We assume metal-containing molecules to be located in a cylinder with a radius of $\rho_o = 1$ mm, the argon pressure is $p = 1$ atm, and the concentration of metal-containing molecules is $c_M = 10\%$ with respect to argon atoms. In addition, we take the gas-kinetic cross section to be $\sigma_{gas} = 3 \cdot 10^{-15}\,\text{cm}^2$ and use this value in the rate constant of destruction of metal-containing molecules and their radicals in collisions with argon atoms. Along with the parameters of cluster evolution given in Table 9.2, we represent in Figure 9.8 typical temperatures for the generation of tungsten clusters from molecules WCl_6 in argon at pressure of 1 atm and the tungsten atom concentration of 10% with respect to argon atoms.

Table 9.2 Parameters of evolution of metal-containing molecules in hot argon ($p = 1$ atm, $c_M = 10\%$, $\rho_o = 1$mm).

Compound	MoF_6	IrF_6	WF_6	WCl_6
ρ, g/cm^3	2.6	6.0	3.4	3.5
T_m, K	290	317	276	548
T_b, K	310	326	291	620
ε_X, eV	4.3	2.5	4.9	3.6
ε_M, eV	6.3	6.5	8.4	8.4
T_1, K	2200	1200	2500	1700
T_2, K	4100	4000	5200	5200
δT, 10^3 K	12	3.7	13	4.8
T_o, K	3700	6000	5000	5800
T_*, K	3600	2900	4600	4200
τ_{eq}, 10^{-5} s	12	17	8	9
τ_o, 10^{-3} s	5	0.1	1	0.1
N_m, 10^{17}cm^{-3}	2.0	2.5	1.6	1.8

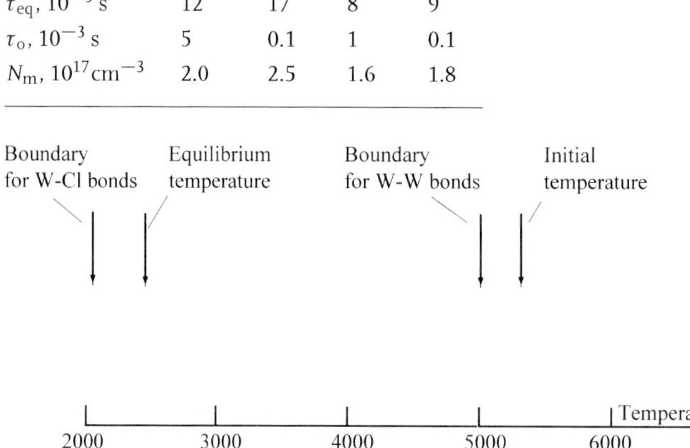

Fig. 9.8 Typical temperatures for growth of tungsten clusters from WCl_6 molecules.

Thus, the heat balance of a mixture consisting of a dense hot buffer gas and metal-containing molecules includes heat absorption due to the destruction of metal-containing molecules and heat release as a result of the joining of metal atoms in clusters, which are compensated by heat transport from surrounding regions.

Problem 9.38

Analyze nucleation in a hot buffer gas with metal-containing molecules within the framework of a model where the rate of formation of free metal atoms Q is approximated by an appropriate dependence.

If criterion (9.54) is fulfilled, thermodynamic equilibrium of a buffer gas with an admixture of metal-containing molecules includes the destruction of molecules and the formation of a condensed metal in the form of large clusters. This leads to the following scheme of basic processes:

$$MX_k \to M + kX\,, \quad M + M_n \to M_{n+1}\,. \qquad (9.63)$$

Note that in contrast to nucleation in a pure metal vapor that proceeds according to the scheme (9.20), one can ignore now slow three body processes, and the first stage of cluster growth processes due to the formation of diatomic molecules proceeds as

$$MX + M \to M_2 + X\,.$$

Hence, in this case a buffer gas does not partake in the nucleation process, so the nucleation rate does not depend on the number density of buffer gas atoms. But a buffer gas determines the destruction of metal-containing molecules and their radicals, and thus the specific rate of formation of free metal atoms as a result of molecule destruction is $Q = N_a \nu_d = N_a N_b k_d$, where N_a is the number density of buffer gas atoms, N_b is the number density of bound atoms at the end of the destruction process, and the rate ν_d of this process is given by formula (9.56).

Note that the nucleation process proceeds in a restricted spatial region where metal-containing molecules are located from the beginning. Free halogen atoms and molecules leave the cluster region, whereas clusters remain in this region because of low mobility. Thus, we neglect the role of halogen atoms in cluster growth, and though the nucleation process is considered in a uniform system, for other processes it is not uniform.

Under the above conditions, we obtain the following balance equations for the number density of free metal atoms N_m, the number density of clusters N_{cl}, and the cluster size n (number of cluster atoms):

$$\frac{dN_m}{dt} = Q - N_{cl} k_n N_m\,, \quad \frac{dN_{cl}}{dt} = k_{ch} N_m^2\,, \quad \frac{dn}{dt} = k_n N_m\,. \qquad (9.64)$$

Here $k_n = k_o n^{2/3}$ is the rate constant of atom attachment to a cluster and k_{ch} is the rate constant of the process of formation of a diatomic metal molecule M_2. We consider a diatomic metal molecule that is transformed later into a growing cluster as a nucleus of condensation, and attachment of atoms to it takes place in pairwise collisions as well for larger molecules and clusters.

An analysis of this set of balance equations shows that the cluster number density N_{cl} and cluster size n grow in time, whereas the number density of free atoms N_m grows in the first stage of the cluster growth process and drops in its second stage. Roughly, one can divide the time into three ranges, as shown in Figure 9.9, so at $t \le \tau_{max}$ the number density of free atoms grows, and at time $\tau \sim 1/\nu_d$ metal-containing molecules are destroyed.

For the second stage of the nucleation process ($\tau > t > \tau_{max}$) we have

$$Q = N_a N_b k_d \sim N_{cl} \frac{dn}{dt} \sim k_{ch} N^2 n\,,$$

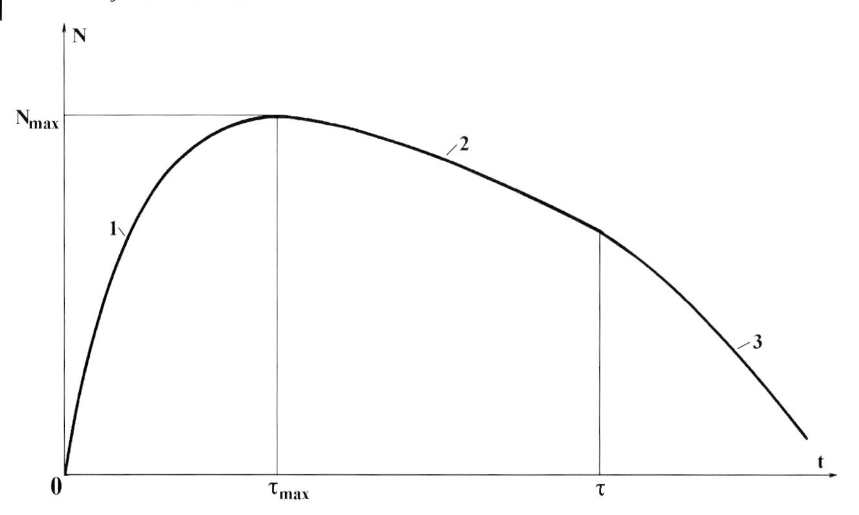

Fig. 9.9 Time dependence of the number density of free metal atoms in the course of formation of metal clusters from metal-containing molecules in a buffer gas.

and assuming $k_\mathrm{d} \sim k_\mathrm{ch}$, we obtain in the second stage of the nucleation process

$$N_\mathrm{m} \sim \sqrt{\frac{N_\mathrm{a} N_\mathrm{b}}{n}} \,. \tag{9.65}$$

Note that from the first equation of the set (9.64), the maximum number density N_max of free metal atoms is attained at time

$$\tau_\mathrm{max} \sim \frac{N_\mathrm{max}}{Q} \sim \frac{N_\mathrm{m}}{N_\mathrm{a} N_\mathrm{b} k_\mathrm{d}} \,.$$

Since a typical time of decay of metal-containing molecules is

$$\tau \sim \frac{1}{N_\mathrm{b} k_\mathrm{d}} \,,$$

this gives

$$\tau \gg \tau_\mathrm{max} \,.$$

This means that the time dependence for the number density of free atoms N_m has a form in accordance with Figure 9.9, and the maximum of this value is achieved before total destruction of metal-containing molecules.

In addition, we have from the last equation of the set (9.64) and the relation (9.65) for the number density of free atoms the following estimate for the cluster size n at the end of the nucleation process:

$$n \sim (k_\mathrm{o} N \tau)^3 \sim (k_\mathrm{o} \tau)^3 \left(\frac{N_\mathrm{a} N_\mathrm{b}}{n}\right)^{3/2} \,,$$

which gives

$$n \sim (k_o \tau)^{6/5} (N_a N_b)^{3/5} \sim \left(\frac{k_o}{k_d}\right)^{6/5} \left(\frac{N_a}{N_b}\right)^{3/5} ,$$

and each factor of the last product is large.

In the same manner, using the above formulas for τ_{max} and estimation (9.65) for the number density of free metal atoms at this time, we obtain for a cluster size at this time

$$n \sim \left(\frac{k_o}{n k_d}\right)^3 ,$$

which gives

$$n \sim \left(\frac{k_o}{k_d}\right)^{3/4} \gg 1 ,$$

and since $k_o \gg k_d$, we deal with large clusters in this stage of the nucleation process. Thus, the cluster growth process through decomposition of metal-containing molecules in a buffer gas proceeds mostly at large cluster sizes.

Problem 9.39

Determine the temperature of charged clusters in an arc plasma where the gas temperature T differs from the electron temperature T_e, and the cluster temperature results from collisions of clusters with atoms and electrons.

The cluster temperature T_{cl} results from collisions of atoms and electrons with clusters. Let us use a simple model of collisions such that an atomic particle after collision obtains the cluster's thermal energy on average. This means for $T_e > T$ that an atom obtains on average the energy $\frac{3}{2}(T_{cl} - T)$ from the cluster, and an electron gives to the cluster the energy $\frac{3}{2}(T_e - T_{cl})$ on average. Then the power that a cluster takes from electrons is $\frac{3}{2}(T_e - T_{cl}) \cdot v_e N_e \sigma_e$, where v_e is the average electron velocity, N_e is the number density of electrons, and σ_e is the cross section of electron–cluster collisions. The power that atoms obtain from the cluster is equal to $\frac{3}{2}(T_{cl} - T) \cdot v_a N_a \sigma_a$, where v_a is the average velocity of atoms, N_a is the number density of atoms, and σ_a is the cross section of atom–cluster collisions. The stationary condition, using formulas (2.3) and (4.1) for the rate constants of collisions of atoms and electrons with charged clusters, leads to the following expression for the cluster temperature T_{cl} [86, 210]:

$$T_{cl} = \frac{T + \zeta T_e}{1 + \zeta} ; \quad \zeta = \sqrt{\frac{T_e m_a}{T m_e}} \cdot \left(1 + \frac{Z e^2}{r_o T_e}\right) \cdot \frac{N_e}{N_a} , \tag{9.66}$$

where Z is the cluster charge, and m_e and m are, respectively, the electron and atomic masses. As is seen, the cluster temperature can depend on both the cluster's size and charge.

Part III Complex Plasma

We call a complex plasma or an ionized gas with a dispersed phase a plasma where the presence of particles or clusters influences significantly the plasma properties. One form of complex plasma is a dusty plasma, which contains stable particles. Usually a dusty plasma is a gas discharge plasma with micron-sized dielectric particles. In contrast, a cluster plasma is an ionized gas with clusters that may grow or evaporate in the course of plasma evolution. In practice, a cluster plasma contains nanometer-sized clusters and may be a source of cluster beams as well as a light source or a catalyst for some chemical processes. An atmospheric ionized gas with aerosols is called an aerosol plasma, which is really a type of a dusty plasma with a low concentration of charged particles. We will work with these definitions below.

10
Dusty Plasma

A complex plasma contains macroscopic particles along with electrons, ions, and atoms. A dusty plasma contains micron-sized dust and solid particles. Along with an astrophysical dusty plasma where there are dust particles in a rare ionized gas, a laboratory dusty plasma is formed when solid micron-sized particles are introduced into a gas discharge plasma. If the electric potential of a gas discharge has local minima in some regions, as occurs in the case where ionization waves (striations) may be formed in a gas discharge plasma, these regions may be used as traps for charged dust particles. Then dust particles may be located in these traps for a very long time. A dusty plasma is a specific physical object [211–216] where the interaction of charged particles with the surrounding plasma creates a self-consistent field that governs the properties of this plasma. Interaction between charged dust particles that lie in closest proximity to one another may lead to the formation of structures of these particles in the electric traps of the gas discharge. Therefore, the behavior of dust particles in electric traps of gas discharge may be used to model the order–disorder phase transition and other phenomena for a system of interacting particles.

10.1
Particles in the Positive Column of Glow Discharge

Problem 10.1

Spherical particles are introduced into a gas discharge tube of the positive column of gas discharge with the Schottky ionization balance. Determine the number density of particles when recombination processes on their surface give the same contribution to the plasma ionization balance as the attachment of electrons and ions to walls. Apply this to the case where a cylinder tube with a radius of $R_o = 1\,\text{cm}$ is filled with neon or argon at a pressure of $p = 10\,\text{Torr}$ and a temperature of $T = 300\,\text{K}$. The radius of the dust particles is $r_o = 1\,\mu\text{m}$.

The Schottky regime corresponds to low discharge currents, so the gas discharge current does not heat the gas. The ionization balance in this regime is determined by ionization of atoms with electrons accelerated by the discharge electric field and

Cluster Processes in Gases and Plasmas. Boris M. Smirnov

ISBN: 978-3-527-40943-3

the attachment of electrons and ions to walls of the discharge tube. The balance equation for the number density N_e of electrons has the form

$$\frac{D_a}{\rho}\frac{d}{d\rho}\left(\rho\frac{dN_e}{d\rho}\right) + k_{ion}N_eN_a = 0\,, \qquad (10.1)$$

where ρ is the distance from the axis center, N_a is the number density of atoms, D_a is the ambipolar diffusion coefficient for this plasma, and k_{ion} is the rate constant of electron–atom ionization in an electric field. This equation is valid for low electron number densities with weak gas heating, and therefore the rate of ionization is proportional to the number density of electrons. The boundary condition $N_e(R_o) = 0$, where R_o is the tube radius, accounts for the recombination of electrons and ions at the walls of the discharge. The solution of equation (10.1) leads to the following spatial electron distribution:

$$N_e(\rho) = N_o J_0\left(\rho\sqrt{\frac{k_{ion}N_a}{D_a}}\right)\,, \qquad (10.2)$$

where N_o is the electron number density at the axis and $J_0(x)$ is the Bessel function. The above boundary condition, $N_e(R_o) = 0$, gives the following ionization balance:

$$k_{ion}N_a = \frac{5.78\,D_a}{R_o^2}\,. \qquad (10.3)$$

Here the left-hand side of this balance equation is the rate of ionization per electron and the right-hand side is the rate of recombination at walls.

It is convenient to represent the rate constant k_{ion} of atom ionization by electron impact in the form

$$k_{ion} = \frac{\alpha w_e}{N_a}\,,$$

where α is the first Townsend coefficient and w_e is the electron drift velocity. The dependence of the first Townsend coefficient α on the reduced electric field strength is sharp and is given in Figures 10.1 and 10.2 [217], where electrons are located in neon and argon, respectively, in an electric field.

The solution of equation (10.3) gives the electric field strength E that provides this ionization equilibrium. In particular, under given conditions in the absence of particles in a gas discharge plasma we have approximately $N_a = 3.2 \cdot 10^{17}\,\text{cm}^{-3}$ and

$$\frac{k_{ion}}{D_a} = 1.8 \cdot 10^{-17}\,\text{cm}\,.$$

In addition, the ambipolar diffusion coefficient is determined mostly by the electric field of electrons acting on ions in the course of their joint drift, and the ambipolar diffusion coefficient is

$$D_a = D_i\frac{T_e}{T}\,,$$

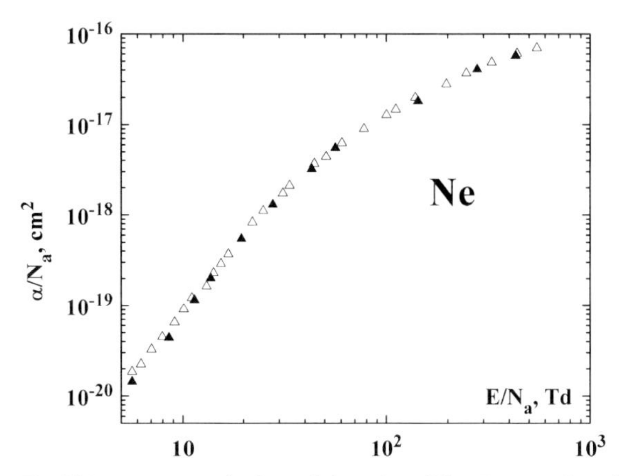

Fig. 10.1 Experimental values of the reduced first Townsend coefficient for neon [217].

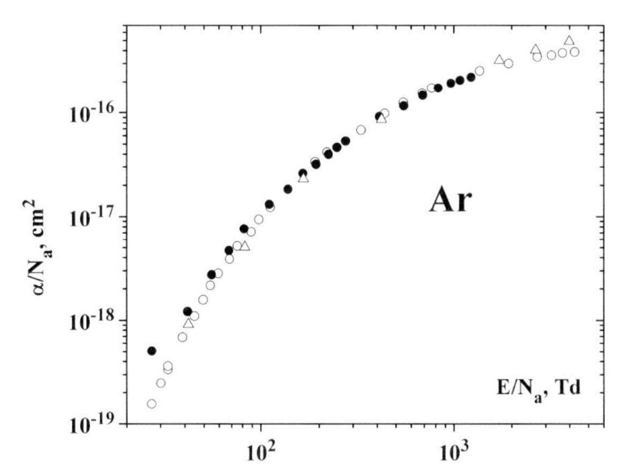

Fig. 10.2 Experimental values of the reduced first Townsend coefficient for argon [217].

where D_i is the ion diffusion coefficient and the T_e is the effective electron temperature. For definiteness, we assume atomic ions Ne^+ or Ar^+ to be located in a gas, so the diffusion coefficient of atomic ions in the parent gas is $D_i = 0.1\,\text{cm}^2/\text{s}$ in the neon case and $D_i = 0.022\,\text{cm}^2/\text{s}$ in the argon case under normal conditions, and under given conditions we have $D_i = 8.4\,\text{cm}^2/\text{s}$ in the neon case and $D_i = 1.8\,\text{cm}^2/\text{s}$ in the argon case under a given pressure. Using experimental data for electron drift in a gas in an external electric field, we obtain the reduced electric field strength $E/N_a \approx 6\,\text{Td}$ in the neon case ($D_a \approx 2 \cdot 10^3\,\text{cm}^2/\text{s}$, $\alpha/N_a \approx 2 \cdot 10^{-20}\,\text{cm}^2$, $w_e = 2 \cdot 10^6\,\text{cm/s}$). In the argon case the reduced electric field strength is $E/N_a \approx 12\,\text{Td}$ ($D_a \approx 6 \cdot 10^2\,\text{cm}^2/\text{s}$, $\alpha/N_a \approx 1 \cdot 10^{-20}\,\text{cm}^2$, $w_e = 1.3 \cdot 10^6\,\text{cm/s}$). Similarly, we obtain a rate constant of atom ionization of approximately $k_{\text{ion}} = 2 \cdot 10^{-14}\,\text{cm}^3/\text{s}$ in the neon case and $k_{\text{ion}} = 9 \cdot 10^{-15}\,\text{cm}^3/s$ in the argon case. From this it follows for the rate of ionization per electron that $\nu_{\text{ion}} = 6.4 \cdot 10^3\,\text{s}^{-1}$ in the neon case and $\nu_{\text{ion}} = 2.9 \cdot 10^3\,\text{s}^{-1}$ in the argon case.

We now consider the charging of particles in this plasma for the limiting case $\lambda \gg r_o$, where $\lambda \sim 10^{-3}$ cm is the mean free path of ions in the parent gas and r_o is the particle radius. The cluster charge follows from equation (7.41) for the reduced cluster charge (7.30), which has the form

$$x = \frac{T_e}{T} \ln \frac{X}{1+x}, \quad x = \frac{|Z|e^2}{r_o T}, \quad X = \sqrt{\frac{T_e m_i}{T m_e}}, \tag{10.4}$$

where m_e, m_i are the electron and the ion masses, respectively.

Solving this equation, we find for a particle radius of $r_o = 1\,\mu$m, the equilibrium cluster charge $|Z| = 790$ in the neon case ($T_e \approx 6$ eV) and $|Z| = 1100$ in the argon case ($T_e \approx 8$ eV). Correspondingly, on the basis of formula (7.31) the recombination coefficient of electrons and ions on a particle is $k_{rec} = 0.77\,\text{cm}^3/\text{s}$ in the neon case and $k_{rec} = 1.0\,\text{cm}^3/\text{s}$ in the argon case. Note that formulas (7.30) and (7.31) give an identical value for the recombination rate constant, and the rate of recombination of electrons and ions due to the presence of particles in a gas discharge plasma is equal to $k_{rec} N_p$, where N_p is the number density of particles.

These evaluations show that there is the same contribution to electron–ion recombination in a gas discharge plasma owing to the attachment of electrons and ions and owing to their recombination on particles, which is governed by the relation $\nu_{ion} = k_{rec} N_p$, at the number density of particles $N_p = 8.3 \cdot 10^3\,\text{cm}^{-3}$ in the neon case and $N_p = 2.9 \cdot 10^3\,\text{cm}^{-3}$ in the argon case. Note that if the particle material is a glass, these values correspond to its average density $\rho = 7.3 \cdot 10^{-8}\,\text{g/cm}^3$ in the neon case and $\rho = 2.6 \cdot 10^{-8}\,\text{g/cm}^3$ in the argon case. The densities of neon and argon in the gas discharge tube are $\rho = 1.1 \cdot 10^{-5}\,\text{g/cm}^3$ and $\rho = 2.1 \cdot 10^{-5}\,\text{g/cm}^3$, respectively.

Problem 10.2

Under the conditions of the previous problem determine the reduced electric field strength when attachment of electrons and ions to particles injected into a plasma of the positive column of a low-current gas discharge leads to the same contribution to electron and ion losses as their attachment to the walls.

Accounting for the recombination of electrons and ions on the particle surface, we obtain the following equation of ionization balance instead of formula (10.3):

$$\alpha w_e = \frac{5.78\,D_a}{R_o^2} + k_{rec} N_p\,. \tag{10.5}$$

A new channel of loss of electrons and ions leads to an increase in the electric field strength in comparison with the balance equation (10.3), but because of a sharp dependence of the first Townsend coefficient α on the reduced electric field strength E/N_a, this gives an insignificant increase in the electric field strength. In particular, under the conditions of the previous problem we have an increase in the reduced electric field strength from 6 Td to 7 Td in the neon case, and from 12 Td to 15 Td in the argon case, when particles are injected into the positive column such that the contribution to the recombination of electrons and ions on walls and on particles is identical. This is obtained on the basis of the data in Figures 10.1 and 10.2.

Problem 10.3

Determine the character of the variation of the electric field strength in the positive column of gas discharge with variation of the gas pressure or a tube radius.

According to the ionization balance equation (10.3), an increase in the tube radius R_o decreases the rate of electron and ion recombination on walls, which leads to a decrease in the reduced electric field strength E/N_a. In the same manner, an increase in the number density of atoms N_a leads to an increase in the ionization rate, which corresponds to a decrease in the reduced electric field strength. One can see that E/N_a is conserved if a variation in these values proceeds in such a way that the combination $N_a R_o^2$ is conserved.

Problem 10.4

Compare the charge density concentrated on particles injected into the positive column of gas discharge and on electrons under the conditions of Problem 10.1. Find its dependence on the particle radius.

When losses of electrons and ions on the walls of a discharge tube and those due to the attachment to particles injected into a plasma are equal, the charge number density under the conditions of Problem 10.1 is $6.6 \cdot 10^6\ \mathrm{cm}^{-3}$ in the neon case and $3.2 \cdot 10^6\ \mathrm{cm}^{-3}$ in the argon case, as follows from the results of Problem 10.1. These values are small compared with the number density of electrons N_e or ions in the positive column of glow discharge, which is $N_e \gg 10^8\ \mathrm{cm}^{-3}$.

If the particle radius r_o varies, according to formula (10.4), then the particle charge varies proportionally to the radius. According to formula (7.31) the recombination coefficient of electrons and ions on the particle surface is also proportional to the particle radius. Hence, if the contribution to the total recombination rate is conserved owing to the attachment of electrons and ions to the particle surface, whereas the particle radius varies, then we obtain for the number density of particles under such conditions $N_p \backsim 1/r_o$. From this it follows that the charge number density is conserved in this case.

Problem 10.5

Assuming the losses of electrons and ions in the positive column of a gas discharge due to their attachment to particles to be small compared with the losses due to the attachment of electrons and ions to the walls, that is, particles do not influence the parameters of the gas discharge plasma, analyze the behavior of the charged particles in this plasma.

We base our analysis on the character of processes in the positive column a gas discharge plasma. The number density of electrons and ions N_e according to formulas (10.2) and (10.3) has the form

$$N_e(\rho) = N_o J_0 \left(2.4 \frac{\rho}{R_o}\right) ,$$

where ρ is the distance from the center and R_o is the tube radius. Next, in the main region this plasma is quasineutral, that is, the fluxes of electrons and ions to the walls are equal, so the flux of electrons j_e to the walls on the scale of electron parameters is zero,

$$j_e = -D_e \nabla N_e + K_e E N_e = 0 ,$$

and the electric field strength **E** is directed to the tube center, that is, the electric field slows down the motion of electrons to walls. Hence, this field compels negatively charged particles to move to the center. One can expect these particles to be concentrated at the tube center.

We now determine the potential energy of particles U in the region of the plasma quasineutrality and the electric field strength E through the plasma electric-potential φ

$$U = Ze\varphi , \quad E = -\frac{d\varphi}{d\rho} .$$

From the expressions for the electron flux and the electron number density we have

$$\varphi = T_e \ln J_0 \left(2.4 \frac{\rho}{R_o}\right) ,$$

where we introduce the electron temperature T_e and use the Einstein relation (6.4). Expanding the Bessel function $J_0(x) = 1 - x^2/4$, $x \ll 1$, we obtain near the tube center

$$U = 1.45 Z T_e \frac{\rho^2}{R_o^2} . \tag{10.6}$$

Since according to formulas (7.40) and (7.41) $Z \gg 1$, even close to the tube center the potential energy for negatively charged particles significantly exceeds a thermal energy.

10.2
Particles in Traps of Gas Discharge

Problem 10.6

Small particles are locked in an electric trap whose size is on the order of R_o and the difference between the electric potentials for the trap center and its walls is φ_o. Estimate the maximum number of charged particles that can be locked in this trap.

Let us assume that charged particles do not change the distribution of the plasma electric potential. Then the maximum number of locked particles results from the relation that the electric potential of charged particles is equal to the trap electric potential φ_o. Taking a typical particle charge to be Z (in units of electron charges), we estimate the maximum number of locked particles n from the relation

$$\varphi_o \sim Zen/R_o\,,$$

which gives

$$n \sim \varphi_o R_o/(Ze)\,.$$

In reality, the number of locked charged particles exceeds this estimate, since the field of charged particles is screened as a result of displacement of plasma electrons and ions.

Problem 10.7

Consider the equilibrium of a charged particle in an electric trap that is determined by the equality of the gravitational force and the electric force of the trap.

Small particles are kept in an electric trap as a result of the equality of the above forces, and the corresponding relation has the form

$$\frac{4}{3}\pi r_o^3 \rho g = ZeE\,, \tag{10.7}$$

where r_o is the particle radius, ρ is the density of the particle material, E is the electric field strength, Z is the particle charge, and g is the free-fall acceleration. Assuming $r_o \gg \lambda$, where λ is the mean free path of gas atoms, and $r_o \gg e^2/T$, where T is the temperature of ions or electrons attaching to a small particle, we have according to formula (7.7) that $Z \sim r_o$. Thus, formula (10.7) allows us to evaluate the radius and charge of the particle in a trap.

Along with formula (10.7), an additional condition must be fulfilled. Let us assume that as a result of a fluctuation the particle charge becomes $Z-1$. Then at this point relation (10.7) is violated, and the particle drops under the action of gravity. This drop will stop if the electric field strength increases when we move down. In this case the particle will be stopped at a lower point, and when its charge becomes the initial one, it returns to the initial point. From this it follows that equilibrium occurs if

$$\frac{dE}{dz} > 0\,, \quad E < 0\,, \tag{10.8}$$

where the z axis is directed up from the Earth's surface. This result is demonstrated in Figure 10.3, where the region of possible capture of a small particle by an electric trap is marked.

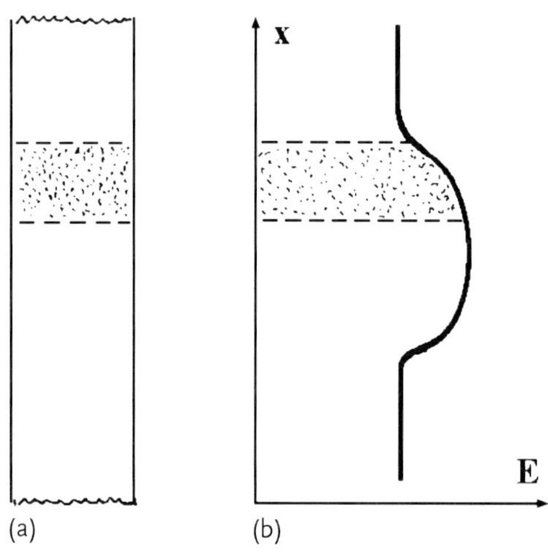

Fig. 10.3 Gas glow discharge in a vertical discharge tube with an electric trap. (a) Tube position; (b) distribution of the electric field strength along the tube. Particles can be captured in the hatched discharge region.

Problem 10.8

Small particles are locked in the positive column of glow discharge. These particles are a sink for electrons and ions of the gas discharge plasma. To compensate the loss of electrons and ions, the electric field strength increases in a region where small particles are located, so additional electrons and ions are formed in the plasma. Determine a typical value of the increase in the electric field strength due to this effect.

The equilibrium electric field strength of the positive column of glow discharge follows from the equality of the number of charged particles (electrons and ions) that are transported to the walls and the corresponding number of electrons and ions resulting from the ionization of atoms by electron impact. For gas discharge of low power, where heat release is low and the gas temperature is constant over the cross section of the discharge tube, the above equality has the form of the Schottky condition (10.3):

$$\nu_{\text{ion}} = 5.8 \frac{D_{\text{a}}}{R_{\text{o}}^2}\,, \tag{10.9}$$

where the ionization frequency is given by $\nu_{\text{ion}} = N k_{\text{ion}}$ (N is the number density of atoms, k_{ion} is the rate constant of ionization of an atom by electron impact, which depends on the electric field strength E of the positive column), R_{o} is the discharge tube radius, and D_{a} is the ambipolar diffusion coefficient for ions and electrons.

The presence of small particles in the positive column leads to an increase in the loss of charged particles and must be compensated by an increase in the ionization rate of atoms in the positive column. The increase in the electric field strength of the discharge leads to the following additional number of charged particles (electrons and ions) produced per unit time:

$$I = e \int N_e \frac{d\nu_{ion}}{dE}(E - E_o) 2\pi\rho d\rho dz \,,$$

where E_o is the electric field strength of the positive column without small particles. On the other hand, the number of electrons and ions that are lost per unit time as a result of attachment to small particles is given by

$$I = 4\pi K_i Z e^2 n \int N_e \cdot \frac{2\pi\rho d\rho}{\pi R_o^2} \,,$$

where we use the Langevin formula (6.45) for the rate constant of attachment of ions to a small particle, Z is the mean charge of a small particle, n is the number of small particles located in the positive column, and the average is taken in this equation over positions of small particles in the cross section of the discharge tube. Equalizing the above expressions, we obtain the following relation:

$$Zn = \frac{d\nu_{ion}}{dE} \frac{\Delta\varphi R_o^2}{4K_i} \,, \tag{10.10}$$

where $\Delta\varphi = \int (E - E_o) dz$ is the increase of the discharge voltage.

Problem 10.9

A gas cylinder discharge tube has a vertical axis such that the anode is located above the cathode. Several small particles that are negatively charged are located in some layer of the positive column of gas discharge. Find the character of the distribution of positive charges near the particles.

Because small particles are locked in some region of gas discharge, they are captured by an electric trap. This means that the electric field strength increases as we move down from the particles. The introduction of small charged particles in the plasma changes the distribution of the charge in this plasma. Below we analyze this effect. The electric field strength in the plasma satisfies the Poisson equation

$$\operatorname{div} E = \frac{\partial E}{\partial z} = 4\pi e(N_+ - N_-) \,,$$

where N_+, N_- are the number densities of positively and negatively charged plasma particles. From the Poisson equation it follows that insertion of small charged particles into a plasma leads to a variation in the electric field strength by the value $\Delta E = 4\pi Zen/(\pi R_o^2) = 4Zen/R_o^2$, where we average over the positions of the small particles in the cross section of the discharge tube. To return to the initial distribution of the electric field far from the small particles, it is necessary to introduce an additional positive charge into the plasma that compensates the negative charge of the small particles. If this charge is in the form of a positive layer located at a distance l from the small particles, it leads to a jump in the discharge electric voltage as a result of the introduction of the small charged particles into the plasma

$$\Delta\varphi = \frac{4Zenl}{R_o^2} . \tag{10.11}$$

Comparing this with formula (10.10), one can find this distance

$$l = \frac{\Delta\varphi\rho_o^2}{4Zen} = \frac{K_i}{\frac{d v_{ion}}{d E}} . \tag{10.12}$$

Note that the accuracy of the positions of the layers in the above consideration is on the order of the distance between nearest small particles. The above assumption of the uniform distribution of charges over a layer is valid within this accuracy.

Problem 10.10

Ascertain the possibility of small particles located in the positive column of gas discharge being locked in an electric trap resulting from interaction of these particles with the discharge plasma. Particles are located in a cylindrical discharge tube where the cathode is located below.

Let us take a simple model of the discharge under consideration, such that the cathode is a positively charged plate and the anode is a negatively charged plate. Then there is an electric field between the electrodes that compels ions to move down to the anode, and electrons cannot move up to the cathode. An additional electric field due to the electric trap in the gas discharge compensates the loss of plasma electrons and ions due to their absorption by the small particles. At the same time, a new electric field also arises under the action of a layer of small negatively charged particles. This additional field is strong near the layer of locked small particles and falls off far from this layer owing to shielding by plasma electrons and ions.

Because of symmetry, the total electric field has a maximum in a region of the particle layer. Since the total electric field increases in a region of this layer, within the framework of the model under consideration we get that the layer of positive ions is located higher than the layer of small particles. This means that the variation of the discharge electric field resulting from interaction of small particles with a gas discharge plasma takes place at a position higher than where these particles are located.

The additional electric field due to the charge of the locked small particles changes sign during the transition through the layer of charged particles. Therefore, according to criterion (7.2), the self-consistent field of the layer, which occurs in a uniform gas discharge, cannot keep small particles of this charge in the plasma of the positive column, though this field can create a trap for particles of another charge. Hence, to keep small particles in a gas discharge, it is necessary to have an initial electric field configuration that creates a trap for the small particles.

Problem 10.11

Analyze the shielding of the field of a small negatively charged particle by ions and electrons of a plasma of a gas glow discharge.

It is known that the spatial distribution of electrons and ions in a plasma is such that they screen plasma electric fields. First we analyze this shielding in a plasma for the field of a charged plasma particle. The number density of electrons N_e and ions N_i of the plasma are given by the Boltzmann law:

$$N_e(R) = N_o \exp\left(\frac{e\varphi}{T_e}\right), \quad N_i(R) = N_o \exp\left(-\frac{e\varphi}{T}\right), \tag{10.13}$$

where N_o is the number density of electrons or ions of the plasma far from the test particle, $\varphi = -Ze/R$ is the electric potential of the charged particle at a distance R from it, and T, T_e are the ion and effective electron temperatures, respectively. Since the energy distribution function of electrons differs from the Maxwell distribution in glow discharge, we introduce the effective electron temperature as $T_e = eD_e/K_e$, where D_e, K_e are, respectively, the diffusion coefficient and mobility of electrons in the gas of glow discharge.

The electric field strength of a test plasma particle satisfies the Poisson equation

$$\operatorname{div}\mathbf{E} = -\Delta\varphi = 4\pi e(N_i - N_e) ,$$

where φ is the electric potential of the test plasma particle. For small values of φ one can expand expressions (10.13) over a small parameter, and this equation takes the form

$$\Delta\varphi = \varphi/r_D^2 ,$$

where the Debye–Hückel radius for plasma particles is given by [155]

$$r_D = \left[4\pi N_o e^2\left(\frac{1}{T_i} + \frac{1}{T_e}\right)\right]^{-1/2} . \tag{10.14}$$

In particular, the electric potential of a plasma particle of charge Z, which is $\varphi = Z/R$ in a vacuum, in a plasma has the form

$$\varphi = \frac{Ze}{R}\exp\left(-\frac{R}{r_D}\right) .$$

Now let us evaluate the electric field of a small charged particle in a plasma of glow discharge. The difference compared to the field of plasma particles consists in the absorption of plasma electrons and ions by the small particle. This process can violate the Boltzmann distribution (10.13) near the particle for plasma electrons and ions. Hence, let us determine the distribution of plasma electrons and ions near the particle.

Under these conditions, electrons and ions transfer their charge to the charged particle as a result of contact with it. The distribution of plasma particles near the charged particle in this case is considered in Problem 7.1 . In particular, the number density of positive ions of a plasma in the field of a negatively charged particle of charge Z at a distance R from the particle follows from the expression for the ion current toward the surface of the charged particle, which has the form

$$I_{\mathrm{i}} = -4\pi R^2 e D_{\mathrm{i}} \left(\frac{dN_{\mathrm{i}}}{dR} + \frac{Ze^2}{TR^2} N_{\mathrm{i}} \right) .$$

Accounting for $J_{\mathrm{i}} = \mathrm{const}$ in space, that is, the absorption of electrons and ions is absent in the space outside the particle, one can consider this relation to be an equation for the ion number density. Solving this equation with the boundary condition $N_{\mathrm{i}}(r_{\mathrm{o}}) = 0$, where r_{o} is the particle radius, we obtain

$$N_{\mathrm{i}}(R) = N_{\mathrm{o}} \frac{1 - \exp\left(-\frac{Ze^2}{Tr_{\mathrm{o}}} + \frac{Ze^2}{TR}\right)}{1 - \exp\left(-\frac{Ze^2}{Tr_{\mathrm{o}}}\right)} ,$$

where N_{o} is the number density of ions far from the particle. From formula (7.7) it follows that $Ze^2/(Tr_{\mathrm{o}}) \gg 1$, so from this equation for the number density of ions we have $N_{\mathrm{i}}(R) = 0$ near the particle and $N_{\mathrm{i}}(R) = N_{\mathrm{o}}$ far from it. The transition region is found close to the particle's surface, $Ze^2(R - r_{\mathrm{o}})/r_{\mathrm{o}}^2 \sim T$, that is, where $R - r_{\mathrm{o}} \ll r_{\mathrm{o}}$. Thus, absorption of positive ions by a small particle leads to a change in the spatial distribution of positive ions near the small particle compared with the Boltzmann distribution (10.13). As for electrons, their absorption by a small particle is relatively small owing to the high mobility of electrons, so that the electron distribution in the field of a small charged particle is given by the Boltzmann equation. As a result, the Poisson equation for the particle potential in a plasma has the form

$$\Delta\varphi = 4\pi N_{\mathrm{o}} e \left[\exp\left(\frac{e\varphi}{T_{\mathrm{e}}}\right) - 1 \right] , \tag{10.15}$$

and near the particle its electric potential is given by

$$\varphi = -\frac{Ze}{R} .$$

The solution of this equation in the region $e|\varphi| \ll T_{\mathrm{e}}$ has the form

$$\varphi = -\frac{Ze}{R} \exp\left(-\frac{R}{R_{\mathrm{D}}}\right) , \quad R_{\mathrm{D}} = \sqrt{\frac{4\pi N_{\mathrm{o}} e^2}{T_{\mathrm{e}}}} . \tag{10.16}$$

From this formula it follows that the shielding of the field of a small charged particle by plasma electrons and ions is similar to the shielding of fields of charged plasma particles. But the Debye–Hückel radius R_{D} for the field of a small particle (10.16) differs from that of charged plasma particles because of absorption of positive ions by the small particle.

Problem 10.12

Analyze the behavior of the field of a small charged particle in the plasma of glow discharge depending on the plasma's parameters.

It follows from the analysis of the previous problem that the absorption of positive ions by a small negatively charged particle located in the plasma of glow gas discharge leads to the Poisson equation in the form (10.15). The solution of this equation is as follows:

$$\varphi = -\frac{Ze}{R}\exp(-R/R_D)\,, \quad e|\varphi| \ll T_e\,,$$
$$\varphi = -\frac{Ze}{R}\left[1 - R^3/\left(2r_W^3\right)\right]\,, \quad R \ll r_W\,, \quad \frac{Ze^2}{R} \gg T_e\,, \tag{10.17}$$

where $r_W = [3Z/(4\pi N_o)]^{1/3}$ is the Wigner–Seitz radius.

As is seen, the behavior of the particle's electric potential is determined by the parameter $Ze^2/(R_D T_e)$. If this parameter is small, Debye shielding of the particle's electric field takes place. This shielding is similar to that for fields of electrons and ions in a plasma, but the expressions for the Debye–Hückel radii (10.14) and (10.16) are different because of the absorption of positive ions by a small particle. In the other limiting case, $Ze^2/(R_D T_e) \gg 1$, the particle's field is screened by positive ions of a plasma at distances on the order of $Z^{1/3} N_o^{-1/3}$ in accordance with formula (7.11). Let us consider a typical example of an argon plasma of glow discharge with parameters $r_o = 1\mu\text{m}$, $T = 400\,\text{K}$, $T_e = 4\,\text{eV}$, $N_o = 1 \cdot 10^{10}\,\text{cm}^{-3}$. For this example we have $Z = 2 \cdot 10^5$, $r_D = 10\mu\text{m}$, $R_D = 100\mu\text{m}$, and $r_W = 200\mu\text{m}$, $Ze^2/(R_D T_e) = 1$.

Thus, we have two parameters of the size dimensionality in the case of interaction of a small particle with a plasma. Small particles become negatively charged in the plasma because the electron mobility exceeds remarkably that of the positive ions. The action of the particle field leads to a redistribution of electrons and ions near the particle, so that far from the particle its field is shielded. Two parameters of the size dimensionality characterize this phenomenon; the first one is the Debye–Hückel radius R_D, and the second one is the Wigner–Seitz radius $r_W = [3Z/(4\pi N_o)]^{1/3}$. In the region where the particle's electric potential is small $e|\varphi| \ll T_e$, electrons do not partake in shielding the particle field, and because the number density of ions is constant over the region, the electric field at a distance R from the particle is, according to the Gauss theorem, given by

$$E(R) = \frac{Ze}{R^2}\left(\frac{R^3}{r_W^3} - 1\right)\,.$$

Let us call a plasma near a small particle ideal if

$$\frac{r_W}{R_D} = (36\pi)^{1/6}\left(\frac{Z^2 e^6 N_o}{T_e^3}\right)^{1/6} = 2\left(\frac{Z^2 e^6 N_o}{T_e^3}\right)^{1/6} \ll 1\,.$$

If we return to the above example of a particle in an argon plasma, we can conclude that a real plasma near a small particle can be both ideal and nonideal (or a plasma with a strong coupling [214]).

Problem 10.13

Determine the increase in the temperature of a small particle captured by a gas discharge plasma as the result of a current of ions and electrons toward this small charged particle.

Assume that each recombination of an ion and an electron on the surface of a small particle leads to the release of energy I, the ionization potential of the atoms of the discharge gas. Then, on the basis of the Langevin formula (7.5) for charge transfer from a positive ion to a small particle, we have the following heat balance equation in the gas near the particle:

$$-\kappa \nabla T \cdot 4\pi R^2 = J_{\rm i} I/e = 4\pi Z e K_{\rm i} I N_{\rm i} ,$$

where $N_{\rm i}$ is the ion number density far from the particle. The left-hand side of this equation is the total heat power transported a distance R from the particle, and the right-hand side is the power absorbed by the particle. Above we accounted for the equality of the electron and ion currents toward the particle. We assume that the particle is spherical, and we take its average charge to be Ze and its radius to be $r_{\rm o}$. Let us introduce the function $S = \int_{T_{\rm o}}^{T} \kappa \, dT$, where $T_{\rm o}$ is the plasma temperature far from the particle. Taking the temperature dependence of the thermal conductivity coefficient in the form $\kappa(T) \sim T^{\gamma}$, we get

$$S(T) = \frac{T\kappa(T) - T_{\rm o}\kappa(T_{\rm o})}{1+\gamma} .$$

Solving the above equation, we obtain

$$S(T) = Z e K_{\rm i} N_{\rm i} I/R . \tag{10.18}$$

Formula (10.16) gives the temperature distribution near the particle. In particular, from this we obtain for the particle temperature $T_{\rm p}$

$$S(T_{\rm p}) = Z e K_{\rm i} N_{\rm i} I/r_{\rm o} . \tag{10.19}$$

Let us consider an example of an argon plasma ($I = 16\,{\rm eV}$) of glow discharge with typical values of the number densities of atoms $N = 1 \cdot 10^{17}\,{\rm cm}^{-3}$ and charged atomic particles $N_{\rm i} = 1 \cdot 10^{10}\,{\rm cm}^{-3}$. Then we have $K_{\rm i} = 500\,{\rm cm}^2/({\rm V \cdot s})$, $Z/r_{\rm o} = 2 \cdot 10^9 {\rm cm}^{-1}$, and formula (10.19) gives $S = 4 \cdot 10^{-3}\,{\rm W/cm}$, which corresponds to heating the particle by approximately 10 K.

10.3
Structures of Particles in Dusty Plasma

When charged clusters or small particles are captured by traps of gas discharge, they are located together in a plasma of this trap. The interaction of these particles with ions and electrons of the plasma and with the walls of the trap leads to the formation of structures in the positions of small particles in the trap. The structure of captured particles is called a plasma crystal. This crystal is stable and can be considered as a specific physical object. It has properties of common crystals, and, like them, the plasma crystal can be melted or it can decay as a result of the action of an external field, but these processes proceed in a special way.

Problem 10.14

Analyze the structure of the positions of charged particles located in an electric trap of gas discharge.

Charged particles captured by an electric trap of a gas discharge can form structures. We assume the electric trap of the gas discharge to have cylindrical symmetry when the gas discharge is located in a cylindrical tube. In addition, the trap has mirror symmetry with respect to its central plane, which is perpendicular to the trap's axis.

The positions of particles in layers depend on the character of the interaction of the particles. First we consider the case where Coulomb repulsion acts between particles, and all the particles have an identical charge. Let us consider a layer of these particles. From the symmetry consideration, this layer is located in a plane perpendicular to the discharge axis. The electric field of this layer of charged particles in the plane of the layer is directed along the radius that connects a given point and the layer center. Hence, in the case of Coulomb interaction of particles, they are located on circles, and within the limits of one circle they form regular polygons. Evidently, the same result for the positions of charged particles relates to the case of several layers of particles.

If the Debye–Hückel radius of the gas discharge plasma is smaller than the distance between the nearest particles, they can form a different structure. If we take any particle of the layer, we can see that its nearest neighbors form a regular polygon, so that the force acting on a particle from the neighboring particles is zero. The number of polygon vertices is $2k$, where k is an integer. Then the total force acting on each particle is zero. This corresponds to a regular structure consisting of regular polygons. The positions of peripheral particles are also determined by the interaction of the particles with a field of trap walls. Hence, the regular structure of particles is violated near the walls of the trap, and the boundary conditions for the particles near the walls are of importance for the symmetry of the structure that the particles form.

Problem 10.15

Small particles are captured by an electric trap of glow gas discharge and form there a horizontal layer. Estimate the average charge of particles of this layer.

The electric field strength near a layer of small charged particles is $E = 2\pi\sigma$ according to Gauss's theorem, where σ is the layer charge per unit area, that is,

$$E = \frac{2\pi Zen}{\pi R_o^2},$$

where Z is the average particle charge in units of electron charge, n is the number of particles, and R_o is the discharge tube radius. From formula (10.10) it follows that

$$E = \frac{\Delta\varphi}{2K_i} \cdot \frac{dv_{ion}}{dE}, \tag{10.20}$$

where $\Delta\varphi$ is the additional electric voltage of the electric trap that supports the particle layer. The electric field (10.20) of this layer creates an additional voltage of a trap, which we denote by $\delta\varphi$. This field separates electrons and ions near the layer. As a result, the currents of electrons and ions toward the charged particles are equal to each other.

Let us assume that ions and electrons cannot penetrate through the layer of charged particles because they are absorbed by these particles. Then, neglecting inelastic collisions of electrons with atoms, one can estimate the jump in the electric voltage near the layer on the basis of the equation

$$e\delta\varphi = T_e \ln\left(\frac{K_e}{K_i}\right), \tag{10.21}$$

where T_e is the effective electron temperature and K_e, K_i are, respectively, the mobilities of electrons and ions.

In this case ions that are absorbed by particles move to the particle layer only from the cathode, that is, these ions are located above the layer, and the distribution of the electric field along the tube's axis is given in Figure 10.4 in this case.

Problem 10.16

Small particles are captured by an electric trap of glow discharge under the conditions of the previous problem and form there a horizontal layer. Estimate the portion of the positive ions from the plasma of gas discharge that penetrate through this layer.

Above we assumed that structures of small particles in glow discharge absorb all the ions that are located near them. Then the process is characterized as follows. Positive ions move up to the cathode, so ions formed below the particle layer are absorbed by them. Positive ions that are formed above the layer go outside it.

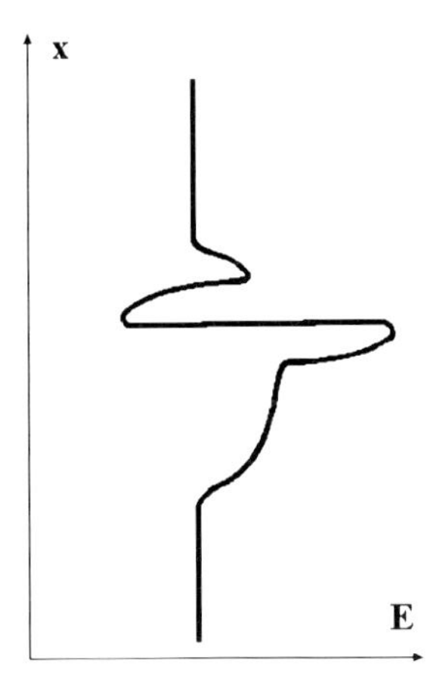

Fig. 10.4 Distribution of the electric field strength along a vertical gas discharge tube with captured small particles that are negatively charged and form a horizontal layer in the discharge tube.

Electrons penetrate partially through the particle layer, so that the electron and ion currents absorbed by the particle layer are equalized.

Let us consider this picture in detail. First, the decrease in the electric field strength due to the charged particles is limited. Evidently, the electric field strength from the layer of charged particles E must be smaller than the electric field strength E_o in the gas discharge $E < E_o$; otherwise, the layer becomes the cathode and the gas discharge disconnects above this layer. Second, the electric field strength is constant over the discharge cross section far from the layer if the distance from it exceeds by a significant amount the distance between the nearest particles of the layer. At small distances from the layer, the structure of the particle positions in the layer must be taken into account.

Below, on the basis of the above remarks, we estimate the probability that positive ions will penetrate through the net of small particles forming the layer. The current of positive ions is proportional to their drift velocity, and the ion drift velocity is proportional to the electric field strength. Then the probability P of penetration of an ion through a given point of the layer is

$$P = \frac{E_o}{\sqrt{E_o^2 + E_l^2}} ,$$

where E_o is the discharge electric field strength and E_l is the electric field created by the particles of the layer.

For simplicity, let us take the square net of particles with a distance of $2l$ between the nearest particles, and account for the interaction of a test ion with the four nearest particles of the net where these particles form a square. Using a rough estimation, we obtain the electric field strength acting on a test ion near the center of the square:

$$\mathbf{E}_l = \frac{Ze\rho}{l^3\sqrt{2}} ,$$

where ρ is the distance of a test ion in the layer from the square's center and Ze is the charge of each particle of the square. Averaging the probability of ion penetration through the layer over the circle inscribed in the square, we get

$$\langle P \rangle = \int_0^l \frac{2\pi\rho d\rho}{\pi l^2} E_o \left(E_o^2 + \frac{Z^2 e^2 \rho^2}{2l^6} \right)^{-1/2} = 2\left(\sqrt{1+\delta^2} - \delta\right) ,$$

where the parameter δ is equal to $E_o l^3 \sqrt{2}/Ze$. Let us introduce the electric field strength of the layer far from the layer, that is, at distances large compared with l. At such distances the layer structure is not essential, but we consider distances where the layer field does not screen the plasma. We have $E_l = 2\pi Ze/l^2$, which leads to the estimate

$$w \sim \frac{E_o}{E_l} , \tag{10.22}$$

for the probability that ions will penetrate through the layer of small negatively charged particles, so that this probability is comparable to unity.

Note that the assumptions applied are not valid between particles in the main region. But this leads to the conclusion that because $E_o > E_l$, the majority of positive ions can penetrate through the layer. This means that in one electric trap of gas discharge there can exist several layers of captured small particles. Thus, small particles that are captured by an electric trap in gas discharge and are charged there negatively can form space structures in the gas discharge plasma.

11
Aerosol Plasma

According to the definition, an aerosol plasma is weakly ionized atmospheric air located at various altitudes of the Earth's atmosphere with an admixture of aerosol particles – solid and liquid particles. Aerosol particles, or aerosols, include water drops, ice crystallites, and dusty particles on the order of 0.01 to 100 μm in size that exist in atmospheric air. The kind and size of aerosols in the Earth's atmosphere depend both on the atmospheric altitude and on local conditions there. In particular, water aerosols are located close to the Earth's surface, and tropospheric aerosol is formed from SO_2, NH_3, and NO_x molecules, as well as mineral salts and other compounds that result from natural processes on the Earth's surface and may reach high altitudes.

The smallest aerosols, Aitken particles [6, 7], are formed as a result of photochemical reactions at altitudes above the cloud boundary. They have a size of 10–100 nm and a number density of 10^2–$10^4\,\mathrm{cm}^{-3}$ [218]. Aitken particles scatter the Sun's radiation and are responsible for the blue color of the sky. Aitken particles partake in atmospheric chemical reactions and charge transfers from initially formed electrons and simple ions to large aerosols.

The processes involving aerosols are different at different altitudes and are responsible for the interaction of atmospheric air with solar radiation [219, 220]. Having the general properties of small particles in neutral and ionized gases [151, 152, 221], water aerosols located at altitudes of a few kilometers determine electrical phenomena in atmospheric air, including lightning [219, 220, 222]. The process of charging the Earth that is the basis of atmospheric electric processes is complicated and includes many stages. But the falling of water-charged aerosols under the action of gravitational forces is the only way to create an electric current that charges the Earth's surface. In turn, the Earth's electric field is important for the charging of aerosols, and such processes will be considered below. Aerosol particles in the Earth's atmosphere may be formed by natural processes and as a result of human activity [223].

When solid clusters are joined as a result of contact with one another in atmospheric air, porous structures can be formed in which solid clusters preserve their individuality. These structures are fractal aggregates if their joining takes place in the absence of external fields. This name results from the dependence of the matter density in fractal aggregates on their size. The specifics of the growth of fractal

Cluster Processes in Gases and Plasmas. Boris M. Smirnov

ISBN: 978-3-527-40943-3

aggregates in a buffer gas are connected to the diffusion character of their motion. This can lead to a nonlinear dependence of the nucleation rate on the aggregate number density. Because of the micron size of fractal aggregates, they interact effectively with external fields. Even small electric fields can cause effective interaction between fractal aggregates owing the interaction of induced charges. As a result of nucleation in an external electric field, so-called fractal fibers are formed, which are elongated fractal structures.

11.1 Growth and Charging of Aerosol Particles in an External Electric Field

Problem 11.1

Determine the rate of association of two neutral spherical aerosols in a uniform electric field as a result of interaction of induced dipole moments.

The action of the atmospheric electric field on aerosols is of importance for aerosol behavior. It takes place in the Earth's atmosphere and consists in the separation of charges inside aerosols, which leads to additional interactions between aerosols. Such interactions between aerosols may determine the character of association of aerosols. Below we consider this mechanism of aerosol association when the approach of aerosols results from the interaction of dipoles induced by an external electric field. Then the interaction potential of aerosols that is responsible for their approach at large distances R between them is given by

$$U(\mathbf{R}) = \frac{\left[\mathbf{D}_1\mathbf{D}_2 - 3(\mathbf{D}_1\mathbf{n})(\mathbf{D}_2\mathbf{n})\right]}{R^3} , \qquad (11.1)$$

where $\mathbf{R} = R\mathbf{n}$, and $\mathbf{n}$ is the unit vector directed along $\mathbf{R}$. The dipole moments of aerosols induced by an electric field of strength $\mathbf{E}$ are

$$\mathbf{D}_1 = \alpha_1 \mathbf{E} , \quad \mathbf{D}_2 = \alpha_2 \mathbf{E} .$$

where α_1, α_2 are the aerosol polarizabilities. Hence, the interaction potential of aerosols is

$$U = -\frac{\alpha_1\alpha_2 E^2}{R^3} \cdot (3\cos^2\theta - 1) , \qquad (11.2)$$

where θ is the angle between vectors $\mathbf{R}$ and $\mathbf{E}$. As is seen, the attraction of aerosols takes place for some angle range $0 < \theta < \arccos(1/\sqrt{3})$, and this angle range is responsible for the aerosol association. The normal F_n and F_τ tangential components of the force between aerosols owing to the interaction of induced dipole moments are

$$F_n = \frac{3\alpha_1\alpha_2 E^2}{R^4} \cdot (3\cos^2\theta - 1) , \quad F_\tau = \frac{3\alpha_1\alpha_2 E^2}{R^4} \cdot \sin 2\theta .$$

For simplicity, below we ignore the tangential force for small angles θ, which are responsible for aerosol association. Under this assumption, the angle θ does not vary during cluster approach. Then, solving the motion equations for aerosols $v_R = dR/dt = F_n(K_1 + K_2)/e$, we obtain by analogy with formula (6.48)

$$\tau = \frac{eR^5}{15\alpha_1\alpha_2 E^2(K_1 + K_2)(3\cos^2\theta - 1)} .$$

To determine the average association time we take a volume around a test aerosol in the center-of-mass frame of reference in such a way that the time of approach to the center is the same for all points of the surface. Then if R_o is the distance from this surface for $\theta = 0$, the distance for other angles $\cos\theta > 1/\sqrt{3}$ is equal to $R_o(3\cos^2\theta - 1)^{1/5}$. This elemental volume is given by

$$V = \int_0^{R_o} R^2 dR \int_{1/\sqrt{3}}^{1} 2\pi d\cos\theta \cdot [(3\cos^2\theta - 1)/2]^{3/5} = 0.518 R_o^3 .$$

Hence, in averaging the association time over the initial relative positions of the aerosols, we will use in the expression for the probability of the location of the nearest aerosol $P = \exp(-NV)$ in the volume $V = 0.518R^3$ instead of the volume $V = 4\pi R^3/3$ for an isotropic distribution of aerosols. Averaging in this manner over initial relative positions of the aerosols, we get for the average association time for the test aerosol

$$\overline{\tau} = \int \tau NV dP = \frac{0.14e}{\alpha_1\alpha_2 E^2 N^{5/3}(K_1 + K_2)} ,$$

and the effective rate constant of this process is

$$k_{\text{ind}} = \frac{1}{N\overline{\tau}} = \frac{7.1\alpha_1\alpha_2 E^2 N^{2/3}(K_1 + K_2)}{e} . \tag{11.3}$$

The dependence of this rate constant on the number density N of aerosols testifies to the nonlinear character of aerosol approach in a buffer gas. Note that formula (11.3) relates to a rare buffer gas where the mean free path λ of atoms in this gas is large compared with the radius of spherical particles r_o.

Problem 11.2

Compare the rate constants of association of two neutral clusters if the cluster approach results from their diffusion motion in a buffer gas or from the interaction of induced dipole moments under the action of an external electric field.

In comparing formulas (6.43) and (11.3) for the rate constants of cluster (or aerosol) association, we take the cluster radii to be $r_1 = r_2 = r_o$ and their mobilities to be $K_1 = K_2 = K$ in a buffer gas. Taking joining clusters to be identical, we also equalize their polarizabilities $\alpha_1 = \alpha_2 = \alpha$. The ratio of the association rate constants for the above mechanisms, accounting for the Einstein relation (6.4) for cluster mobility and the diffusion coefficient in a buffer gas, takes the form

$$\frac{k_{\text{ind}}}{k_{\text{as}}} = \frac{0.28\alpha^2 E^2 N^{2/3}}{r_o T} .$$

We simplify this expression assuming the cluster to be a metal particle, and its polarizability in accordance with formula (5.3) is $\alpha = r_o^3$. This leads to the following ratio of the association rate constants in a rare buffer gas:

$$\frac{k_{\text{ind}}}{k_{\text{as}}} = \frac{0.28 r_o^5 E^2 N^{2/3}}{T} . \tag{11.4}$$

As is seen, cluster association due to the interaction of induced electric moments is preferable at high electric field strengths and large cluster sizes.

To demonstrate the reality of the mechanism of induced association of clusters, we consider an example of water drops in atmospheric air. Take a high degree of air humidity such that the water content is 0.1 g/g (grams of water per gram of air). This corresponds to the density of water drops under typical atmospheric conditions to the water density of $1.3 \cdot 10^{-4}\,\text{g/cm}^3$, which gives $N r_o^3 = 3 \cdot 10^{-5}$. Under these conditions formula (11.4) gives the following electric field strength when the ratio of the above rate constants equals one at temperature $T = 300$ K:

$$E = \frac{E_o}{r_o^{3/2}} ,$$

where the cluster radius is expressed in microns and $E_o = 3.7\,\text{kV/cm}$. As is seen, for large water aerosols the induced mechanism of association is realistic.

Problem 11.3

Determine the rate of association of a small spherical particle with a large chain aggregate as a result of interaction of induced charges in a uniform electric field in the limit of a dense buffer gas when the mean free path λ for atomic motion is small compared with the radius of the spherical particles r_o.

Let us model the chain aggregate by a cylindrical metal particle. If its length is sufficiently large, the axis of this cylindrical particle is directed along the electric field axis (Figure 5.1). The charge distribution under the action of an external electric field along the length of a cylindrical particle and for a spherical particle of the same radius is given in Figure 11.1, when a spherical particle is located near the end of the cylindrical particle. Then contact of these particles leads to a charge redistribution, as shown in Figure 11.2. Such behavior of an induced charge is the basis for aggregation of the particles under consideration, and we will use this character of the charge distributions below.

The induced electric charge per unit length of the aggregate is $\sigma(z) = Cz$, where the z axis is directed along the aggregate axis and the origin is located at the middle of the aggregate. Taking the aggregate length to be $2l$, we have

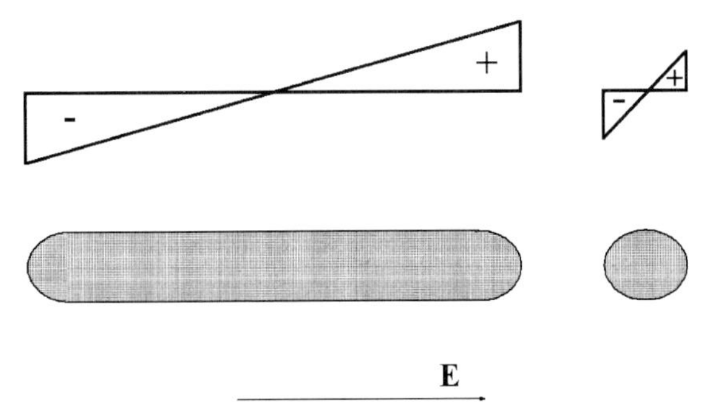

Fig. 11.1 Distribution of charges induced under the action of a uniform electric field for separated cylindrical and spherical particles.

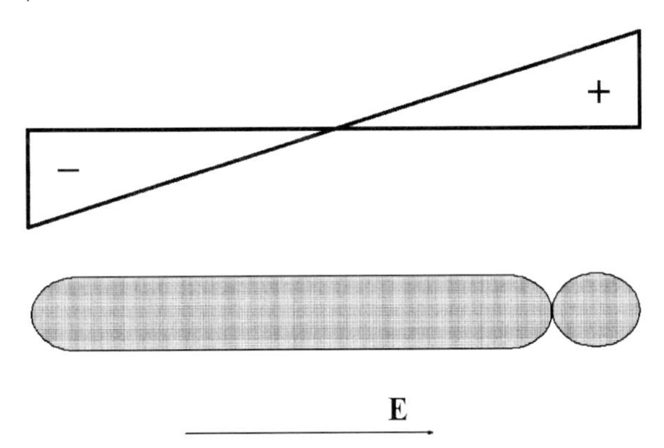

Fig. 11.2 Distribution of charges induced under the action of a uniform electric field for cylindrical and spherical particles in contact with each another.

$-l \leq z \leq l$. Hence, the aggregate dipole moment is $D_2 = \int_{-l}^{l} Cz^2 dz$, that is, $C = 3D_2/(2l^3) = 3\alpha_2 E/(2l^3)$, where the aggregate polarizability α_2 along its axis is given by formula (5.11), $\alpha_2 = l^3/[3\ln(l/r_o)]$, and r_o is the aggregate radius. The interaction potential of a charge e and a dipole moment $\mathbf{D}$ is given by $U(R) = e\mathbf{Dn}/R^2$, where R is the distance between these objects and $\mathbf{n}$ is the unit vector along $\mathbf{R}$. We have $R = \sqrt{(z-z')^2 + \rho^2}$, where z' is the charge coordinate along the polar axis and z, ρ are coordinates of the dipole. From this we obtain for the interaction potential of the induced dipole moments between cylindrical and spherical particles of an identical radius, given by formula (5.16),

$$U(R) = \int_{-l}^{l} Czdz \frac{\mathbf{D}_1\mathbf{n}}{R^2} = \frac{3D_2}{2l^3} \int_{-l}^{l} zdz \frac{\mathbf{D}_1\mathbf{n}}{R^2} .$$

In particular, in the limiting case $R \gg l$ it follows that the interaction potential of the two dipole moments is given by

$$U(R) = \frac{\mathbf{D}_1\mathbf{D}_2 - 3(\mathbf{D}_1\mathbf{n})(\mathbf{D}_2\mathbf{n})}{R^3} .$$

Let us consider the other limiting case $l \gg R$ in the attraction region near the end of a cylindrical particle, where $z > l$. Then we get the following expression for the interaction potential:

$$U(R) = \frac{3D_2}{2l^3} \int_{-l}^{l} z' dz' (\mathbf{D}_1 \mathbf{n})/R^2 = \frac{3D_1 D_2}{2l^3} \int_{-l}^{l} z'(z' - z) dz'/R^3 .$$

In this limiting case the interaction potential is determined by the region of the cylindrical particle near its end. This allows us to replace z' under the integral by l, so we get

$$U(R_o) = -\frac{3D_1 D_2}{2l^2 R_o} ,$$

where $R_o = \sqrt{z^2 + \rho^2}$ is the distance of the spherical particle from the end of the cylindrical one, and this equation is valid for $z > l$. From this for the force that acts on the spherical particle from the cylindrical one, we get

$$\mathbf{F} = -\frac{3D_1 D_2}{2l^2 R_o^2}\mathbf{n} = -\frac{3\alpha_1\alpha_2 E^2}{2l^2 R_o^2}\mathbf{n} = -\frac{r_o^3 l E^2}{2R_o^2 \ln(l/r)}\mathbf{n} .$$

Now we use the method used for the derivation of formulas (6.48) and (11.3). Take the cylindrical particle to be motionless. Then the velocity v of the motion of the spherical particle to the end of the cylindrical particle follows from the equation

$$F = 6\pi r_o v\eta = -\frac{r_o^3 l E^2}{2R_o^2 \ln(l/r_o)} ,$$

so for the velocity of a spherical particle we have

$$v = \frac{dR_o}{dt} = \frac{F}{6\pi\eta r_o} = \frac{r_o^2 l E^2}{12\pi\eta R_o^2 \ln(l/r_o)} .$$

The solution of this equation gives

$$\tau = \frac{4\pi\eta R_o^3 \ln(l/r_o)}{r_o^2 l E^2} , \tag{11.5}$$

for a time of attachment of the spherical particle to the end of the cylindrical particle if the initial distance from this particle is R_o. Averaging this time on the basis of the averaging method used in the derivation of formulas (6.50) and (11.3), we get by using the probability of a given distance of the nearest spherical particle from the end of the cylindrical one

$$\overline{\tau} = \frac{3\eta \ln(l/r_o)}{r_o^2 l E^2 N} ,$$

where N is the number density of spherical clusters. This expression is valid if the following criterion is fulfilled:

$$Nl^3 \gg 1 ,$$

and gives for the rate constant of cluster association

$$k_{as} = \frac{1}{N\overline{\tau}} = \frac{r_o^2 l E^2 N}{3\eta \ln(l/r_o)} . \tag{11.6}$$

Problem 11.4

Find the criterion when the growth of chain aggregates dominates in the growth of spherical clusters located in a buffer gas in an external electric field and aggregation results from the interaction of induced moments of associating clusters.

Let us analyze the association of spherical clusters and chain aggregates in a gas in a strong electric field from the standpoint of spherical clusters. Comparing the rate constants of association in accordance with formulas (11.3) and (11.6) in the diffusion regime of atom–cluster collisions, we find that the attachment of spherical clusters to chain aggregates proceeds more effectively than the joining of spherical clusters if the following criterion holds true:

$$N_{cyl} \gg \frac{N^{5/3} r_o^3}{l} \ln\left(\frac{l}{r_o}\right) ,$$

where N is the number density of spherical clusters, N_{cyl} is the number density of chain aggregates, and $N r_o^3 \ll 1$. From this one can find the number density of cylindrical particles when the rate of attachment of a test spherical cluster to the ends of cylindrical particles exceeds remarkably that for association with a spherical cluster.

Problem 11.5

Determine the rate of association of a small spherical particle with a large chain aggregate as a result of interaction of induced charges in a uniform electric field.

Modeling the chain aggregate by a cylindrical metal particle, we use formula (11.6) for the attachment of a spherical particle to a cylindrical metal particle of the same radius r_o when charges are induced on the ends of the cylindrical particle under the action of an electric field. Then the equation of growth of a cylindrical particle has the form

$$\frac{dl}{dt} \sim r_o k_{as} N \sim \frac{r_o^2 l E^2 N^2}{\eta \ln(l/r_o)} ,$$

because the attachment of each spherical particle to the end of the cylindrical one increases its length by $2r_o$. Note that the rate of growth for a cylindrical particle is independent of its length.

Let us compare this with a time of association of spherical particles in an external electric field due to the interaction of induced charges. We then use formula (11.3) for the rate constant of association of spherical particles in an external electric field and use expression (6.8) for the particle mobility in formula (11.3). Then we have for the typical time of association of spherical particles

$$\overline{\tau} = \frac{1.4\eta}{r_o^5 E^2 N^{5/3}} .$$

From this we have the criterion for which spherical particles attach to cylindrical ones:

$$N_{cyl} \gg \frac{N^{5/3} r_o^3}{l} \ln\left(\frac{l}{r_o}\right) ,$$

where N and N_{cyl} are, respectively, the number densities of spherical and cylindrical particles. Since spherical particles form a gas in a buffer gas, we have $N r_o^3 \ll 1$. One can see that at the corresponding number density of cylindrical particles the rate of attachment of a test spherical particle to the ends of cylindrical particles exceeds remarkably that of its association with a spherical particle. A typical time of growth of a cylindrical particle in a gas of spherical ones is determined by formula (11.5). From the equation $dl/dt \sim r_o/\overline{\tau}$ we estimate a typical time τ_l of the growth of a cylindrical particle with an accuracy up to logarithmic terms to be

$$\tau_l \sim \frac{\eta}{r_o^3 E^2 N} .$$

Note that this time does not depend on the length l of the cylindrical particle.

Problem 11.6

The charging of two aerosols results from their collision in an electric field. Determine a range of particle sizes and fields where this charging mechanism holds true.

When two neutral spherical particles with radius r_o are located in an external electric field with strength E, they accept an induced dipole moment. If they have contact with each other such that the axis joining the centers of the particles is directed along the field, one of these particle obtains a negative charge, and the other becomes positively charged. Then the charge of each particle is estimated as

$$q \sim \frac{\alpha E}{r_o} \sim E r_o^2 .$$

The removal of the charged particles requires the condition that the energy of the Coulomb attraction of the particles is comparable to a thermal energy of the particles (proportional to T), that is,

$$U \sim \frac{q^2}{r_o} \sim E^2 r_o^3 \sim T .$$

The removal of charged particles requires a restricted electric field strength

$$E \leq E_{cr} , \quad E_{cr}^2 = \frac{T}{r_o^3} . \tag{11.7}$$

In particular, at room temperature, $T = 300$ K, and for a particle radius of $r_o = 1\ \mu$m, the critical electric field strength E_{cr} according to formula (11.7) is $E_{cr} \sim 60$ V/cm. The particle charge is given by

$$q \sim E r_o^2 \sim \sqrt{T r_o}\,. \tag{11.8}$$

For this numerical case with a particle radius of $r_o = 1\ \mu$m and a critical electric field strength of E_{cr} the particle charge is $q \sim 4e$.

Problem 11.7

Determine the minimum particle size when the above charging mechanism is possible, so that spherical particles collide in an external electric field and conserve an induced charge in the course of removal due to thermal energy.

This mechanism of particle charging as a result of collision of particles in an external electric field holds true under the condition

$$q \geq e\,,$$

where e is the electron charge. Thus, we have the following relations in this limiting case:

$$T \sim E_{cr}^2 r_o^3\,, \quad e \sim E_{cr} r_o^2\,.$$

From this it follows that

$$r_o \sim \frac{e^2}{T}\,, \quad E_{cr} \sim \frac{e}{r_o^2} \sim \frac{T^2}{e^3}\,.$$

These formulas give the minimum particle size $r_o \sim 0.06\ \mu$m, and the critical electric field strength $E_{cr} \sim 4$ kV/cm corresponds to this particle size.

Problem 11.8

After the joining of a small spherical particle and a cylindrical particle (Figure 11.2), the induced charge of the cylindrical particle is transferred partially to the spherical particle. Estimate the charge of the spherical particle in an electric field assuming the radii of the spherical and cylindrical particles to be equal.

The charge per unit length induced on a cylindrical particle according to formula (5.10) is

$$\frac{dq}{dz} = \frac{Ez}{2\ln\left(\frac{l}{r_o}\right)}\,.$$

We assume the cylindrical axis to be directed along the electric field of an electric field strength E; $2l$ is the particle length and z is the coordinate along the cylinder with the origin at its middle. When a neutral spherical particle attaches to an end of the cylindrical one, the induced charge flows partially to the spherical particle. Then the charge q_o of the spherical particle is estimated as

$$q_o \sim \frac{Elr_o}{\ln\left(\frac{l}{r_o}\right)}, \quad l \gg r_o. \tag{11.9}$$

Let us apply these formulas to an atmospheric cloud, taking $E \sim 10^3$ V/cm, $r_o \sim 5$ μm, and $l \sim 50$ μm. Then we obtain

$$q_o \sim 5 \cdot 10^4 e.$$

This mechanism may be responsible for the formation of clouds at altitudes of several kilometers, where the temperature is close to 0 °C. Then cylindrical ice particles may exist along with spherical liquid water drops.

Problem 11.9

Estimate the interaction potential between a cylindrical particle and a spherical particle of the same radius that attaches to the cylindrical one.

The interaction potential between the cylindrical and spherical particles is estimated as

$$U = \int dz \frac{q_o \frac{dq}{dz}}{l - z} \sim Elq_o \sim \frac{E^2 l^2 r_o}{\ln\left(\frac{l}{r_o}\right)}.$$

For the above numerical case, which is typical for an atmospheric cloud, $E \sim 10^3$ V/cm, $r_o \sim 5$ μm, $l \sim 50$ μm, we obtain $U \sim 10^5$ eV. Let us suppose that binding between the cylindrical particle (ice) and the spherical particle (water drop) proceeds through the breaking of bonds between molecules of the solid and liquid particles. Taking the energy of breaking of one bond to be approximately 1 eV, we find that the spherical particle is removed if the number of contacts between particles is approximately 10^5.

Problem 11.10

Compare typical charges of a spherical particle when it is located in an ionized gas or has acquired a charge as a result of collision with a cylindrical particle in an external electric field.

The charge of a spherical particle with radius r_o located in an ionized gas is given by formula (7.7):

$$Z \sim \frac{r_o T}{e^2},$$

and the charge sign corresponds to electrons or ions with greater mobility in a gas.

In the case of charging in a collision with a cylindrical particle in an electric field, a spherical particle may obtain a charge of a different sign depending on which end of the cylindrical particle it attaches. The charge of the spherical particle after the collision expressed in electron charges e is estimated on the basis of formula (11.9):

$$Z = \frac{q_o}{e} \sim \frac{E l r_o}{e \ln(l/r_o)} .$$

From this it follows that collision with a cylindrical particle in an electric field leads to a larger charge if the potential of the electric field over the particle length exceeds significantly a typical thermal energy

$$e E l \gg T \ln(l/r_o) .$$

11.2 Electrical Processes in Aerosol Plasma

An aerosol plasma consists of negative and positive ions of atmospheric air and aerosols, liquid drops, or micron-sized dust particles that can carry a positive or a negative charge. The behavior of this plasma determines some electrical properties of the Earth's atmosphere and processes in it. Negative and positive ions of the atmosphere can attach to aerosols, and aerosols acquire a charge whose sign depends on the types of ions in the atmosphere. Next, falling down under the action of the gravitational force, charged aerosols create an electric current in the atmosphere, and this electric current charges clouds. Therefore, the transport of charged aerosols in the atmosphere is the basis of atmospheric electric phenomena.

For these processes it is important that the Earth's surface is charged negatively, and we use the following average parameters. The electric field strength near the Earth is $E = 130\,\mathrm{V/m}$ and the Earth's charge is $Q = 5.8 \cdot 10^5\,\mathrm{C}$ (or the Earth's electric potential is $U = 300\,\mathrm{kV}$) [224, 225]. One can model the Earth as a capacitor, one of whose electrodes is its surface, and the other is a sphere located at a distance of 6 km from the Earth's surface. From this we conclude that all the electric processes of the Earth's atmosphere take place at altitudes below approximately 10 km, that is, they proceed in lower atmosphere layers.

One can be based on the following simple scheme of the electric machine of the Earth's atmosphere [226]. Discharging of the Earth results from currents of positive and negative atmospheric ions that are formed under the action of cosmic radiation and radioactivity from the Earth's surface. Charging of the Earth's surface starts from clouds where the positive and negative charges are separated, and a lower cloud edge has a high electric potential with respect to the Earth's surface, 20–100 MV [227]. Subsequent lightning activity between the lower cloud edge and the Earth's surface leads to charging of the Earth (Figure 11.3) [228].

Note that the character of charge separation in clouds as a result of falling of water aerosols depends on the sort of atmospheric ions. That may be different depending on small admixtures to atmospheric air in a given region. Therefore, the lower cloud edge may be charged both negatively and positively. Nevertheless, the negative charge of the lower cloud edge is preferable. In reality, the number of lightning discharges from clouds to the ground with transfer of the negative

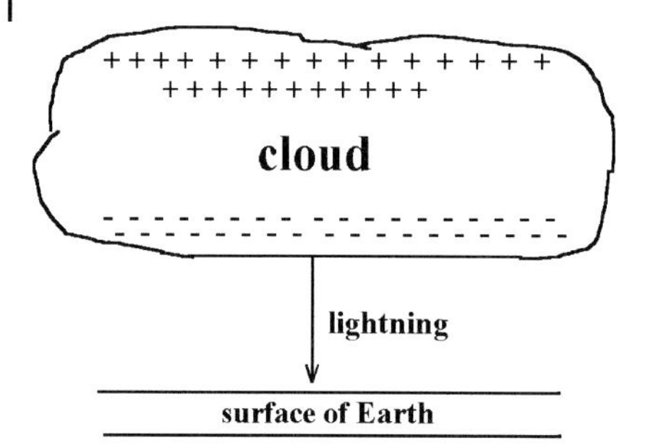

Fig. 11.3 Character of charging of the Earth in a cloud when charge separation results from falling negatively charged water drops and charge transfer from a cloud to the Earth happens through lightning [228, 237].

charge to the Earth, exceeds the number with a positive charge by 2.1 ± 0.5 times on average [229]. The ratio of transferred negative charge to the transferred positive charge in this manner is 3.2 ± 1.2 on average [229]. In addition, there are 100 negative flashes per second over the entire planet on average [227].

Problem 11.11

Determine the average charge of small particles in a weakly ionized gas if the particle number density is high enough, so that small particles collect a significant part of the plasma charge. Assume a typical particle radius to be large compared with the mean free path of ions and take the diffusion coefficients of positive and negative ions to be nearby.

We assume a large size of aerosols compared with the mean free path of ions in air. Because the mean free path of ions in air at atmospheric pressure is $\lambda \sim 0.1\,\mu$m, this corresponds to the following criterion for an aerosol particle radius r_o:

$$r_o \gg 0.1\,\mu\text{m}\,.$$

Note that the relation between the diffusion coefficient of positive D_+ and negative D_- atmospheric ions may be arbitrary depending on the ion types, which in turn depend on the presence of admixtures in the atmosphere. Below we take for definiteness $D_+ > D_-$, so that aerosol particles obtain a positive charge as a result of ion attachment. Then repeating the deduction of formula (7.7) and accounting for the difference in the density of positive N_+ and negative N_- ions, we obtain from the Fuks formula (7.2) for the charge of aerosol particles instead of formula (7.7)

$$\exp\left(\frac{Ze^2}{r_o T}\right) = \frac{D_+ N_+}{D_- N_-}\,.$$

To this relation must be added the condition of quasineutrality of the atmospheric plasma, which gives

$$Z N_p + N_+ = N_-\,,$$

where N_p is the number density of aerosol particles.

Usually the values of the diffusion coefficients D_+ and D_- are close, so that $Ze^2/(r_o T)$ is a small parameter. Expanding the above relation over this small parameter, we obtain for the equilibrium aerosol particle charge

$$Z = \frac{(D_+/D_-) - 1}{e^2/(Tr_o) + N_p/N_-} .$$

From this it follows that the charge of aerosol particles does not influence the plasma charge if

$$N_p \sim \frac{N_- e^2}{Tr_o} \ll N_- .$$

In the other limiting case, we have the following charge of an aerosol particle:

$$Z = \frac{N_-}{N_p} \cdot \left(\frac{D_+}{D_-} - 1\right) . \qquad (11.10)$$

This formula does not require closeness of D_- and D_+ because $Ze^2/(Tr_o) \ll 1$ for $D_- \sim D_+$. This allows one to expand the initial relation over a small parameter.

Problem 11.12

Determine the average charge of a particle in a rare weakly ionized gas if the particle number density is high enough such that small particles collect a significant part of the plasma charge. Assume a typical particle radius to be small compared with the mean free path of ions.

Taking the relation between ion masses $m_+ > m_-$, we get that the aerosol particles have a negative charge $-q = -|Z|e$. On the basis of formula (7.35) we obtain for this charge

$$(1 + x)\sqrt{m_-} N_+ = e^{-x}\sqrt{m_+} N_- ; \quad x = \frac{qe}{r_o T} = \frac{Ze^2}{r_o T}$$

or

$$x = \frac{Ze^2}{r_o T} = \ln\left(\frac{N_-}{N_+}\sqrt{\frac{m_+}{m_-}}\right) . \qquad (11.11)$$

This formula, together with the plasma quasineutrality condition

$$ZN_p + N_+ = N_- ,$$

allows us to determine the average charge of particles and to compare the total particle charge with that of the positive ions. Note that in this case the influence of the particles on the plasma charge is weaker than under the conditions of the previous problem.

Problem 11.13

Aerosol particles in the Earth's atmosphere catch ions that recombine on their surface, that is, the attachment of ions to aerosol particles leads to the decay of atmospheric plasma. Estimate a typical number density of aerosol particles when the rate of ion recombination resulting from ion attachment to aerosol particles is comparable to the rate of pairwise recombination of ions in a given space.

Atmospheric ions result from the action of cosmic rays and radioactivity from the Earth's surface. Subsequently positive and negative ions in the Earth's atmosphere perish as a result of pairwise recombination and attachment of ions to aerosol particles. The balance of ions due to these processes is described by the following balance equations for the number density of positive N_+ and negative N_- ions of atmospheric plasma:

$$\frac{dN_+}{dt} = G - \alpha N_+ N_- - \frac{I_+ N_p}{e} , \quad \frac{dN_-}{dt} = G - \alpha N_+ N_- \frac{I_- N_p}{e} , \tag{11.12}$$

where G is the ionization rate by an external source, α is the ion–ion recombination coefficient, N_p is the particle number density, and I_+, I_- are ion currents directed toward the aerosol particle that are given by the Fuks formula (7.2). In the case $|Z| < r_o T/e^2$, the ion current is described by the Smoluchowski formula (4.22). Then $N_+ = N_- = N_i$, and the balance equation in the stationary case has the form

$$G - \alpha N_i^2 - k N_i N_p = 0 , \tag{11.13}$$

where $k = 4\pi D r_o$, and $D = (D_+ + D_-)/2$ is the average diffusion coefficient of ions in air. This formula gives that the recombination of ions on the surface of aerosol particles is stronger than in a given space under the following condition:

$$N_p \gg \frac{\sqrt{G\alpha}}{k} . \tag{11.14}$$

Then the number density of ions is given by

$$N_i = \frac{G}{k N_p} . \tag{11.15}$$

Let us make numerical estimates for an atmospheric plasma. Then $G \sim 10\,\text{cm}^{-3}\text{s}^{-1}$, $\alpha \sim 10^{-6}\,\text{cm}^3/\text{s}$, and for ions with a mobility of $K = 1\,\text{cm}^2/(\text{V} \cdot \text{s})$ we have $k/r_o \sim 0.3\,\text{cm}^2/\text{s}$. Criterion (11.14) has the form $N_p r_o \gg 0.01\,\text{cm}^{-2}$. In particular, for particles of size $r_o \sim 10\,\mu\text{m}$ this gives $N_i \gg 100\,\text{cm}^{-3}$ (for water vapor this corresponds to the density $\rho \gg 4 \cdot 10^{-7}\,\text{g/cm}^3$) and, according to the above balance equations for ions, this leads to the estimate $N_i N_p \sim 3 \cdot 10^4\,\text{cm}^{-6}$.

Problem 11.14

Evaluate the rate of decay of atmospheric plasma if this is determined by ion attachment to small aerosol particles of size $r_o \ll e^2/T$.

Small aerosol particles of size $r_o \ll e^2/T$ are mostly neutral, and charged particles have a single charge. Indeed, the number of aerosol particles with charge $\pm 2e$ is exponentially small because the Coulomb barrier for a charged particle exceeds remarkably a the thermal ion energy. In this regime, recombination of ions is relatively weak under the condition $k_{1,0} N_1 N_i \gg \alpha N_i^2$. In accordance with the above expression, this criterion can be rewritten in the form

$$N_p \gg \sqrt{\frac{G}{\alpha}} \cdot \frac{e^2}{r_o T} . \tag{11.16}$$

In particular, for water vapor at room temperature this gives $N_p \gg 4 \cdot 10^5 \sqrt{G}$, where G is expressed in cubic centimeters per second, and N_p is in reciprocal cubic centimeters.

The rate of plasma decay follows from the balance equation for the number density of ions:

$$\frac{dN_i}{dt} = -k_{1,0} N_1 N_i = -k_{dec} N_p N_i .$$

We define in this way the rate constant of plasma decay:

$$k_{dec} = k_{1,0} N_1 / N_0 = k_{0,1} .$$

We assume the total charge of particles to be small compared with the total ion charge. From this it follows that small particles do not influence plasma decay if $k_{dec} N_p N_i \ll \alpha N_i^2$, or

$$N_p / N_i \ll \alpha / k_{dec} . \tag{11.17}$$

In particular, let us take typical values of the atmospheric parameters near the Earth's surface: $\alpha \sim 10^{-6}\,cm^3/s$, $N_i \sim 300\,cm^{-3}$, and $k_{01} = \pi r_o^2 v_i$, where $v_i \sim 10^4$ cm/s is a typical thermal velocity of ions and r_o is the particle radius. Then criterion (11.17) has the form $N_p r_o^2 \ll 10^{-8}\,cm^{-1}$. From this we obtain that the recombination of ions on the surface of aerosol particles is important even at low number densities of aerosol particles.

Problem 11.15

Assuming that the charging of aerosol particles in the Earth's atmosphere results from the attachment of atmospheric molecular ions to aerosols, determine the sign of the aerosol charge in the Earth's atmosphere.

Positive and negative ions in the Earth's atmosphere are formed under the action of cosmic rays. Chemical processes in the atmosphere are important for certain

Table 11.1 Mobilities of some negative and positive ions in nitrogen [231–234]. The mobility K (cm^2/(V · s)) is reduced to normal conditions (the nitrogen number density $N = 2.69 \cdot 10^{19}$ cm^{-3} corresponds to a temperature of 273 K and a pressure of 1 atm).

Ion	K
NO_2^-	2.5
NO_3^-	2.3
CO_3^-	2.4
$NO_2^- \cdot H_2O$	2.4
$NO_3^- \cdot H_2O$	2.2
$CO_3^- \cdot H_2O$	2.1
$H^+ \cdot H_2O$	2.8
$H^+ \cdot (H_2O)_2$	2.3
$H^+ \cdot (H_2O)_3$	2.1

kinds of ions under certain conditions [230], and this influences some atmospheric properties. These processes start from ionization processes in the atmosphere, and then the electrons that are formed attach to oxygen molecules, which yields negative ions O^- or O_2^- depending on the electron energy. These negative ions in turn partake in chemical processes that lead to the formation of more stable negative ions. In the same manner, positive ions that start from simple ions O^+, O_2^+, N^+, and $N_2{}^+$ are transformed into other ions depending on admixtures present in atmospheric air. Subsequently, more stable positive and negative ions attach to aerosol particles, which are negatively charged if the diffusion coefficient or the mobility of the negative ions is higher than that for the positive ions, and the aerosol particles are charged positively for inverse relation between ion mobilities.

We give in Table 11.1 the measured mobilities of some ions in nitrogen. These ions may be formed in a humid atmosphere. Negative ions are characterized by high electron binding energies. Indeed, the electron affinity is 2.3 eV for an NO_2 molecule, 3.7 eV for NO_3, and 2.8 eV for CO_3. For comparison, the electron affinity to the oxygen molecule O_2 is 0.43 eV, and for the oxygen atom O it is 1.5 eV. These stable negative ions can form complex ions in a humid atmosphere.

We include in Table 11.1 positive complex positive ions that can be present in a moist atmosphere. A comparison of the mobilities of these ions shows that they are nearby. The average mobility of negative ions from Table 11.1 is (2.3 ± 0.1) cm^2/(V · s), whereas for positive ions it is (2.4 ± 0.3) cm^2/(V · s). Though the mobility of negative ions is lower, the mobility difference does not exceed the statistical error. From this we conclude that in principle aerosol particles in the Earth's atmosphere may be charged both negatively and positively, though the negative charge of aerosol particles is preferable. On the basis of the charge of clouds, one can conclude that the negative charge of aerosol particles is realized in roughly 90% of cases.

Problem 11.16

Taking the total current of Earth discharge to be $I = 1700\,\mathrm{A}$ and estimating the number density of atmospheric ions as $N_i \sim 300\,\mathrm{cm}^{-3}$, ascertain the sort of atmospheric ions that are responsible for this charge transfer.

On the basis of the equation for ion current density i,

$$i = \frac{I}{S} = N_i K E\,,$$

where $S = 5.1 \cdot 10^{18}\,\mathrm{cm}^2$ is the area of the Earth's surface, $E = 1.3\,\mathrm{V/cm}$ is the electric field strength near the Earth's surface, and K is the ion mobility. We have from this for the discharge current density $i = 3.3 \cdot 10^{-16}\,\mathrm{cm}^{-2}\mathrm{s}^{-1} = 2100\,\mathrm{cm}^{-2}\mathrm{s}^{-1}$, and this estimate gives for the ion mobility $K \sim 3\mathrm{cm}^2/(\mathrm{V} \cdot \mathrm{s})$, if we account for the contribution of both positive and negative ions and assume the mobilities of positive and negative ions to be nearby. One can see that this mobility relates to molecular ions. We also have the drift velocity of these ions:

$$w = \frac{i}{2N_i} \approx 3\,\mathrm{cm/s}\,.$$

This relates both to negative and positive ions, and positive ions move toward the Earth under the action of the Earth's electric field, whereas negative ions move away from the Earth.

Problem 11.17

The electric field strength in a quiet atmosphere near the Earth's surface that is negatively charged is $E = 130\,\mathrm{V/m}$ [224, 225], and this field repulses negatively charged particles from the Earth's surface. Find a size of singly negatively charged aerosols of water that will fall to the Earth's surface under the action of gravitational force.

For such particle sizes the repulsive force under the action of the gravitational field is stronger than that due to the electric field, that is,

$$Mg \geq eE\,,$$

where $M = 4\pi\rho r_o^3/3$ is the aerosol mass, ρ is the density of its material, and g is the acceleration of free fall. From this we find the aerosol particle radius when it falls to the Earth:

$$r_o \geq \left(\frac{3eE}{4\pi\rho}\right)^{1/3}\,.$$

For water aerosol particles in a quiet atmosphere this gives

$$r_o \geq 0.1\,\mu\mathrm{m}\,. \tag{11.18}$$

Thus in a real atmosphere, positively charged aerosols and negatively charged aerosols with a radius of $r_o > 0.1\,\mu\mathrm{m}$ may reach the Earth's surface.

Problem 11.18

Assuming that the charging of the Earth results from falling of charged aerosol particles in atmospheric air, find the maximum aerosol size if this process proceeds through the falling of water aerosols. The current to the Earth's surface is about 1700 A, and water circulation in the atmosphere is approximately $4 \cdot 10^{20}$ g/year, or $1.3 \cdot 10^{13}$ g/s. Assume water aerosols to be charged negatively and to be of identical size.

According to measurements, the average current density over land amounts to $2.4 \cdot 10^{-16}$ A/cm^2, and over the ocean it is $3.7 \cdot 10^{-16}$ A/cm^2 [224], which corresponds approximately to the total current of the Earth of $I = 1700$ A, as indicated in the condition for the problem. Assuming it is created by water aerosols and dividing the latter by the rate of total water transport in the Earth's atmosphere, we find the specific rate of electricity transport to be $1.4 \cdot 10^{-10}$ C/g water.

Note that the equilibrium charge of a large aerosol located in an ionized gas is proportional to its radius r_o, and the aerosol mass is proportional to r_o^3, that is, the specific rate of electricity transport is proportional to r_o^{-2}, that is, it drops with an increase in the aerosol radius. Hence, the appropriate size of an aerosol particle for electricity transport is restricted.

In determining the aerosol charge we assume the average mobility of atmospheric ions to be $2\,\text{cm}^2/(\text{V} \cdot \text{s})$, and the relative difference of the mobilities of positive and negative ions we take to be 10%, that is, $\Delta K/K = 0.1$. Then formula (7.7) gives

$$|Z| = \frac{r_o T}{e^2}\frac{\Delta K}{K} = 0.1\frac{r_o T}{e^2} \approx 2\frac{r_o}{a}\,, \tag{11.19}$$

where we take $T = 300$ K and $a = 1$ µm. Expressing the mass of a liquid water drop in grams, we have

$$m = 4.2 \cdot 10^{-12}\left(\frac{r_o}{a}\right)^3 .$$

Correspondingly, the ratio of an aerosol particle's charge to its mass is

$$\frac{|Z|e}{m} = 6.9 \cdot 10^{-8}\left(\frac{a}{r_o}\right)^2 \frac{C}{g} .$$

Hence, the observed specific transport of electricity, being connected with water circulation, may be realized at a radius of aerosol particles of

$$r_o < 23\ \mu\text{m}\,, \tag{11.20}$$

that corresponds to the average charge of aerosol particles $|Z| < 40$ under given conditions. If we use an average current density to the Earth's surface of $i = 3.3 \cdot 10^{-16}$ A/cm^2, we obtain an average flux of charged aerosol particles toward the Earth of $j = 50\,\text{cm}^{-2}\text{s}^{-1}$. Because the average falling velocity for water drops of the indicated size is $v_d = 2$ cm/s, this is provided by the average number density of aerosol particles $N_d \approx 30\,\text{cm}^{-3}$, which corresponds to a water content in the atmosphere of 1 g water per kilogram of air. In addition, the charge number density when the charge relates to charged water aerosols is $N_Z \sim 10^3\,e/\text{cm}^3$ in this case.

The analysis above is based on the charging of an aerosol particle as a result of the attachment of positive and negative molecular ions, which leads to formula (11.19) for the aerosol charge. In reality, charging takes place in collisions between water drops and ice crystallites when charge transfer takes place owing to the surface effects in interactions of their surfaces. Let us assume that the collision of these aerosols leads to a charge change of Z such that interaction of these particles is on the order of the thermal energy T:

$$\frac{Z^2 e^2}{r_o} \sim T\,, \tag{11.21}$$

From this we obtain for the charge-to-mass ratio for falling aerosols

$$\frac{|Z|e}{m} = 1.6 \cdot 10^{-7} \left(\frac{a}{r_o}\right)^{5/2}\,, \quad \frac{\text{C}}{\text{g}}\,,$$

where $a = 1\,\mu\text{m}$. This gives that the observed charge-to-mass ratio is satisfied under the criterion

$$r_o < 17\,\mu\text{m}\,, \tag{11.22}$$

and the upper limit corresponds to the aerosol charge $Z \approx 18$.

Problem 11.19

Determine the size range of charged aerosol particles when the electric current in atmospheric air is created by falling aerosol particles.

We take the electric current in atmospheric air as a sum of the current of ions under the action of the Earth's electric field and the current of charged aerosol particles due to their falling. Then the current density is

$$i = eE(K_- N_- - K_+ N_+) + e|Z|N_p v\,,$$

where E is the Earth's electric field strength, K_-, K_+ are the mobilities of negative and positive ions, N_-, N_+ are the number densities of negative and positive ions, N_p is the number density of aerosol particles, and v is the velocity of aerosols falling under the action of gravitational force. The latter is given by formula (6.13) and for water aerosols in air at temperature $T = 300\,\text{K}$ may be represented in the form

$$v = 0.012 \left(\frac{r_o}{a}\right)^2\,,$$

where $a = 1\,\mu\text{m}$, and the velocity of falling v is expressed in centimeters per second.

The above formula for the current density is written for the case where aerosol particles are charged negatively, that is, $K_- > K_+$, and the current is directed to the Earth's surface. We assume the ion mobilities to be similar and take $\Delta K/K = 0.1$ and $K = (K_- + K_+)/2$. Next, the condition of quasineutrality of air gives $N_p + N_- = N_+$. As a result, the expression for the density of an atmospheric current takes the form

$$i = eN_p(|Z|v - EK)\,. \tag{11.23}$$

Taking the average ion mobility $K = 2\,\mathrm{cm^2/(V\cdot s)}$, we find that the electric current in atmospheric air is determined by aerosol particles if

$$r_o > 15\,\mu\mathrm{m}\,. \tag{11.24}$$

Under these conditions, the drift velocity of positive molecular ions and water drops with radius $r_o = 15\,\mu\mathrm{m}$ is approximately $3\,\mathrm{cm/s}$, which corresponds to a time of transport at a distance $h = 3\,\mathrm{km}$ of $\tau \sim 10^3\,\mathrm{s}$.

It is interesting to compare this time with the time of attachment of positive ions to negatively charged water drops under typical atmospheric conditions when the number density of molecular ions is $N_i \sim 300\,\mathrm{cm^{-3}}$. We have from formula (7.11) for the typical time of recombination

$$\tau_{rec} = \frac{T}{4\pi N_i e^2 D_i} = \frac{1}{4\pi N_i e K_i}\,,$$

where D_i, K_i are the diffusion coefficient and the mobility for positive molecular ions. Taking $K_i \sim 2\,\mathrm{cm^2/(V\cdot s)}$, we obtain $\tau_{rec} \sim 10^3\,\mathrm{s}$, which is comparable to a falling time. In addition, note that the typical charge of a drop with radius $r_o = 15\,\mu\mathrm{m}$ due to the attachment of positive and negative molecular ions of different mobilities is $Z \approx 30$.

Problem 11.20

Analyze the possibility of the charging of the Earth due to falling negatively charged water drops in atmospheric air toward the Earth's surface if aerosols of liquid water are present in the atmosphere only and their charging results from the attachment of positive and negative molecular ions, which in turn are generated through the ionization of air molecules.

Let us consider the scheme of charging of the Earth when atmospheric ions result from the action of cosmic rays and radioactivity of the Earth's surface and then attach to aerosol particles; we assume that water drops play this role. If the mobility of negatively charged atmospheric ions exceeds that of positively charged atmospheric ions, then water drops are charged negatively. These negative water drops fall to the Earth's surface against the electric force, and this leads to charging of the Earth.

We note that in the limit of a large number density of aerosols, when the charge number density of aerosols exceeds that of positive molecular ions, this process is not realized because all the charge is transferred to aerosols, and we do not have the mechanism to give a positive charge to small water drops, and large water drops will carry a negative charge. Because of the absence of such selectivity, the separation of charges is absent in this limiting case.

In the other limiting case, we have negatively charged aerosol particles – water drops and positive molecular ions – and then this mechanism is possible for large water drops. Indeed, criterion (11.24) gives the size of aerosol particles when the drift velocity of their falling under the action of the gravitational force toward the Earth's surface exceeds the drift velocity of positive molecular ions toward the Earth's surface under the action of the Earth's electric field. Criterion (11.20) requires that the transferred charge is connected with the transport of water in the Earth's atmosphere. We have that if criteria (11.20) and (11.24) are fulfilled, this model, including positive molecular ions and negatively charged water drops in the Earth's atmosphere, provides a charging of the Earth, so that this supports the observed electric field strength near the Earth's surface, $E = 1.3\,\mathrm{V/cm}$, and the observed amount of water transport if the Earth's charging is a secondary phenomenon of water transport. Moreover, this model allows one to have positive charging if the mobility of positive molecular ions exceeds that of negative molecular ions that are realized in some cases.

But this model cannot satisfy us for several reasons. First, criteria (11.20) and (11.24) give a narrow range of aerosol sizes that allow for charging of the Earth on the basis of this model. But these estimations do not take into account the flux of positive ions toward the Earth's surface that decreases cluster size in criterion (11.20) and increases cluster size in criterion (11.24). Taking this into account may violate the simultaneous fulfillment of both criteria (11.20) and (11.24). Second, the lifetime of positive ions is limited. Third, the separation of charge occurs in clouds rather than in a quiet atmosphere as the electric field strength in clouds is two orders of magnitude higher on average than in a quiet atmosphere. This increases the lower limit in criterion (11.24), that is, this model is not valid for clouds.

Thus, the electric model of the Earth's atmosphere, which accounts for the presence of molecular ions and water drops, may be valid in principle, but it cannot describe electric processes in a real atmosphere.

Problem 11.21

On the basis of the above results, formulate the character of the working of the Earth's electric machine where the separation of charges in the Earth's atmosphere results from falling of charged water drops and discharge of the Earth's surface is caused by ion currents under the action of the Earth's electric field.

On the basis of the above analysis one can represent the following scheme of an atmospheric electric machine [226]. The atmospheric electric processes start from ionization processes in the atmosphere under the action of cosmic rays and the

Earth's radioactivity. The evolution of forming positive and negative ions with the participation of aerosol particles – small water drops – leads to the separation of the atmospheric charge due to falling of charged water drops. Discharge processes result from the drift of small atmospheric ions under the action of the Earth's electric field. An effective charge separation occurs in clouds where the water concentration is relatively high [235–239].

The types of charged particles in the atmosphere and the character of aerosol charging are of importance in this scheme. Above we considered the case where only water drops are aerosol particles, and then their charging results from the attachment of molecular ions to them. Electric processes in the atmosphere occur when other aerosol particles are present in the atmosphere. This takes place in clouds at altitudes of several kilometers where the air temperature passes through 0 °C as the altitude varies. In this region ice particles are formed along with water drops, and this leads to additional mechanisms of aerosol charging. In particular, above we considered the charging of spherical particles with elongated solid particles when these particles are located in an external electric field. Then interaction between colliding particles is determined by induced charges, and collisions of spherical particles with endings of an elongated particle lead to an effective charge exchange between them. But this mechanism is symmetric with respect to the charge sign, that is, it leads to the formation of both positive and negative spherical particles. Nevertheless, this mechanism may contribute to particle charging because it enforces the rate of aerosol charging.

Another mechanism of aerosol charging results from the collision of a water drop and an ice particle that leads to negative charging of a water drop and positive charging of an ice crystallite [240–246]. The reason for the charge transfer in these collisions consists in different electric properties of the surfaces of water and ice, so that the contact between these surfaces leads to a voltage between them, and the charge transfer compensates this voltage. Assuming colliding particles are of equal size, we estimate the charge transfer on the basis of formula (11.21):

$$Z \sim \sqrt{\frac{r_o T}{e^2}} ,$$

so the interaction potential of particles is on the order of their thermal energy, which allows their separation after the collision.

Thus, we conclude that charge separation in the Earth's atmosphere takes place in clouds with a high water concentration. At altitudes of several kilometers where the atmospheric temperature is close to 0 °C, the melting temperature of water, water drops and ice crystallites are present simultaneously. Heavier water drops are charged negatively and fall, which leads to negative charging of the lower cloud edge. Subsequent accumulation of negative charges on the lower cloud edge and thunderstorms transfer the negative charge to the Earth's surface through lightning flashes. Note that in this two-component aerosol model one can neglect the transport of positive ions to the Earth's surface under the action of the Earth's electric field, and criterion (11.24) does not work in this case.

Problem 11.22

Taking a typical electric field strength in an electrically active cloud to be $E \approx 10\,\mathrm{kV/m}$ [227], determine the minimum size of negatively charged aerosols that can fall under the action of the gravitational force.

Let us repeat the analysis of Problem 11.17 for a new electric field strength. Then we obtain, instead of formula (11.18),

$$r_o \geq 0.4\,\mu\mathrm{m}\,. \tag{11.25}$$

Problem 11.23

The content of water in a cloud is 7 g per kilogram of air [225], which corresponds to the saturated vapor pressure of water at 10 °C. Taking the charge-to-mass ratio to be $1.3 \cdot 10^{-10}$ C/g, which allows for charging of the Earth, and a typical drop size of $r_o = 4\,\mu\mathrm{m}$, determine the rate of charging of the lower cloud edge if the water content given above relates to water drops that become negatively charged. Determine a typical charging time up to a typical electric field strength $E \approx 10\,\mathrm{kV/m}$ [227] in an electrically active cloud if the distance between cloud edges is $h \approx 2\,\mathrm{km}$.

We assume that half of the indicated content of atmospheric water (7 g of water per kilogram of air) exists in the form of water drops, and for the radius of water drops $r_o = 5\,\mu\mathrm{m}$ this corresponds to their density $N_d = 900\,\mathrm{cm}^{-3}$. In air at a temperature of $T = 300$ K the velocity of falling water drops with radius r_o expressed in centimeters per second is

$$v = 0.012 \left(\frac{r_o}{a}\right)^2 ,$$

where $a = 1\,\mu\mathrm{m}$, and for $r_o = 5\,\mu\mathrm{m}$ we have $v_d = 0.3\,\mathrm{cm/s}$, so that the flux of falling drops is $j = N_d v = 260\,\mathrm{cm}^{-2}\mathrm{s}^{-1}$. Assuming that the charging of water drops results from their collisions with ice crystallites, and their charge is given by formula (11.21), we have for the drop charge $Z \approx 10$, which leads to the current density of charging of a lower cloud edge $i = 4 \cdot 10^{-16}\,\mathrm{cm}^{-2}\mathrm{s}^{-1}$; this is larger than the average discharge current density in a quiet atmosphere.

We also estimate a typical time of charging of water drops in collisions with ice crystallites, assuming they are of equal size. Formula (2.7) gives for the rate of collision of a water drop with an ice crystallite $k \sim 3.5 \cdot 10^{-7}\,\mathrm{cm}^3/\mathrm{s}$, which gives a typical time of particle charging as $\tau \sim 1\,\mathrm{h}$. Note that subsequent collisions of charged water drops and ice crystallites do not lead to charge recombination, that is, this air with positive ice particles and negative water drops is an equilibrium state of this system.

Assuming that at the end of the process of cloud charging the negative and positive charges are concentrated in narrow layers, we have for the charge density

$$\sigma = \frac{E}{4\pi} = 5.5 \cdot 10^7 \frac{\mathrm{e}}{\mathrm{cm}^2} ,$$

and the cloud charging process lasts $\sigma / i \sim 1$ day.

Note that in the analysis of the electric phenomena in the Earth's atmosphere we account for processes in the atmosphere involving aerosol particles. In reality, transport processes and heat balance processes in the atmosphere are important for the accumulation and separation of electric charges in the atmosphere. Therefore, the above numerical results may be considered as estimations.

11.3
Growth of Fractal Structures Involving Solid Clusters

The fractal structure of physical objects is a specific porous structure [130–133, 247, 248]. We consider below geometric structures consisting of identical particles, fractal aggregates (Figure 11.4), and bounds between neighboring particles of a fractal aggregate are established at the point of their contact. Pores inside fractal aggregates are determined by the character of their formation, and the relative volume of pores increases with an increase in aggregate size. Below we consider a fractal aggregate consisting of identical spherical elemental particles with radius a and mass m_0. Characterizing a fractal aggregate by its radius R and mass M, we have the following relation between these parameters and the parameters of elemental particles according to formula (5.50) [130]:

$$M = m_0 \left(\frac{R}{a}\right)^{\beta} . \tag{11.26}$$

The parameter β, the fractal dimensionality of a fractal aggregate, is the principal characteristic of a fractal system. In the case of three-dimensional space, we have $\beta < 3$, that is, fractal systems are not compact. The fractal dimensionality is determined by the character of the aggregate growth.

We consider two limiting mechanisms of this process, which correspond to two growth models. Diffusion-limited aggregation (DLA) takes place when elemental particles with diffusional motion in a buffer gas attach to a motionless aggregate. The fractal dimensionality of this aggregate is given by $\beta = 2.45$ in three-dimensional space [249]. Cluster–cluster diffusion-limited aggregation (CCA) takes place when aggregates join to form a larger aggregate as a result of their contact,

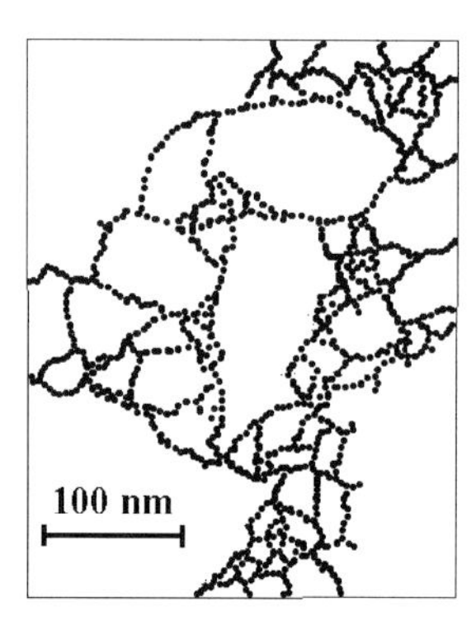

Fig. 11.4 Fragment of a fractal aggregate.

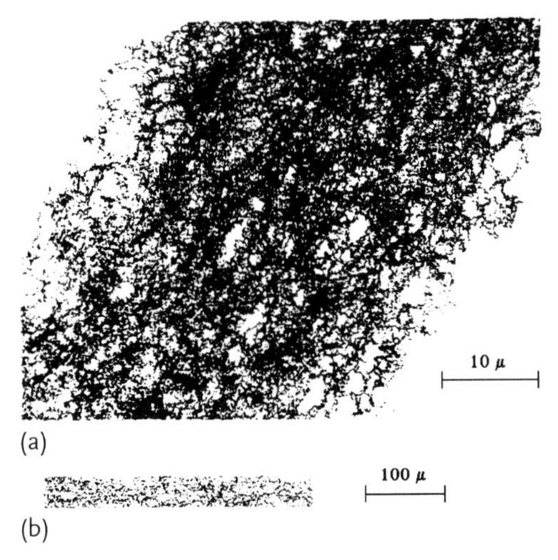

Fig. 11.5 Fragment of a fractal fiber that is cut off (a) across and (b) along the fiber [251].

and the motion of the aggregates in the buffer gas has diffusional character. The fractal dimensionality for the CCA mechanism of aggregation in three-dimensional space is $\beta = 1.77$ [250]. We will use these two mechanisms of growth of fractal aggregates below.

Similar to compact small particles, fractal aggregates may be joined to each other. If joining proceeds in the absence of external fields, this process corresponds to the CCA mechanism of fractal aggregate growth. Because of the relatively large size of fractal aggregates, their interaction in external fields through induced charges is important for the growth of structures. As a result, fractal fibers (Figure 11.5) are formed [251, 252] that have a fractal structure at small distances and are elongated along an external electric field. This structure may be responsible for the formation of the skeleton of ball lightning [253, 254], which is a tangle of entangled fractal fibers. Owing to this structure, the skeleton of ball lightning has simultaneously properties of gases, liquids, and solids.

Problem 11.24

Spherical clusters with radius a are joined to a fractal aggregate in a buffer gas in the diffusion regime of atom–cluster collisions. Assuming the probability of cluster sticking as a result of their contact with the surface of a fractal aggregate to be one, determine the rate of the aggregate growth.

Fractal aggregates are rare systems with a low density, and their density decreases as the aggregate size increases. This determines the specifics of the growth of aggregates from the elemental particles located in the buffer gas. According to formula (11.26), we have for the number of small elemental particles contained in the fractal aggregate with radius R

$$n = \left(\frac{R}{a}\right)^{\beta} . \tag{11.27}$$

The rate of aggregate growth is given by the Smoluchowski formula (4.22):

$$\frac{dn}{dt} = 4\pi R D ,$$

where D is the diffusion coefficient of the aggregate in the buffer gas, which is expressed through the aggregate radius by formula (6.12). From the above relations it follows that

$$R = \left(R_o^{\beta-1} + 4\pi D \frac{\beta-1}{\beta} a^{\beta} t\right)^{1/(\beta-1)} , \tag{11.28}$$

where R_o is the initial aggregate radius. This formula gives in the limit of large time periods $R \sim t^{1/(\beta-1)}$. A typical value of the fractal dimensionality for this process in the three-dimensional case is $\beta = 2.45$, so that the law of the growth of the large aggregate is $R \sim t^{0.69}$ for the DLA regime of the aggregation process.

Note that an aggregate with radius R collects particles during its growth and forms in average a gaseous ball with radius

$$R' = \left(\frac{3}{4\pi N}\right)^{1/3} \cdot \left(\frac{R}{a}\right)^{\beta} .$$

From this it follows that the number density of small particles in a gas drops if

$$\frac{R'}{R} = \left(\frac{3}{4\pi N}\right)^{1/3} R^{(\beta-3)/3} a^{-\beta/3} > 1 .$$

The growth proceeds up to an aggregate radius of

$$R_c = \left(\frac{3}{4\pi N}\right)^{1/(3-\beta)} a^{-\beta/(3-\beta)} , \tag{11.29}$$

when the atom number density in the fractal aggregate coincides with the number density of free atoms in the gas. Here R_c is the correlation radius of this system. In particular, for the case of DLA ($\beta = 2.45$), this formula has the form

$$R_c = 0.074 N^{-1.82} a^{-4.45} .$$

At larger radii the density of small particles in the aggregate becomes lower than that in the gas.

Problem 11.25

Determine the rate of cluster–cluster aggregation in a uniform electric field due to the interaction of induced dipole moments of joining aggregates. Compare it with the rate of the growth of fractal aggregates according to cluster–cluster aggregation in the absence of an electric field.

We assume fractal aggregates to be similar to spherical metallic particles. Then the effective rate constant of their joining is given by formula (6.50):

$$k_{ef} \sim \frac{\alpha_1 \alpha_2 E^2 N^{2/3}}{\eta R} ,$$

where α_1, α_2 are the polarizabilities of joining aggregates, E is the electric field strength, N is the number density of aggregates, and η is the viscosity coefficient of the buffer gas in which the process occurs. Considering aggregates as metal particles, we have $\alpha_1 \sim \alpha_2 \sim R^3$, where R is a typical radius of the aggregates at a given time.

Let us use the definition of the effective rate constant of the process

$$\frac{dn}{dt} = k_{ef} N n ,$$

where n is the average number of elemental particles constituting a fractal aggregate. We have $N \sim N_o/n$, where N_o is the total number density of elemental particles in the aggregates. On the basis of formula (11.26) we obtain

$$\frac{dn}{dt} \sim R^5 E^2 N_o^{2/3}/\eta \sim a^5 E^2 N_o^{2/3} n^{5/\beta - 2/3}/\eta ,$$

where a is the typical radius of an elemental particle and β is the fractal dimensionality of the aggregates.

Comparing this with the rate of cluster–cluster aggregation in the absence of electric fields (11.28), we obtain for the ratio of typical times of these processes

$$\frac{\tau_{CCA}}{\tau_{el}} \sim \frac{E^2}{T} N_o^{2/3} a^5 n^{5/\beta - 2/3} , \tag{11.30}$$

where τ_{el} is the typical time of aggregation resulting from the interaction of induced dipole moments and τ_{CCA} is the typical time of DLA in the absence of an electric field. As is seen, aggregate growth due to the interaction of induced dipole moments is essential for large aggregates. In particular, taking the fractal dimensionality of an aggregate $\beta = 1.77$, we have the following criterion for the mechanism of aggregation due to the interaction of induced dipole moments to determine the rate of aggregate growth:

$$E \gg \frac{T^{1/2}}{N_o^{1/3} a^{5/2} n^{1.1}} . \tag{11.31}$$

Problem 11.26

Determine the rate of growth of a fractal fiber by means of attachment of fractal aggregates to its ends as a result of interaction of induced dipole moments of joining aggregates in a uniform electric field.

A fractal fiber (Figure 11.5) [251, 252] is a specific physical object that resembles a chain aggregate, where fractal aggregates are its elemental particles instead of solid clusters in chain aggregates. The aggregation process proceeds similar to the scheme for Problem 11.3 until the induced dipole moment of the forming aggregate D is relatively small, that is, $DE \ll T$, where E is the electric field strength and T is a typical thermal energy. Under this condition, a forming fractal aggregate can be considered spherical. If the opposite criterion is fulfilled, a forming fractal aggregate resembles a cylindrical particle whose axis is directed along the electric field. In this case spherical aggregates attach to a fractal fiber near its ends.

A typical time τ_l of the growth of a fractal fiber as a result of attachment of fractal aggregates to the ends of the fractal fiber follows from formula (11.5):

$$\tau_l \sim \frac{\eta \ln(l/r_o)}{r_o^2 l E^2 N} .$$

Note one more feature of the growth of a fractal fiber compared with a continuous cylindrical particle. We assume that the conductivity of the forming fiber is great enough, so that the electric potential of its surface is constant over the fiber when it is located in an external electric field. But the formation of strong bonds between fractal aggregates inside a fractal fiber requires a large time, so that this scenario can be valid only for a slow process.

Problem 11.27

Assuming the diffusion regime of collision between gas atoms and fractal aggregates located in a gas, determine the typical time of aggregate growth up to a given size for the CCA mechanism of cluster aggregation. The fractal dimensionality of the forming aggregates is β.

Using the rate constant (6.44) for the aggregation process, assume the sizes of joining aggregates to be similar, so that the rate constant (6.44) of aggregate association does not depend on the radii of colliding aggregates:

$$k_{\text{dif}} = \frac{8T}{3\eta} ,$$

where η is the viscosity coefficient of the buffer gas. Then the evolution of the aggregate radius corresponds to the diffusion-limited mechanism, and the average number of particles in a forming fractal aggregate is given by

$$n = k_{\text{dif}} N_o t ,$$

where N_o is the initial number density of small particles constituting the fractal aggregate.

On the basis of formula (11.27), we have from this for the evolution of the mean radius of a fractal aggregate

$$R = a\left(\frac{8T}{3\eta}N_o t\right)^{1/\beta} . \quad (11.32)$$

For the aggregation of solid particles in a dense gas, that corresponds to the CCA mechanism and is characterized by the fractal dimensionality $\beta = 1.7 - 1.8$, formula (11.32) gives

$$R = a\left(\frac{8T}{3\eta}N_o t\right)^{0.57\pm0.02} .$$

Problem 11.28

Small spherical particles join in fractal aggregates on the basis of the CCA mechanism of aggragation. These particles are located in a buffer gas, and their initial number density is N_o. At the end of the process all the particles form a bound system similar to an aerogel that occupies all the gaseous volume. Find a time of formation of this system.

The aggregate growth finishes when the number density of particles in the fractal aggregate coincides with the initial number density N_o of small particles in the buffer gas. Then the system of free small particles in the buffer gas is transformed into an aerogel – a bulk system of bound particles. Figure 11.6 exhibits an element of an aerogel as a porous bulk system of bound particles. According to formula (11.29), the aggregate radius that corresponds to the formation of an aerogel is

$$R_c = \left(\frac{3}{4\pi N}\right)^{1/(3-\beta)} a^{-\beta/(3-\beta)} ,$$

and a typical time t of transition from free particles in a buffer gas to an aerogel through the formation of fractal aggregates is given by

$$t = \frac{3\eta}{8T}\left(\frac{3}{4\pi}\right)^{1/(3-\beta)}\left(N_o a^{\beta}\right)^{-\frac{4-\beta}{3-\beta}} . \quad (11.33)$$

Taking $\beta = 1.77$, the typical fractal dimensionality for fractal aggregates resulting from the CCA mechanism of this process, we transform formula (11.33) into

$$t = \frac{3\eta}{8T}\left(\frac{3}{4\pi}\right)^{0.75}\left(N_o a^{1.77}\right)^{-1.75} = \frac{0.13\eta}{T N_o^{1.75} a^{1.01}} . \quad (11.34)$$

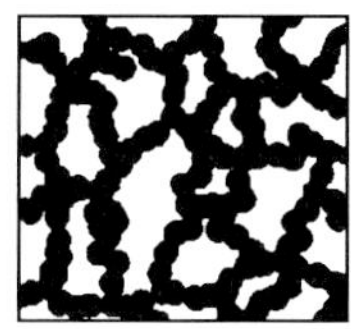

Fig. 11.6 Fragment of a fractal aerogel.

Note the features of the formed system of bound particles, an aerogel. It is uniform for large sizes $R \gg R_c$ and has a fractal structure for $R \ll R_c$. This means that if we cut off a sphere with radius R from the aerogel, it will have fractal properties when $R \ll R_c$. The value R_c is called the correlation radius of the aerogel.

Problem 11.29

The growth of a fractal aggregate results from its fall under the influence of gravity in a buffer gas containing small particles that attach to the fractal aggregate. Find a typical time of transformation of the fractal aggregate into an aerogel.

Consider the character of growth of a falling aggregate when the gravitational force determines its motion. The aggregate growth occurs as result of two mechanisms. First, a falling aggregate collects small particles that are located in its path. Second, particles go to the surface of the falling aggregate as a result of diffusion in the buffer gas and attach to the aggregate. Then the balance equation for the number of particles n constituting the aggregate is given by the equation

$$\frac{dn}{dt} = v_1 + v_2 ,$$

where v_1, v_2 are the rates of the corresponding mechanisms of aggregation.

The rate of the first process is

$$\nu_1 = \pi R^2 v N ,$$

where R is the aggregate radius, v is the velocity of gravitational fall, and N is the number density of particles in the buffer gas. The velocity of a falling aggregate can be determined by analogy with formula (6.13) from the equality of the gravitational force $P = mgn$ of the aggregate and the resistance force $F = 6\pi R\eta v$, where m is the mass of an individual particle, g is the free-fall acceleration, n is the number of elemental particles of the aggregate, and η is the viscosity coefficient of the buffer gas. Note that for simplicity we consider the aggregate as a spherical symmetric system, while this process of aggregation stretches it in the vertical direction. From this it follows that for the drop velocity we have

$$v = \frac{mgR^{\beta-1}}{6\pi\eta a^{\beta}} ,$$

so the rate of this process is given by

$$\nu_1 = \frac{mgR^{\beta+1}N}{6\eta a^{\beta}} = \frac{mgNan^{1+1/\beta}}{6\eta} .$$

The rate of the second aggregation process is given by the Smoluchowski formula (4.22):

$$\nu_2 = 4\pi DRN = 4\pi DNan^{1/\beta} ,$$

where D is the diffusion coefficient of individual particles in the buffer gas. Thus, the balance equation for the number of particles of the aggregate has the form

$$\frac{dn}{dt} = mgNan^{1+1/\beta}/(6\eta) + 4\pi DNan^{1/\beta} = Nk_1n^{1+1/\beta} + Nk_2n^{1/\beta} ,$$

where $k_1 = mga/(6\eta)$, $k_2 = 4\pi Da$. As is seen, the aggregation process is accelerated when the first mechanism of aggregation becomes the principal one. This means a finite time of transformation of the aggregate into the aerogel, which is given by

$$t = \frac{1}{N}\int_0^\infty \frac{dn}{k_1n^{1+1/\beta} + k_2n^{1/\beta}} = \frac{\pi}{Nk_1^{1-1/\beta}k_2^{1/\beta}\sin\frac{\pi}{\beta}} . \tag{11.35}$$

The main contribution to the integral over time is determined by aggregate sizes $n \sim k_2/k_1$. We assume that the number of particles in the aerogel formed exceeds this value remarkably. For a typical fractal dimensionality of this aggregate, $\beta = 2.45$, the time of aerogel formation is

$$t = \frac{3.3}{Nk_1^{0.59}k_2^{0.41}} . \tag{11.36}$$

Problem 11.30

The character of formation of an aerogel in a buffer gas containing small aggregating particles consists in the joining of fractal aggregates as a result of their collision and their fall in a gravitational field. Determine the time of aerogel formation.

This process is analogous to that of the previous problem, so

$$\nu_1 = \pi(R_1 + R_2)^2|v_1 - v_2|N_a , \quad \nu_2 = 4\pi(D_1 + D_2)(R_1 + R_2)N_a ,$$

where N_a is the parameters density of aggregates at a given time and subscripts in the other terms denote the parameters of colliding aggregates. Let us use the simple size distribution function of aggregates $f \sim \exp(-n/\overline{n})$, where $\overline{n}$ is the average aggregate size. Averaging over aggregate sizes on the basis of this distribution function, we get the balance equation for the mean number of aggregate particles:

$$\frac{d\overline{n}}{dt} = 1.4N_o k_1\overline{n}^{1+1/\beta} + (\pi + 2)N_o k_2 ,$$

where $N_o = \overline{n}N_a$ is the initial number density of particles, $k_1 = mga/(6\eta)$, and $k_2 = 2T/(3\eta)$.

From this for the time of aerogel formation we get

$$t = \int_0^\infty \frac{d\overline{n}}{1.4 N_o k_1 \overline{n}^{1+1/\beta} + (\pi + 2) N_o k_2}$$

$$= \frac{\pi\beta}{\left[(\beta + 1) N_o \sin \frac{\pi}{\beta+1} \cdot (1.4 k_1)^{\beta/(\beta+1)} [(\pi + 2) k_2]^{1/(\beta+1)}\right]} .$$

For a typical fractal dimensionality for the CCA mechanism of aggregation we take $\beta = 1.77$, so this formula takes the form

$$t = N_o^{-1} k_1^{-0.64} k_2^{-0.36} . \tag{11.37}$$

12
Cluster Plasma

A cluster plasma is an ionized gas containing clusters [36, 86, 255]. This system as a physical object is similar in principle to a dusty plasma, but there are two differences between these systems. First, clusters are usually on the nanometer scale, whereas micron sizes are typical for dust particles. Second, clusters grow and evaporate in the course of evolution of a cluster plasma, whereas the size of dust particles does not vary in a dusty plasma. An example of a cluster plasma that we will use for illustration is an ionized buffer gas (e.g., an inert gas or nitrogen) and a metal vapor of low concentration. Under these conditions, metal clusters and metal atoms are important for cluster processes, whereas a buffer gas serves as a stabilizing component. Below we analyze several such systems that may be used for the generation of metal clusters.

The properties of a cluster plasma are determined by the kinetics of processes involving clusters. There are various methods to generate metal clusters through a cluster plasma. The simplest and most reliable method is based on a dense plasma such as a high-pressure arc plasma. Metal atoms are introduced into a flow of such a plasma and then are converted into clusters. The advantage of such a system is that metal clusters formed at a certain point of a buffer gas flow do not vary their position in a flow because of the small diffusion coefficient of clusters. One can see that the best method to obtain high-intensity cluster beams is to inject a metal into the dense plasma flow in the form of a metal-containing chemical compound (e.g., halogen–metal compounds). After decay of these molecules in a dense plasma, metal atoms join in metal clusters, whereas another compound component propagates over the entire plasma volume. This scenario is possible under certain conditions because of the large binding energy of metal atoms in clusters.

There are other methods for generation of metal clusters, but in any case clusters are formed as a result of joining of metal atoms. This process takes place in a chamber of generators of metal clusters, and to obtain a cluster beam it is necessary to eject clusters from the aggregation chamber through an orifice (or nozzle), and the character of cluster motion in gas flows shows that because of their inertia, clusters do not follow the gas flow in an orifice region and can attach to chamber walls. Hence, the behavior of clusters near the orifice requires a special analysis and then a specific construction of the aggregation chamber near the orifice to avoid cluster losses.

Cluster Processes in Gases and Plasmas. Boris M. Smirnov

ISBN: 978-3-527-40943-3

If a cluster plasma is nonuniform in a given space, and there is an equilibrium between metal atoms and clusters in each part of the cluster plasma, a cluster instability develops. Indeed, in this case the equilibrium between some parts of a nonuniform cluster plasma occurs through the diffusion of metal atoms to a cold region and the diffusion of clusters to a hot plasma region. But when clusters grow, the mean free path of metal atoms with respect to their attachment to clusters decreases, and the diffusion coefficient of the clusters decreases. This leads to an instability that violates the equilibrium in the weakly nonuniform cluster plasma. As a result, a specific character of cluster transport and cluster growth takes place in a nonuniform cluster plasma.

12.1
Clusters in a Dense Arc Plasma

Problem 12.1

Analyze the general features of cluster plasma generation in an plasma flow of a high-pressure arc.

A general scheme for generation a cluster plasma on the basis of arc discharge is given in Figure 12.1 [36, 256, 257]. A high-pressure arc plasma flow is created after arc discharge. Metal-containing molecules are injected into the central part of this flow in the form of liquid drops. Owing to a high temperature of this plasma flow, molecules are decomposed into atoms, and then metal atoms join to form clusters.

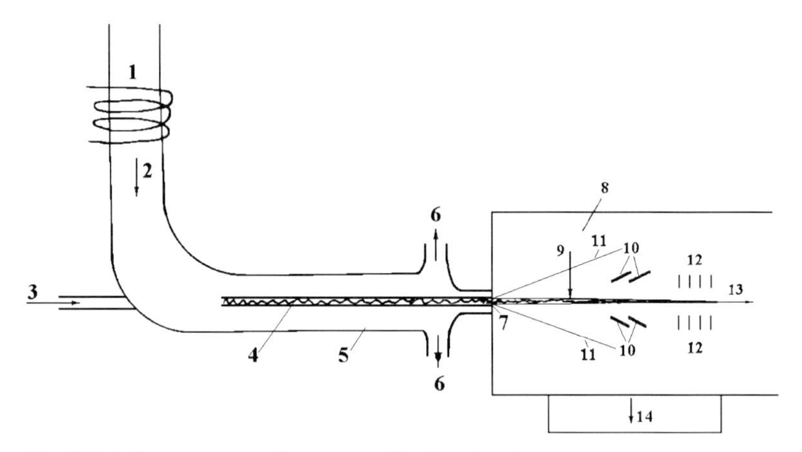

Fig. 12.1 A cluster generator from an arc afterglow: 1 – plasma generator; 2 – plasma flow; 3 – insertion of a gaseous compound of a heat-resistant metal; 4 – cluster beam in plasma flow; 5 – afterglow tube; 6 – output plasma flow; 7 – nozzle for plasma expansion in a vacuum; 8 – vacuum camera; 9 – electron beam; 10 – skimmers; 11 – expanding buffer gas; 12 – electric field optics; 13 – cluster beam of a buffer gas; 14 – pumps.

In this manner, the metal property is used that the binding energy between metal atoms is stronger than that in metal-containing molecules. For definiteness, we will be guided below by metal compounds MX_6, where M is a metal atom and X is a halogen atom. Of course, this scheme may be realized for a restricted number of metal compounds, and we analyze it in detail below.

According to this scheme, the first stage of the cluster generation process is the formation of a vapor of metal atoms, and subsequently this vapor is transformed into a gas of clusters within the framework of the standard scheme analyzed above. It is necessary to conserve metal atoms inside a gas discharge chamber, and this is achieved owing to a high pressure of the buffer gas. It is convenient to use a cylindrical tube as a chamber with a gas discharge plasma flow.

Note one more problem of this scheme of generation of cluster beams. The attachment of atoms and clusters to the walls of the tube with a plasma flow is absent because of a high gas pressure. A subsequent injection of this flow into a vacuum (or a low-pressure gas) proceeds through an orifice with a small diameter to conserve a high gas pressure inside the tube. Therefore, the attachment of metal clusters to the walls near the exit orifice becomes significant, and this is an additional problem with this scheme. Below we consider various elements of this scheme of generation of metal clusters.

Problem 12.2

Metal-containing molecules MoF_6 are inserted in the center of an arc argon plasma that flows through a cylindrical tube in the form of a cylinder jet with a radius of 0.1 mm and pressure of 10 atm. The argon pressure is 1 atm, and the initial temperature is 5000 K. Assuming the tube radius to be large compared with that occupied by metal atoms and clusters, estimate a radius of the region occupied by clusters at the end of nucleation.

The processes in a cluster plasma take place according to scheme (9.50), which leads to the transformation of gaseous MoF_6 into Mo clusters. In estimations we assume the rate of decomposition of MoF_6 molecules given by formula (9.56) to be large and the supersaturation rate (2.58) of a formed metal vapor to be large, so that a nucleation time is determined by formula (9.27). Taking the reduced rate constant in this formula using the data in Table 2.1 to be $k_o \sim 2 \cdot 10^{-10}\,\mathrm{cm^3/s}$, we get for the parameter (9.21) $G \sim 10^5$.

Next, after injection of a jet of molecules MoF_6 into argon, the pressure of these molecules and metal atoms after molecular decomposition will correspond to the argon pressure. Then the nucleation time of transformation of metal atoms into clusters is $\tau_{nuc} \sim 3 \cdot 10^{-7}$ s, in accordance with formula (9.27). The distance ρ of propagation of metal atoms during this time as a result of diffusion in a buffer gas is

$$\rho^2 = 4D\tau_{nuc} ,$$

where D is the diffusion coefficient of metal atoms in this gas. Estimating the diffusion coefficient of metal atoms on the basis of the data in formula (6.16) and reducing it to the conditions under consideration, we obtain $D \sim 0.2\,\text{cm}^2/\text{s}$. From this we have for a radius of jet expansion $\rho \sim 5\,\mu\text{m}$, which is small or comparable with the initial jet radius. Thus, neglecting the diffusion of clusters, we obtain that the width of a cluster region corresponds to the initial width of the molecular jet, and according to formula (9.27) these clusters contain 10^3–10^4 molybdenum atoms.

Problem 12.3

Analyze the character of the pulse method of metal cluster generation.

A general scheme of this method of cluster generation is given in Figure 12.2 [186, 258]. A liquid drop consisting of metal-containing molecules with a lightly ionized admixture is injected into a dense buffer gas and heated by pulsed gas discharge. This drop absorbs an electromagnetic wave of high-frequency discharge because of the higher degree of plasma ionization there. This leads to strong heating of the drop and surrounding gas up to temperatures of several thousand Kelvin during a short time and as a result of consumption of a small electromagnetic energy. Under this action, the compounds constituting the drop are decomposed into atoms, the metal atoms join to form clusters, and the other components propagate over the buffer gas as a result of diffusion in the form of atoms and molecules, while the diffusion coefficient of forming clusters is small and the clusters remain in the initial region. Cluster growth proceeds until the buffer gas with clusters expands through a nozzle in a vacuum. As a result of standard methods [259–261], atoms of a buffer gas are pumped after the nozzle or orifice and clusters are charged by a crossing electron beam and their motion is governed by electric fields. The pulsed beam formed contains large charged metal clusters. This scheme for pulse generation of a metal cluster beam may be realized under certain external conditions, and we consider these below.

We suppose for simplicity that the liquid drop injected into a buffer gas has a spherical shape and conserves this shape after transformation into a gas. The

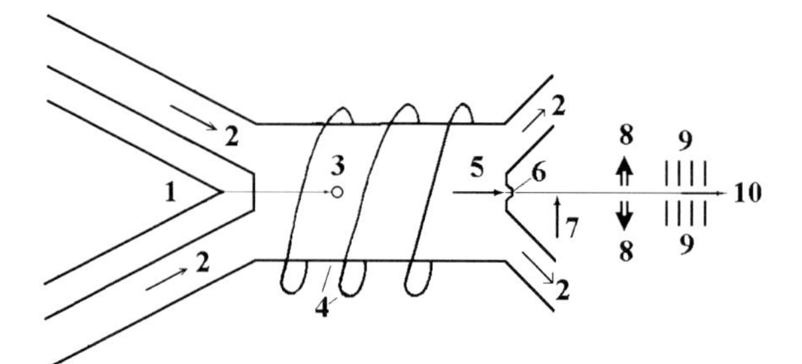

Fig. 12.2 Pulse generation of a cluster beam. 1 – source of liquid drops; 2 – flow of buffer gas; 3 – liquid drop in a buffer gas flow; 4 – waveguide and gas discharge; 5 – flow of a buffer gas with metal clusters; 6 – nozzle or orifice; 7 – electron beam; 8 – pumping; 9 – electric field optics; 10 – beam of charged clusters.

buffer gas is located in a waveguide, and a pulsed gas discharge causes the fast heating of the gas up to a certain temperature T_o. As a result, the drop is transformed into a gas consisting of molecules, the molecules are decomposed into atoms, and the atomic metal vapor is transformed into a gas consisting of clusters. For cluster generation criterion (9.54)

$$T_1 < T_o < T_2$$

must be fulfilled.

Problem 12.4

Find the parameters of pulse generation of metal clusters if drops of compounds MoF_6, IrF_6, WF_6, and WCl_6 with a radius of 10 μm are injected into argon at a pressure of 50 atm and are excited by pulsed gas discharge. The drops contain a lightly ionized admixture, and the input energy is consumed in drop excitation to a great extent. Parameters of the molecules indicated are given in Table 9.2.

Drops of the indicated compounds are excited by a short pulse (about 10^{-6} s) of an electromagnetic wave that acts mostly on the drops owing to the presence of a lightly ionized admixture. As a result of absorption, the drop is evaporated and expands, and argon penetrates inside the drop. For definiteness, we assume the concentration of argon atoms inside the drop in the course of the process to be $c = 50\%$. We give in Table 12.1 the total mass of metal atoms M in a drop and the total number of metal atoms n_M in the drop. Note that in this case the input energy for heating and destruction of the drop is not large; in particular, the energy that is consumed for the transformation of this drop into metal and halogen atoms is about 10^{-4} J, that is, various forms of gas discharges are available for this.

We give in Table 12.1 typical temperatures T_1, T_2 that characterize the decomposition of the above-mentioned molecules and the joining of metal atoms to form clusters. We take the temperature T_o of argon outside a drop such that $N_{sat}(T_o) = 0.1 N_m$, and its values are given in Table 12.1. Note that since the temperature T_* at the drop center is lower than T_o, the number density of metal atoms at this temperature $N_{sat}(T_*)$ is significantly less (Table 12.1).

Problem 12.5

Analyze the character of heat transport and metal transport for the pulse method of generation of metal clusters.

The heat regime of this drop is similar to that in the stationary regime. Indeed, the energy for the destruction of metal-containing molecules is taken from the surrounding buffer gas, which does not contain metal atoms. Halogen atoms formed as a result of the destruction of metal-containing molecules and their radicals leave the drop region, whereas metal atoms join to form clusters and remain in this region. Such a regime is established during a typical time of

$$\tau_{eq} = \frac{r^2}{3\chi(T_*)} ,$$

Table 12.1 Parameters of cluster formation as a result of the destruction of a drop with radius 10 μm. This drop consists of metal-containing molecules and is located in argon at a pressure of 50 atm [186].

	MoF_6	IrF_6	WF_6	WCl_6
r, μm	45	48	46	40
M, 10^{-9} g	4.9	16	8.9	6.8
n_M, 10^{13}	3.1	5.0	2.9	2.2
T_2, 10^3 K	5.9	5.7	7.3	7.3
T_o, 10^3 K	5.0	4.9	6.2	6.2
T_*, 10^3 K	4.3	2.8	5.2	4.4
N_m, 10^{19} cm^{-3}	4.2	6.6	3.5	4.2
$N_{sat}(T_*)$, 10^{17} cm^{-3}	0.05	$5 * 10^{-4}$	1	0.05
τ_{eq}, 10^{-6} s	0.3	0.8	0.3	0.3
τ_o, 10^{-6} s	15	4	10	2
$\overline{n(\tau_o)}$, 10^5	10	2	4	0.6
$n_M/\overline{n}$, 10^7	3	26	8	35
Δx, μm	1	1	1	1

where χ is the thermal diffusivity coefficient of argon, whose values are taken from [148, 262], as well as the values of the diffusion coefficients for halogen atoms (they are modeled by inert gas atoms) and argon. The forming halogen atoms leave the drop region freely, and the inverse process of their attachment to metal-containing radicals is not significant. The total time of transformation of metal-containing molecules into metal clusters given by formula (9.62) is determined mostly by the destruction of these molecules and their radicals

$$\tau_o = \frac{6}{\nu_d(T_*)} ,$$

where $\nu_d(T_*)$ is the rate of destruction of molecules and radicals in collisions with argon atoms. Under these conditions, cluster growth is determined by the coagulation process, and the average number of cluster atoms $\overline{n}$ with time t is given by formula (9.32):

$$\overline{n} = 6.3 \cdot (k_o N_m t)^{1.2} ,$$

where N_m is the total number density of bound metal atoms and k_o is the reduced rate constant of cluster collisions (Tables 2.1 and 2.2). The longer the drop is located in the buffer gas, the larger the clusters that are formed in this process. We give in Table 12.1 the minimal value of the average cluster size $\overline{n}(\tau_o)$ if a time of cluster location in argon corresponds to the total time of destruction of metal-containing molecules. In reality, this process leads to larger cluster sizes. The ratio $n_M/\overline{n}$ is the maximum number of clusters formed in this process.

The decay of metal-containing molecules with the release of halogen atoms leads to an increase in the gas pressure, whereas the joining of radicals and clusters causes a decrease in the pressure. We assume a soft regime of these processes, so that atoms of a buffer gas penetrate into this region, compensating the variations in the gas pressure. In addition, we assume that the volume of the region where bound metal atoms are located is preserved in the course of formation and growth of clusters; the concentration of bound metal atoms in the volume where they are located with respect to atoms of a buffer gas is taken as $c_M = 50\%$.

Problem 12.6

Determine the displacement of a metal component in the expansion of a drop with a metal-containing compound at the end of the evolution of the metal component for the pulse regime of cluster generation.

As a result of these processes, released fluorine atoms spread over a given space owing to their diffusion in argon. Let us estimate the average displacement Δx of metal clusters as a result of their diffusion motion, which is given by the formula

$$\Delta x^2 = 2 \int D dt \, ,$$

where the diffusion coefficient D of large clusters in a buffer gas is given by formula (6.5) in the case where the cluster size is small compared with the mean free path of atoms in the gas. According to formula (6.15) we represent this dependence on the cluster size in the form $D = d_o / n^{2/3}$. On the basis of the balance equation for a number n of cluster atoms

$$\frac{dn}{dt} = k_o n^{2/3} N_m \, ,$$

we obtain for the average displacement Δx of metal clusters as a result of their diffusion

$$\Delta x^2 \approx \frac{3.7 d_o n^{1/6}}{k_o N_o} \, ,$$

where N_o is a typical number density of metal atoms. Taking the parameters of the injected drops of metal-containing molecules in accordance with the data in Table 12.1, one can estimate $\Delta x \approx 1\,\mu\text{m}$ for the regimes under consideration. This value is less than a size of injected drops, that is, propagation of clusters over a volume by their diffusion in a buffer gas is not significant.

Analyzing the results of all the tasks for the pulse generation of metal clusters, one can conclude that this method of generation of metal clusters from a plasma allows one to obtain intense pulsed beams of clusters. The number density of bound atoms in cluster beams obtained by the pulse plasma method is two orders of magnitude higher than that for the stationary plasma regime and is comparable with that in intense beams of clusters consisting of gas atoms or molecules. But in intense beams of clusters consisting of gas atoms or molecules, free atoms or molecules are also present. They are found in equilibrium with clusters and can influence subsequent processes of the application of cluster beams. In pumped beams of metal clusters, free metal atoms are practically absent.

12.2
Laser Generation of Metal Clusters

Problem 12.7

A flux of metal atoms is emitted by a metallic surface as a result of its irradiation by a narrow pulsed laser beam [263, 264]. Analyze the character of cluster beam generation in this method of atom formation.

The scheme of this method of cluster generation is given in Figure 12.3 [265, 266]. A focused pulsed laser beam is directed to a metallic surface and partially absorbed by it. Heating of a metallic surface is compensated by the evaporation of atoms, which creates a flux of evaporated metal atoms. This beam of evaporated atoms is cooled owing to expansion, and when the number density of metal atoms exceeds significantly that at the saturated vapor pressure at a current temperature, the nucleation processes start and an atomic beam is later transformed into a beam of metal clusters.

Of course, this regime of cluster generation when the energy balance in the irradiated spot of a metal surface consists in transformation the energy of absorbed laser radiation into the heating of a surface and subsequent evaporation of atoms corresponds to a certain range of laser intensity and pulse duration since there are different channels of consumption of the laser energy. On the other hand, because in this regime of the energy balance the absorbed energy is converted into the energy of evaporated atoms, which in turn are converted into metal clusters, this method of cluster generation is characterized by high efficiency.

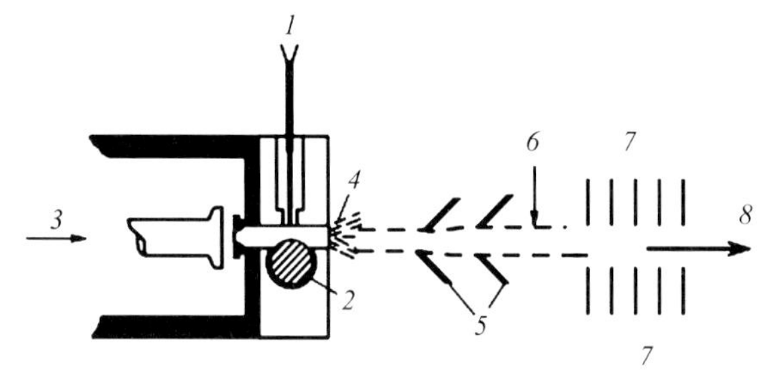

Fig. 12.3 Laser generation of a cluster beam. 1 – laser beam; 2 – rod; 3 – flow of a buffer gas; 4 – beam of a buffer gas and clusters; 5 – skimmers; 6 – electron beam; 7 – ion optics and accelerator; 8 – final beam.

Problem 12.8

Laser generation of metal clusters results from the interaction of a short laser pulse with a metal surface in a regime where the absorbed energy is consumed in the evaporation of metal atoms. For copper, molybdenum, and tungsten surfaces, determine the temperature T of the evaporated surface if the power of the absorbed laser radiation is transformed into the power of atom evaporation and ranges from 10^4 to $10^6\ \mathrm{W/cm^2}$.

In the regime under consideration the balance equation for the power per unit surface has the form

$$J_{\mathrm{las}} = \varepsilon_o j_{\mathrm{ev}} ,$$

where J_{las} is the energy flux of laser radiation absorbed by the metallic surface, ε_o is the binding energy of metal atoms, and j_{ev} is the flux of evaporated atoms. We neglect a thermal energy of atoms compared to the atom binding energy in bulk metal and use formula (4.13) for the flux of evaporated atoms. Taking the basic temperature dependence for this flux to be approximately $\exp(-\varepsilon_o/T)$ and assuming the surface temperature to be on the order of the boiling point T_b, we obtain for the above balance equation

$$J_{\mathrm{las}} = J_b \exp\left(\frac{\varepsilon_o}{T_b} - \frac{\varepsilon_o}{T}\right) ,$$

where the energy flux J_b corresponds to the boiling point T_b:

$$J_b = \varepsilon_o \sqrt{\frac{T_b}{2\pi m_a}} N_b .$$

Here m_a is the atomic mass and N_b is the number density of atoms at the boiling point. The values of the parameter J_b are $3.4\cdot 10^4\ \mathrm{W/cm^2}$ for copper, $3.9\cdot 10^4\ \mathrm{W/cm^2}$ for molybdenum, and $3.6\cdot 10^4\ \mathrm{W/cm^2}$ for tungsten.

From the above balance equation for laser absorption and atom evaporation we obtain for the surface temperature

$$T = \frac{\varepsilon_o}{\ln \frac{J_b}{J_{\mathrm{las}}} + \frac{\varepsilon_o}{T_b}} . \qquad (12.1)$$

We give in Table 12.2 the values of the surface temperature determined on the basis of formula (12.1). Note that the vapor pressure near the metal surface at each temperature is proportional to the ratio J_{las}/J_b.

Problem 12.9

Find the criterion for the duration of a laser pulse when the energy of laser radiation absorbed by a metal surface is consumed for the evaporation of metal atoms rather than for heating the surrounding metal.

Table 12.2 Temperature of a metal surface T expressed in 10^3 K depending on the energy flux J_{las} of absorbed laser radiation.

J_{las}, W/cm^2	10^4	$3 \cdot 10^4$	10^5	$3 \cdot 10^5$	10^6
Cu	2.6	2.8	3.1	3.4	3.7
Mo	4.5	4.8	5.2	5.7	6.3
W	5.4	5.8	6.2	6.7	7.3

A rough criterion that metal evaporation gives the main contribution to the consumption of laser radiation has the form

$$J_{las}\tau \gg TNl \sim TN\sqrt{\frac{\chi}{\tau}} .$$

Here N is the number density of metal atoms, l is the heat penetration depth during the pulse duration τ, and χ is the thermal diffusivity coefficient. Using rough estimations $T \sim 10^4$ K and $N \sim 10^{22}\,\text{cm}^{-3}$, and the thermal diffusivity coefficient is $\chi \sim \sqrt{\lambda v}$, where $\lambda \sim 10^{-7}$ cm is the mean free path for electrons in a hot metal and $v \sim 10^7$ cm/s is a thermal velocity of electrons (we assume the electron number density to be on the order of N). On the basis of these estimations we obtain the above criterion in the form

$$J_{las}\sqrt{\tau} \gg 10^3 \frac{\text{J}}{\text{cm}^3\text{s}^{1/2}} . \qquad (12.2)$$

As follows from criterion (12.2), the laser method of cluster generation is favorable for large laser intensities. For example, let us take the pulse duration $\tau \sim 10^{-5}$ s that corresponds to the absorbed laser intensity $J_{las} \gg 3 \cdot 10^5\,\text{W/cm}^2$. In addition, electric breakdown of a gas of evaporated atoms is possible at large laser intensities [267] when the energy of laser radiation is consumed upon multiplication of electrons in a vapor as a result of ionization processes.

Problem 12.10

Analyze additional channels of the energy balance in the absorption of laser radiation by a metal surface when this method is used for generation of metal clusters.

In considering the energy balance of a heated metal surface, we note that it emits radiation. As an estimation we model the heated surface as a blackbody, and we have that at the lowest temperature $T = 2600$ K in Table 12.2 the radiation intensity is 260 W/cm^2 and at the highest temperature, $T = 7300$ K, it is $1.6 \cdot 10^4\,\text{W/cm}^2$. As is seen, radiation of a heated surface gives a small contribution to the energy balance; moreover, the radiation intensity is lower than that for a blackbody of the same temperature.

Another mechanism of laser absorption is due to a plasma formed in the vicinity of an irradiated metal surface. Indeed, the formation of electrons as a secondary effect of laser irradiation causes the subsequent heating of electrons as a result of the interaction of laser radiation with electrons that have collided with evaporated atoms. The ionization of atoms by electron impact is a subsequent process of this interaction that leads to electric breakdown of an evaporated vapor. As a result, subsequent absorption of laser radiation results from the interaction with a forming plasma and does not lead to evaporation of atoms. Therefore, laser breakdown of an evaporated vapor decreases the efficiency of atom evaporation.

Laser breakdown proceeds at high laser intensities and for long laser pulses. Under typical conditions, the threshold of this instability is about 10^7 W/cm^2 for a specific laser power [267] if the laser pulse duration is about 10^{-6}–10^{-5} s and size of the irradiated spot is about 10–100 μm. Summing the results of the above analysis, we find that effective evaporation of a metal surface proceeds in a narrow range of laser intensities and pulse durations.

Problem 12.11

Analyze the character of transformation of an atom flux resulting from the irradiation of a metal surface by a laser pulse into a beam of metal clusters.

Above we considered only the first stage of cluster generation in the laser irradiation of a metal surface that consists in evaporation of metal atoms. The subsequent stages of evolution of a beam of evaporated metal atoms are given in Figure 12.4. The metal atoms that form at the surface are characterized by the semi-Maxwell velocity distribution function of atoms:

$$f(v_x) = \frac{m_a}{T_{sur}} j_{ev} \exp\left(-\frac{m_a v_x^2}{2T_{sur}}\right), \quad v_x > 0,$$

where m_a is the atomic mass, T_{sur} is the temperature near the surface, x is directed perpendicular to the surface, and the atom flux from the metal surface is

$$j_{ev} = \int_0^\infty v_x f(v_x) d v_x = N_{sur}\sqrt{\frac{2T_{sur}}{\pi m_a}}.$$

Correspondingly, in this case the energy flux due to free metal atoms is

$$J_{ev} = \int_0^\infty v_x \frac{m v_x^2}{2} f(v_x) d v_x + T_{sur} j_{ev} = 2T j_{ev}.$$

The first term corresponds to the longitudinal atomic motion, and the second one relates to the transverse direction. If the probability of atom attachment to the surface upon contact is one, the number density of free metal atoms at the surface is equal to half the number density for the saturated vapor pressure because half of the atoms will move away from the surface.

This distribution of free metal atoms is realized at distances from the surface on the order of the mean free path of atoms in a forming vapor and less. After several collisions this atom flux is transformed into a gas-dynamic flux. Characterizing this

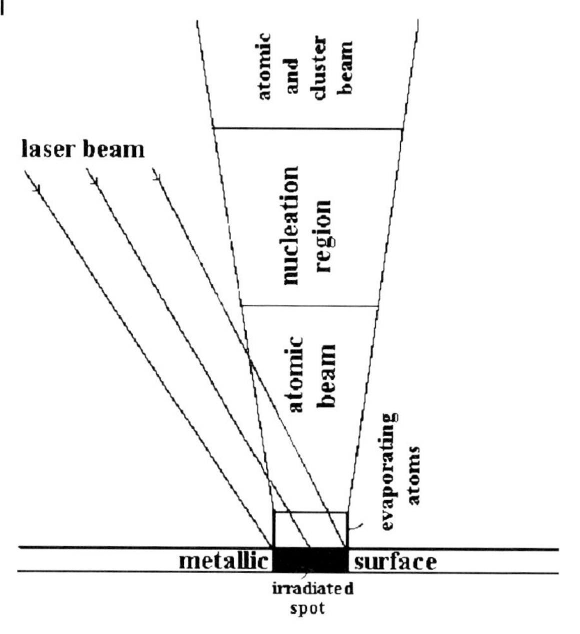

Fig. 12.4 Stages of laser pulse generation of clusters when focused laser radiation is absorbed by a metal surface and is consumed for evaporation of surface metal atoms.

beam by the atom temperature T_b and the number density N_b near the surface, we have that this beam expands the sound speed

$$u = \sqrt{\frac{\gamma T_b}{m_a}} ,$$

and for an atomic gas $\gamma = 5/3$.

If we assume that free metal atoms do not return to the surface as a result of collisions, from the conservation of the atom flux and heat flux in the course of beam formation one can find the connection between the parameters of the fluxes [268–270]:

$$T_b = 0.69 T_{sur} , \quad N_b = 0.25 N_{sat} . \tag{12.3}$$

As the atom flux moves away from the surface, it expands, and for the adiabatic character of the expansion process we have

$$\frac{N(x)}{T^{3/2}(x)} = \text{const} .$$

If the boundary of a gas dynamic beam of atoms forms an angle α with the normal to the surface, and the radius of an irradiated spot at the surface is R_o, then the radius R of the beam boundary at a distance x from the surface is

$$R = R_o + x \tan \alpha ,$$

and the number density of atoms at this distance is

$$N(x) = \frac{N_b}{\left(1 + \frac{x \tan \alpha}{R_o}\right)^2},$$

where N_b is the number density of metal atoms in a beam near the surface. Correspondingly, in the adiabatic regime of beam expansion we have

$$T(x) = \frac{T_b}{\left(1 + \frac{x \tan \alpha}{R_o}\right)^{4/3}},$$

where T_b is the beam temperature near the surface. Because of the sharp temperature dependence of the number density of metal atoms, in the course of removal from the surface the degree of vapor supersaturation (2.58) increases. At some distances the nucleation of an atomic beam starts. But because of fast propagation of the atomic beam, the total nucleation in it is problematic. To increase the time during which nucleation is possible, a buffer gas is added, and an atomic beam is stopped in the buffer gas when the vapor pressure in the beam corresponds to the pressure of the buffer gas. Note that this method of cluster generation under typical conditions leads to the formation of small clusters of sizes below $n \sim 100$.

Table 12.3 lists the parameters of an evaporated metal surface and the atomic and cluster beams formed at the specific laser power of $3 \cdot 10^6\ \mathrm{W/cm^2}$ if 30% of the laser energy is absorbed by the surface and the sticking probability for metal atoms to the surface is one. The assumption that the radius of an irradiating spot on the surface exceeds significantly the mean free path of atoms is also used. The data in Table 12.3 are based on the general character of processes [267, 272]. A laser beam heats the surface up to the temperature T_{sur} that creates the pressure p_{sur} of a metal vapor near the surface. The expansion angle for the atomic beam is taken

Table 12.3 Parameters of atom evaporation from a metal surface under the action of a laser pulse.

Metal	Cu	Mo	Ag	W
T_{sur}, 10^3 K	3.2	5.3	3.3	6.7
p_{sur}, atm	7.4	6.4	68	7.4
T_b, 10^3 K	2.2	3.7	2.3	4.6
N_b, $10^{18}\mathrm{cm}^{-3}$	4.2	2.2	38	2.0
p_b, atm	1.3	1.1	12	1.5
$p_{sat}(T_b)$, atm	0.03	0.01	0.05	0.015
j_{ev}, $\mathrm{g/cm^2 s}$	150	120	1800	180
$\overline{n}$	100	200	60	200
l, μ	50	100	20	100

as $\alpha = 10^\circ$, and $\overline{n}$ is the average cluster size when all atoms are transformed into clusters. Next, T_b, N_b, p_b are, respectively, the temperature, the number density of atoms, and the pressure in a beam formed near the surface, $p_{sat}(T_b)$ is the saturated vapor pressure at the beam temperature T_b, and $\overline{n}$ is the average cluster size after nucleation is complete. This parameter, together with a typical length $l = u\tau$ of nucleation, is found on the basis of formula (9.27). Note that because the binding energy of atoms ε_o in a cluster exceeds significantly a thermal energy of free atoms, the nucleation process leads to a strong gas heating. As a result, a small amount (10–20%) of the beam atoms are converted into clusters after expansion of the beam. After beam expansion in a buffer gas, additional nucleation is possible.

The cluster beam intensity for the laser method of cluster generation is less than that for the plasma method when clusters are formed in an arc cluster plasma, but the mass flux of metal atoms in a beam near the surface j is higher in the laser method.

12.3
Generation of Clusters from a Metal Surface

Problem 12.12

Determine the maximum intensity of bound metal atoms in a cluster beam if the metal atoms are formed on a metal surface, and an equilibrium is established between a hot buffer gas and the metal surface.

The traditional scheme of cluster generation (free jet expansion method) [260, 261, 271] for lightly evaporated metals consists in the accumulation of an atomic vapor in a chamber, and then this vapor passes through a nozzle and is converted into a beam of clusters. There are various methods to generate metal atoms from a metal surface in a neighboring region (Figure 12.5). In the case of equilibrium of these metal atoms with a buffer gas, their number density does not

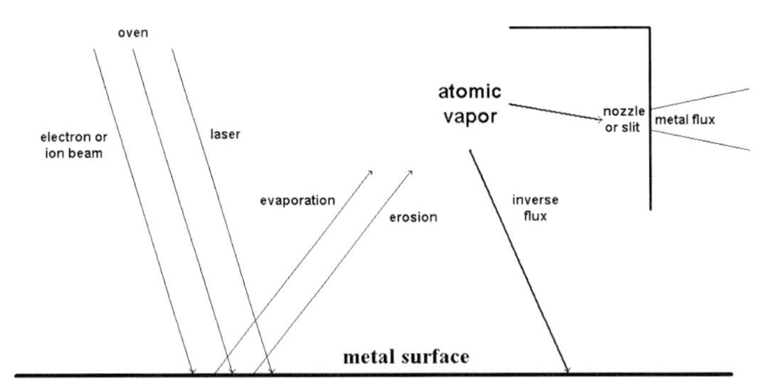

Fig. 12.5 Processes in the course of generation of beams of metal cluster by excitation of a metal surface, accumulation of evaporated atoms in a chamber, and formation of a cluster beam after passage of an atomic beam through a nozzle [145].

exceed that at the saturated vapor pressure since an excess of atoms will attach to the surface. We take the melting point as the upper surface temperature from the requirement of the surface stability. As is seen, the saturated number density of metal atoms at the melting point $N_{sat}(T_m)$ is the upper value for the number density of metal atoms under equilibrium conditions. The values of this parameter $N_{sat}(T_m)$ are given in Table 12.4. Correspondingly, the maximum mass flux in the form of metal atoms when a buffer gas with metal atoms passes though an orifice is

$$J_{max} = m_a \sqrt{\frac{T}{2\pi m_a}} N_{sat}(T_m) . \qquad (12.4)$$

Values of the maximum flux of metal atoms through an orifice J_{max} are given in Table 12.4. Note that the orifice radius exceeds significantly the mean free path of metal atoms in a buffer gas.

Table 12.4 Number density of metal atoms $N_{sat}(T_m)$ at the melting point T_m and saturated vapor pressure, and the metal flux J through a nozzle or an orifice when the equilibrium takes place in a hot buffer gas at the melting point of the metal surface [186, 273].

Metal	T_m, K	$N_{sat}(T_m)$, 10^{13} cm^{-3}	J_{max}, mg/(cm^{2}s)
Ti	1941	2.4	4.4
V	2183	11	22
Fe	1812	17	33
Co	1768	3.5	6.8
Ni	1728	2.1	4.0
Zr	2128	0.005	0.013
Nb	2750	0.39	1.2
Mo	2886	13	41
Rh	2237	1.9	5.5
Pd	1828	22	59
Ta	3290	2.3	8.3
W	3695	13	65
Re	3459	8.1	39
Os	3100	9.6	44
Ir	2819	3.2	14
Pt	2041	0.081	0.31
Au	1337	0.017	0.05
U	1408	$5.4 \cdot 10^{-7}$	$2 \cdot 10^{-3}$

Problem 12.13

The maximum rate of deposition of silver onto a substratum that is attained experimentally by free jet expansion of a silver vapor from an oven is 74 nm/s [274, 275]. Compare this value with the maximum one according to (12.4).

In the silver case we have the melting point for a bulk system as $T_m = 1235$ K and the boiling point as $T_b = 2435$ K, for a bulk system and because the atom binding energy for silver is $\varepsilon_o = 2.87$ eV, the saturated vapor pressure at the melting point is

$$p_{sat}(T_m) = \exp\left[-\frac{(T_b - T_m)\varepsilon_o}{T_m T_b}\right] = 1.7 \cdot 10^{-6}\,\text{atm}\,,$$

that corresponds to the number density of silver atoms $N_{sat}(T_m) = 1.0 \cdot 10^{13}\,\text{cm}^{-3}$. From this, on the basis of formula (12.4), we obtain $J_{max} = 2.2\,\text{mg}/(\text{cm} \cdot \text{s})$. Assuming the density of atoms in a film that forms as a result of the deposition of atoms or clusters is identical to that of bulk silver, $\rho = 10.5\,\text{g/cm}^3$, we find from this

$$\frac{dl}{dt} = \frac{J_{max}}{\rho} \approx 2\frac{\mu\text{m}}{\text{s}}\,.$$

As is seen, the maximum experimental rate of silver deposition onto a target is one order of magnitude lower than that at the melting point of the oven walls and at equilibrium between the vapor inside the oven and its walls.

Problem 12.14

Determine the concentration of bound metal atoms in a cluster beam if metal atoms are formed on a metal surface and equilibrium is established between a hot buffer gas and the metal surface. Find the efficiency of cluster formation in a nonuniform hot buffer gas as a result of atom evaporation from a heated metal surface.

The above analysis shows that the generation of metal cluster beams is problematic in the free jet expansion method when a metal vapor from a metal surface is accumulated in a chamber and then this vapor is converted into a cluster beam by free jet expansion in a vacuum or a low-pressure gas. The reason for this is a low saturated vapor pressure of many metals at the melting point, as well as the equilibrium number density of atoms in a chamber does not exceed that at the saturated vapor pressure and at the surface temperature because an excess of the equilibrium vapor will attach to the surface. Evidently, the surface loses stability at the melting point, so that the vapor pressure in a region near the heated metal surface is restricted by the saturated vapor pressure at the melting point. This value is not usually enough for conversion of a metal vapor into a cluster beam because of the short expansion time, and the standard scheme of cluster beam generation is not suitable for most metals.

One can overcome this problem if we use unequilibrium conditions near the metal surface. A simple scheme to generate metal clusters is represented in Figure 12.6 [273], where free metal atoms are formed as a result of evaporation of

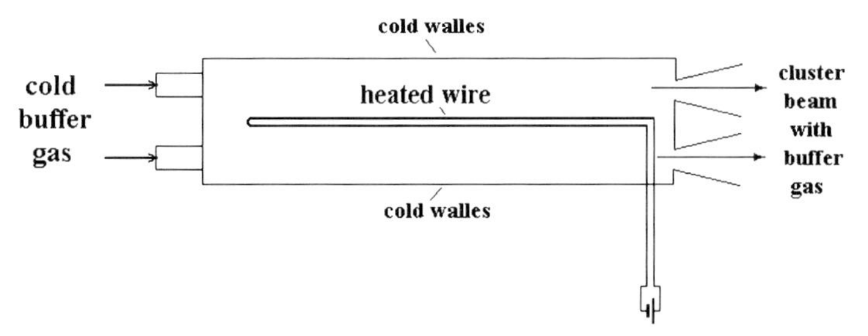

Fig. 12.6 Method of cluster beam generation as a result of evaporation of a metal surface heated by an electric current and subsequent nucleation in a cold region of a buffer gas flow.

a hot metal surface. These atoms are captured by the flow of a buffer gas and are then converted into clusters in a cold region since the flow moves between hot and cold surfaces. Equilibrium between the metal surface and clusters is absent, and some of the evaporating atoms attach to clusters in a cold region, which leads to the transformation of an evaporating metal vapor into metal clusters that are moving in the flow of buffer gas.

Note that the problem of generation of intense beams of metal atoms is important for metal transport. Indeed, the basic applications of cluster beams consist in the deposition of cluster beams on a target with the fabrication of thin films or the creation of new materials with embedded clusters, and such applications result from metal transport in the form of cluster beams between two objects. Because equilibrium between clusters and a source of metal atoms is absent, the number density of metal atoms in cluster beams is not restricted by equilibrium conditions, which must be fulfilled for atomic beams. Therefore, cluster beams are a more convenient form of metal transport compared to atomic beams.

Let us consider this problem for a simple geometry where a hot metal surface and a cold surface are parallel planes, and a dense buffer gas flows between them. We assume the heat balance between these planes to be due to the thermal conductivity of the buffer gas, that is, we have

$$q = -\kappa \frac{dT}{dx} = \text{const} ,$$

where q is the heat flux between planes, κ is the thermal conductivity coefficient for the buffer gas, T is its temperature, and x is the distance from the hot metal surface. We assume for simplicity that the process of atom attachment to clusters does not contribute to the heat balance of the buffer gas. Then if clusters of a typical size n are found in equilibrium with metal atoms in a region of a buffer gas whose temperature is T, the equilibrium number density of free atoms in a cluster region according to formula (4.19) is given by

$$N_{\mathrm{m}}^{\mathrm{eq}} = N_{\mathrm{sat}}(T) \exp\left(-\frac{2A}{3n^{1/3}T}\right) ,$$

where $N_{\mathrm{sat}}(T)$ is the atom number density at the saturated vapor pressure at temperature T of the cluster region and A is the specific cluster surface energy. Thus, the rates of cluster formation and cluster growth are determined by the number density of atoms $N_{sat}(T_m)$ related to the saturated vapor pressure at the surface temperature. These values are given in Table 12.5 for the metals under consideration.

In considering the dynamics of the cluster growth process, we base our analysis on scheme (9.20) for the formation and growth of clusters. Atoms that evaporate from a hot metal surface propagate in a given space in the form of a wave, and, in the absence of the nucleation process, the number density of atoms is $N_{sat}(T_m)$ in a region that is attained by the wave (we take the surface temperature to be equal to the melting point T_m), and the number density $N_{sat}(T)$ corresponds to the saturated vapor pressure at the surface temperature T. According to formula (9.27), the time τ for conversion of a metal vapor into clusters is given by

$$\tau = \frac{(54G)^{1/4}}{k_o N_m} \,.$$

During this time atoms displace at a distance of

$$l \approx \sqrt{2D\tau} \,,$$

where D is the diffusion coefficient of metal atoms in a buffer gas. On the basis of scheme (9.20) for the nucleation process, we neglect the evaporation process, so nucleation takes place in a cold region located far from the hot surface. We take for definiteness the distance l_o from the surface, where the equilibrium number density is low compared with the saturated number density $N_m = N_{sat}(T_m)$, according to formula

$$l_o \approx \frac{2T_m^2}{\varepsilon} \,, \qquad \varepsilon = \varepsilon_o - \frac{2A}{3n^{1/3}T_m} \,,$$

and the criterion

$$l \gg l_o \,,$$

holds true if nucleation takes place in a cold region. Evidently, the optimal pressure of a buffer gas corresponds to

$$l \approx l_o \,,$$

which provides the maximum efficiency for converting evaporating atoms into clusters.

Table 12.5 contains the parameters of the process under consideration for some metals when the temperature of a hot surface is the metal melting point. Argon is taken as a buffer gas at a pressure p that satisfies the relation $l \approx l_o$. We use the fact that the diffusion coefficient of metal atoms in a buffer gas is inversely proportional to the buffer gas pressure, as well as the parameter G defined by formula (9.21). Then the temperature gradient near the surface is taken as T_m/L, where T_m is the

Table 12.5 Parameters of a metal flux from a hot surface in a buffer gas [186].

	l_o, mm	p, atm	$G(T_m)$	D, cm^2/s	j_{dif}, cm^{-2}s^{-1}	j_{dif}, μg/cm^{2}s	j_{ev}, cm^{-2}s^{-1}	ξ, %	c_m
Ti	0.7	10	210	0.45	$1.5 \cdot 10^{14}$	0.012	$9.3 \cdot 10^{17}$	0.03	$6.1 \cdot 10^{-7}$
V	0.8	3.7	570	1.5	$1.6 \cdot 10^{15}$	0.14	$2.0 \cdot 10^{18}$	0.08	$6.8 \cdot 10^{-6}$
Cr	1.1	0.037	49000	150	$2.3 \cdot 10^{19}$	2000	$1.3 \cdot 10^{22}$	0.18	0.14
Fe	0.9	2.1	670	1.9	$2.0 \cdot 10^{15}$	0.18	$2.0 \cdot 10^{18}$	0.10	$1.2 \cdot 10^{-5}$
Co	1.3	6.0	420	1.3	$2.9 \cdot 10^{14}$	0.028	$7.0 \cdot 10^{17}$	0.04	$1.8 \cdot 10^{-6}$
Ni	0.81	50	24	0.072	$2.4 \cdot 10^{12}$	$2.3 \cdot 10^{-4}$	$5.3 \cdot 10^{16}$	0.004	$1.3 \cdot 10^{-8}$
Cu	0.77	50	17	0.046	$1.3 \cdot 10^{12}$	$1.4 \cdot 10^{-4}$	$3.2 \cdot 10^{16}$	0.004	$7.0 \cdot 10^{-9}$
Nb	0.69	86	30	0.085	$3.9 \cdot 10^{12}$	$6.0 \cdot 10^{-4}$	$6.3 \cdot 10^{16}$	0.007	$1.4 \cdot 10^{-8}$
Mo	0.87	20	120	0.39	$6.7 \cdot 10^{14}$	0.11	$3.0 \cdot 10^{18}$	0.022	$3.0 \cdot 10^{-6}$
Rh	1.2	0.12	22000	79	$4.3 \cdot 10^{18}$	730	$1.3 \cdot 10^{20}$	3.3	0.023
Pd	1.0	1.3	920	2.8	$5.1 \cdot 10^{15}$	0.90	$2.7 \cdot 10^{18}$	1.4	$3.4 \cdot 10^{-5}$
Ag	0.85	6.0	120	0.3	$6.0 \cdot 10^{13}$	0.011	$2.1 \cdot 10^{17}$	9.5	$2.9 \cdot 10^{-6}$

melting point and $L = 1\,\text{cm}$. The diffusion flux of atoms from the metal surface in a buffer gas is

$$j_{\text{dif}} \approx D\frac{N_{\text{m}}}{l_{\text{o}}}\,,$$

whereas the flux of evaporating atoms near the surface is

$$j_{\text{ev}} = \sqrt{\frac{T_{\text{m}}}{2\pi m}}N_{\text{sat}}(T_{\text{m}})\,,$$

where m is the atomic mass. Just the diffusion flux of free atoms determines the rate of cluster growth. The portion of the initially evaporating atoms that later attach to clusters is given by

$$\xi = \frac{j_{\text{dif}}}{j_{\text{ev}}}\,,$$

and Table 12.5 contains the above fluxes and the efficiency of conversion of evaporating atoms into clusters.

The heat regime under consideration is such that the nucleation process contributes slightly to the heat balance of a buffer gas, and the corresponding criterion has the form

$$\kappa\frac{dT}{dx} \gg D\frac{N_{\text{m}}}{l_{\text{o}}}\varepsilon_{\text{o}}\,.$$

Introducing the concentration $c_m = N_{\text{m}}/N_{\text{a}}$ of metal atoms, we rewrite this criterion in the form

$$c_m \ll c_{\text{o}} = \frac{l_{\text{o}}}{\varepsilon_{\text{o}}}\frac{dT}{dx}\frac{\kappa}{DN_{\text{a}}}\,,$$

Taking $dT/dx = T_{\text{m}}/L$, where $L = 1\,\text{cm}$, we obtain $c_{\text{o}} \sim 1$ in this formula, whereas the concentration c_m of free metal atoms in a buffer gas is lower by several orders of magnitude (Table 12.5). Hence, the heat balance of a buffer gas is determined by the thermal conductivity of a buffer gas, and a typical specific heat flux is of the order of $100\,\text{W/cm}^2$ for the cases in Table 12.5.

Thus, the character of formation of clusters near a hot metal surface is as follows. Evaporating atoms reach a cold region where the formation of clusters is possible. Clusters are formed first at a distance l_{o} from the hot surface, where the equilibrium number density of free metal atoms is given by formula (4.19). The temperature gradient in the direction outside the hot surface creates the gradient of the equilibrium number density of free metal atoms, which causes the diffusion flux of atoms in the case of equilibrium between free metal atoms and clusters. Therefore, clusters are formed and grow in a narrow region, and the expansion of this region as a result of atom attachment to clusters proceeds slowly in the course of atom evaporation because of a strong temperature dependence (4.19) of the equilibrium number density of free atoms. Next, the rate of cluster formation is very different for different metals (Table 12.5). This value can be increased if a liquid metal surface is used for evaporation.

Problem 12.15

Analyze the efficiency of the cluster generation method through the flow of a buffer gas near a heated metal surface.

The cluster growth under consideration proceeds under nonequilibrium conditions in a cold buffer gas when metal atoms result from the vaporization of a hot metal surface and gas flows near this surface. We assume the heat balance near the surface is determined by the thermal conductivity of the buffer gas, that is, the nucleation process gives a small contribution to the heat balance, and that a typical time of gas flow τ is large in comparison to a time $\tau_{\rm cl}$ of the nucleation process. Taking into account that nucleation proceeds in a cold region of the buffer gas, we ignore cluster evaporation in a nucleation region. Then we have the width of the nucleation region in the first stage of nucleation when evaporating atoms move in a buffer gas as a result of diffusion and join of clusters in a cold region

$$\Delta x \approx \sqrt{2D\tau_{\rm cl}} \approx 2G^{1/8}\left(\frac{D}{k_{\rm o}N}\right)^{1/2} ,$$

where D is the diffusion coefficient for metal atoms in a buffer gas. Later, new evaporating atoms attach to clusters, and the cluster region shrinks. Then the total flux of atoms attaching to clusters is

$$j \approx \frac{D}{l}N ,$$

where l is the distance of a cluster region from the metal surface and N is the number density of free metal atoms near the surface. From this we have at time t

$$n \sim (k_{\rm o}Nt)^3 , \quad N_{\rm cl} \sim KN^2N_{\rm a}t , \quad N_{\rm b} \sim N_{\rm cl}n \sim \frac{N}{G}n^{4/3} ,$$

where $N_{\rm b}$ is the number density of bound metal clusters and n is a typical cluster size. Taking $\Delta x \sim l$, and using the relation $N_{\rm b}\Delta x \sim j\,t$, we find for a typical width of the cluster region

$$\Delta x \sim \left(Dt\frac{N}{N_{\rm b}}\right)^{1/2} \sim \left(\frac{DG}{k_{\rm o}Nn}\right)^{1/2} .$$

As follows from the above equations, the maximum number density of clusters and bound metal atoms increases in time, whereas the distance from a hot surface, where it is reached, decreases. The above consideration holds true if one can neglect the evaporation of clusters in a nucleation region. We take the boundary of a cold region, where cluster evaporation is not important, such that the rate of atom attachment to clusters is half the rate of cluster evaporation. Using these rates for the liquid drop model for clusters [17], we find a distance λ of this boundary from the hot surface:

$$\lambda = \frac{T_m^2}{\varepsilon_o \frac{dT}{dx}} \left(\frac{2A}{3T_m n^{1/3}} + \ln 2 \right) .$$

Here ε_o is the binding energy per atom for a bulk metal, A is the specific surface energy of clusters (the total binding energy of atoms for a cluster consisting of n atoms is $\varepsilon_o n - A n^{2/3}$), and the temperature gradient dT/dx is determined by a buffer gas. As cluster size increases, the boundary moves to the hot surface. Because evaporating atoms are absorbed near this boundary, the maximum flux of absorbed atoms when clusters are large and the surface temperature is near the melting point is given by

$$j_{dif} \sim \frac{D N_{sat}(T_m)}{\lambda} = \frac{D\varepsilon_o N_{sat}(T_m)}{T_m^2 \ln 2} \frac{dT}{dx} . \tag{12.5}$$

Table 12.6 contains some parameters of this nucleation process in addition to those given in Table 12.5. Taking the surface temperature as the melting point T_m of the metal, we have for the flux of evaporating atoms near the surface

$$j_{ev} = \sqrt{\frac{T_m}{2\pi m}} N_{sat}(T_m) , \tag{12.6}$$

where m is the atomic mass. The evaporating atoms attach to the surface, and a small number of them form clusters. These atoms move away from the surface as a result of diffusion, and the diffusion flux of atoms is given by formula (12.5).

Table 12.6 Parameters of a metal flux from a hot surface in a buffer gas at a pressure of 1 atm.

	T_m, K	$p_{sat}(T_m)$, atm	$N_{sat}(T_m)$, 10^{13} cm^{-3}	$G(T_m)$, 10^3	j_{ev}, cm^{-2}s^{-1}	j_{dif}, μg/cm^{2}s	ξ, %
Ti	1941	$6.0 * 10^{-6}$	0.23	2.1	$9.3 \cdot 10^{17}$	0.60	0.80
V	2183	$2.5 * 10^{-5}$	8.4	2.1	$2.0 \cdot 10^{18}$	1.5	0.90
Cr	2180	$5.2 * 10^{-3}$	1700	1.8	$1.3 \cdot 10^{22}$	7600	0.68
Fe	1812	$2.4 * 10^{-5}$	10	1.4	$2.0 \cdot 10^{18}$	1.2	0.67
Co	2750	0.011	2800	2.5	$7.0 \cdot 10^{17}$	0.53	0.77
Ni	1728	$6.3 * 10^{-7}$	0.27	1.2	$5.3 \cdot 10^{16}$	0.037	0.56
Cu	1358	$3.6 * 10^{-7}$	0.20	0.85	$3.2 \cdot 10^{16}$	0.019	0.56
Nb	2750	$1.2 * 10^{-6}$	0.32	2.6	$6.3 \cdot 10^{16}$	0.17	1.7
Mo	2886	$5.9 * 10^{-5}$	15	2.4	$3.0 \cdot 10^{18}$	6.7	1.4
Rh	3237	$2.8 * 10^{-3}$	630	2.6	$1.3 \cdot 10^{20}$	280	1.3
Pd	1828	$4.6 * 10^{-5}$	18	1.2	$2.7 \cdot 10^{18}$	3.7	0.79
Ag	1235	$2.9 * 10^{-6}$	1.7	0.72	$2.1 \cdot 10^{17}$	0.21	0.56

The maximum efficiency of conversion of evaporating atoms into clusters

$$\xi = \frac{j_{\text{dif}}}{j_{\text{ev}}} ,$$

is given in Table 12.5 for $dT/dx = T_m/L$ with $L = 1\,\text{cm}$, where the atom fluxes are determined by formulas (12.5) and (12.6). Note that the flux of nucleating atoms (12.5) reduced to unit area of the hot surface, that is, the total yield of bound atoms is proportional to the area of the heated metal surface. Table 12.6 contains the values of fluxes in accordance with formulas (12.5) and (12.6).

Thus, the character of formation of clusters near a hot metal surface is as follows. Evaporating atoms reach a cold region where the formation of clusters is possible. Clusters are formed first at a distance Δx from the hot surface, where the equilibrium number density of free metal atoms corresponds to that at the saturated vapor pressure for the surface temperature. The clusters that are formed become traps for free metal atoms, and therefore the number density of bound atoms increases in time, and the region of their maximum moves to the hot surface, until the processes of cluster evaporation stop it. This method provides high fluxes for metal transport in the form of clusters, and the rate of cluster formation is strongly different for different metals. This value can be increased if a liquid metal surface is used for evaporation.

Note that use of a weakly ionized buffer gas for cluster generation allows one to control the process of cluster growth. Indeed, clusters obtain a charge in an ionized gas, and then coagulation of clusters resulting from the joining of clusters is absent. Therefore, though the main processes in this system are not connected with plasma processes, in reality use of a plasma allows us to control both the character of cluster growth and the motion of the resultant cluster beam.

12.4 Generation of Metal Clusters in Magnetron Discharge

Problem 12.16

Analyze the character of formation of metal clusters in magnetron discharge.

Magnetron discharge is an effective source of metal atoms and in a regime under heightened pressure may be used for cluster generation [276–280]. Because electrons near the cathode are locked in the magnetic field of this discharge, support of this discharge is provided by the fast ions of a buffer gas that bombard the cathode. The energy of these ions is a few hundred electronvolts, and along with the generation of the secondary electrons for discharge maintenance, the cathode bombardment by ions leads to the formation of fast atoms. These metal atoms collide with atoms of a buffer gas and then may form clusters or attach to walls of the discharge chamber. The forming clusters are captured by a buffer gas flow and are transported to the exit of the magnetron chamber, where the flow passes through a small orifice in a vacuum. After passage through the orifice, the clusters

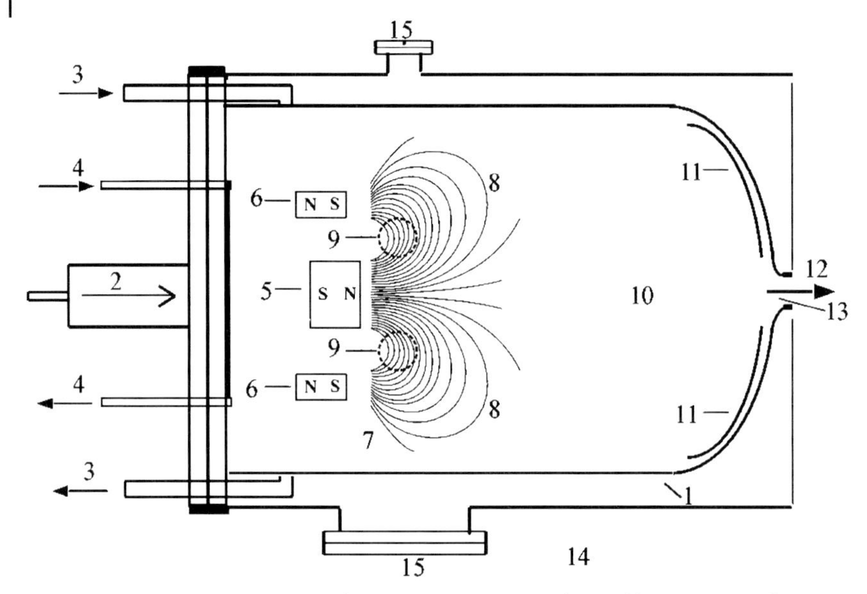

Fig. 12.7 Magnetron (aggregation) chamber [282, 283]. 1 – magnetron (aggregation) chamber; 2 – flow of a buffer gas; 3 – liquid nitrogen for cooling of the chamber; 4 – water for magnetron cooling; 5 – internal cylindrical magnet; 6 – external ring magnet; 7 – cathode; 8 – magnetic lines of force; 9 – ring of captured electrons (race track); 10 – secondary plasma; 11 – electrode for secondary plasma; 12 – flow of buffer gas with clusters; 13 – orifice; 14 – envelope; 15 – pumping.

are charged by a crossed electron beam and are transformed into a cluster beam under the action of specific electric fields, while the buffer gas is pumped after the orifice. A general scheme for the magnetron source of beams of metal clusters is given in Figure 12.7 [281, 282].

An advantage of this method for generation of metal atoms is the high energy of the forming atoms, which is on the order of 10 eV. The mean free path of these atoms in a buffer gas exceeds that for thermal metal atoms by almost an order of magnitude. Therefore, the probability of metal atoms returning to the cathode is not close to one, as it takes place in the case of atom evaporation from the surface. A disadvantage of this method of metal atom generation is a high probability that the metal atoms will attach to walls, that follows from a low pressure of the buffer gas, which is necessary for the maintenance of magnetron gas discharge.

Thus, the efficiency of magnetron discharge as a source of metal clusters results from the competition between the nucleation processes and the transport processes for atom drift to walls. This competition is the subject of our subsequent analysis.

Problem 12.17

Metal atoms are sputtered from a cathode surface as a result of bombardment of the cathode with fast ions of magnetron discharge. Find the probability P that a sputtered atom will return to the cathode before its thermalization.

We assume that the first collision of a sputtered atom with buffer gas atoms leads to the subsequent isotropic motion of a fast atom, and the spatial distribution $F(x)$ of metal atoms over distances x from the cathode has the form

$$F(x) = \frac{1}{\lambda} \exp\left(-\frac{x}{\lambda}\right), \quad \lambda = \frac{M+m}{M} \frac{1}{N_a \sigma^*(\varepsilon)},$$

where M, m are the masses of a metal and buffer gas atoms, respectively, N_a is the number density of buffer gas atoms, and $\sigma^*(\varepsilon)$ is the diffusion cross section of elastic scattering in the collision of a metal atom and a buffer gas atom. This distribution function as the probability is normalized by the condition $\int F(x)dx = 1$, and the probability that a metal atom will remove to a distance above y from the cathode is

$$\int_y^\infty F(x)dx = \exp\left(-\frac{y}{\lambda}\right).$$

As a result of diffusion, in the course of thermalization an atom is displaced, and we have the Gauss formula for the distribution function of atom distances z from the cathode if this distance is x after the first collision involving a sputtered atom:

$$W(x,z) = \frac{1}{\sqrt{2\pi\Delta}} \exp\left[-\frac{(z-x)^2}{2\Delta^2}\right]; \quad \int_0^\infty W(x,z)dz = 1,$$

where

$$\Delta^2 = 2\int_0^t D dt,$$

and D is the diffusion coefficient of a metal atom in a buffer gas, and the upper limit corresponds to the time of atom thermalization.

From this we obtain for the probability of atom attachment to the cathode in the course of its thermalization

$$P = \int_0^\infty dx \int_{-\infty}^0 dy F(x) W(x,y).$$

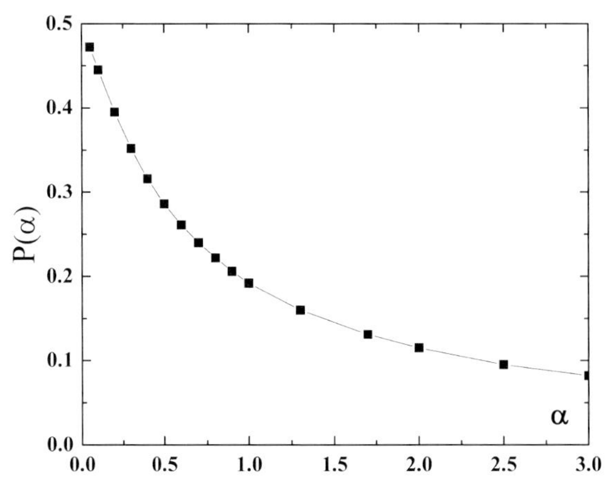

Fig. 12.8 The probability $P(\alpha)$ that a fast sputtered atom will return to the cathode according to formula (12.7).

Thus, the probability that a sputtered atom will attach to the cathode may be represented in the form

$$P(\alpha) = \frac{1}{\sqrt{\pi}} \int_0^\infty dx \int_0^\infty dy \exp[-x - (\alpha x + y)^2] , \quad \alpha = \frac{\lambda}{\sqrt{2}\Delta} . \tag{12.7}$$

Thus, the probability that a sputtered atom will return to the cathode is a result of two processes, atom removal from the cathode as a result of its sputtering, which is described by the parameter λ, and subsequent diffusion of this atom over a given region that is given by the parameter Δ. If the ratio of these parameters α is small, the probability is small. In another limiting case of a large α, propagation of an atom in a space is determined by diffusion, and the probability P is 1/2. Figure 12.8 gives function (12.7) in an intermediate range of α.

Problem 12.18

Compare typical times of isotropization and thermalization for fast metal atoms sputtered from a cathode by ion bombardment. The initial energy of sputtered metal atoms exceeds significantly their thermal energy.

Because of the weak dependence of the collision cross section on the collision energy, one can model colliding atoms by hard spheres, and then the diffusion coefficient for fast atoms with velocity v in a gas is given by

$$D = \left\langle \frac{v^2}{3\nu} \right\rangle ,$$

and the collision rate ν is

$$\nu(\varepsilon) = \frac{v}{\lambda(\varepsilon)} ,$$

where ε is the energy of a fast atom, and the brackets denote the average over the atom velocities. This gives the diffusion coefficient $D(\varepsilon)$ for fast metal atoms of a given energy in a buffer gas:

$$D(\varepsilon) = \frac{v^2}{3\nu(v)} = \frac{v\lambda(\varepsilon)}{3} .$$

Let us analyze the character of collision of a fast metal atom with atoms of a buffer gas. Expressing the velocity of a fast atom $\mathbf{v}$ through the relative velocity of colliding atoms $\mathbf{g}$ and the center-of-mass velocity $\mathbf{V}$ as

$$\mathbf{v} = \mathbf{V} + \frac{M}{M+m}\mathbf{g} ,$$

we obtain for the energy variation as a result of atom collisions

$$\Delta\varepsilon = \frac{Mm}{M+m}\mathbf{V}(\mathbf{g}-\mathbf{g}') = \mu V g(1-\cos\vartheta) = 2\frac{\mu}{M+m}\varepsilon(1-\cos\vartheta) .$$

Here m, M are the masses of buffer gas and metal atoms and ϑ is the scattering angle in the center-of-mass reference frame. Hence, the rate of energy variation for a fast metal atom is given by the equation

$$\frac{d\varepsilon}{dt} = -2\frac{\mu}{M+m}\varepsilon v N\sigma^*(v) , \tag{12.8}$$

where ε is the kinetic energy of a test metal atom and $\sigma^*(v) = \int(1-\cos\vartheta)d\sigma$ is the transport cross section for scattering of a fast metal atom by a motionless buffer gas atom. From this we obtain the time τ_ε of energy variation from E_o to ε:

$$\tau_\varepsilon = \frac{M+m}{\mu}\int_\varepsilon^{E_o}\frac{d\varepsilon}{2\varepsilon v N\sigma^*(v)} . \tag{12.9}$$

Neglecting the energy dependence for the cross section of atom collision, we obtain from this

$$\tau_\varepsilon = \frac{4(M+m)^2}{Mm}\frac{1}{N\sigma^*(v)}\left(\frac{1}{v}-\frac{1}{v_o}\right) , \tag{12.10}$$

where v_o is the initial velocity of a fast atom and v is its final velocity, which is assumed to be large compared with a thermal velocity. One can compare this time with the relaxation time τ_P, which is defined by the character of the balance equation for the velocity component v_x of a fast atom

$$\frac{dv_x}{dt} = -\frac{v_x}{\tau_P} ,$$

where v_x is the component of the mean velocity of a released atom in the direction of the initial velocity, and the expression for the relaxation time has the form

$$\tau_P = \frac{M+m}{M}\frac{1}{Nv\sigma^*(v)} . \tag{12.11}$$

By definition, during the relaxation time τ_P a released metal atom loses its initial direction of motion.

The ratio of the energy relaxation time (12.10) to the momentum relaxation time (12.11) is given by [282]

$$\frac{\tau_\varepsilon}{\tau_P} = \frac{4(M+m)}{m}\left(\frac{v_o}{v} - 1\right), \tag{12.12}$$

and because the initial atom velocity v_o exceeds significantly the final one v, that is, $v_o \gg v$, we obtain that the thermalization of fast atoms lasts significantly longer than the isotropization of fast atom motion.

Problem 12.19

Fast metal atoms are formed in magnetron discharge as a result of bombardment of the cathode by ions of a buffer gas. Assuming that the diffusion cross section of elastic scattering of a metal atom on a buffer gas atom σ^* depends on the relative energy ε of collision as $\sigma^*(\varepsilon) \sim \varepsilon^{-1/4}$ and taking the initial energy of metal atoms to be $E_o = 10\,\text{eV}$, determine the parameter α defined by formula (12.7) for titanium, copper, and silver cathodes. Evaluate the probability that a sputtered atom will return to the cathode under these conditions.

Let us determine the average distance Δ of propagation of a fast metal atom in a buffer gas as a result of diffusion, which is given by

$$\Delta^2 = 2\int D(t)dt = 2\int D(\varepsilon)\frac{d\varepsilon}{d\varepsilon/dt} = \frac{(M+m)^2}{mM}\int D(\varepsilon)\frac{\lambda(\varepsilon)d\varepsilon}{\varepsilon v}$$
$$= \frac{(M+m)^2}{3mM}\int\frac{\lambda(\varepsilon)^2 d\varepsilon}{\varepsilon}. \tag{12.13}$$

Taking the dependence of the interaction potential $U(R)$ on the distance R between metal and buffer gas atoms in a range $U \sim \varepsilon$ as $U(R) \sim R^{-k}$, we have $\sigma^*(\varepsilon) \sim \varepsilon^{-2/k}$, and $\lambda(\varepsilon) \sim \varepsilon^{2/k}$. This gives for a region size of atom propagation Δ, when the atomic energy varies from the initial one E_o to the binding energy ε_b for atoms in a bulk metal,

$$\Delta^2 = \frac{4(M+m)^2}{3mM}\int_{\varepsilon_b}^{E_o}\frac{\lambda^2(\varepsilon)d\varepsilon}{\varepsilon} = \lambda^2(E_o)\frac{k}{12}\frac{(M+m)^2}{mM}\left[1-\left(\frac{\varepsilon_b}{E_o}\right)^{4/k}\right]. \tag{12.14}$$

In the limit $E_o \gg \varepsilon_b$ this formula gives

$$\Delta = \lambda(E_o)\sqrt{\frac{k}{12}\frac{(M+m)^2}{mM}}.$$

Taking $k = 8$ in accordance with the condition for the problem, that is, $\sigma^*(\varepsilon) \sim \varepsilon^{-1/4}$, we obtain from this for the parameter α in formula (12.7)

$$\alpha = \sqrt{\frac{3mM}{4(m+M)^2}}. \tag{12.15}$$

In particular, in the case of titanium, copper, and silver we have $\alpha = 0.43, 0.42$, and 0.38 according to this formula. This gives for the probability that an atom will return to the cathode the values $P(\alpha) = 0.31, 0.31$, and 0.32, respectively, in accordance with formula (12.7).

Problem 12.20

Metal atoms are formed in a stationary magnetron discharge as a result of bombardment of the cathode by ions of a buffer gas. In the course of thermalization and in the next stage of evolution, metal atoms propagate inside the magnetron chamber and then attach to the walls of the magnetron chamber or are transformed into clusters. Determine the dependence of the cluster formation efficiency on the discharge power, the gas pressure, and the chamber radius. We are guided by discharge in argon at a pressure of 1 mbar, a power of 100 W, a chamber radius of 5 cm, and titanium, copper, and silver as the cathode material.

We first note that cluster growth takes place if the cluster temperature is restricted, so cluster evaporation is not essential. Hence, the temperature of the buffer gas is lower than the equilibrium one given by formula (4.19) and represented in Table 4.1. Table 12.7 contains the same values of the lowest number densities of free metal atoms and higher cluster sizes, which correspond to the conditions of magnetron discharge. Thus, the argon temperature in the magnetron chamber must be lower than the values in Table 12.7. The parameters of formula (4.19) are taken from [17].

The temperature of a buffer gas in a magnetron chamber is determined by the specific power that is converted into heat. We take into account that the main part of the magnetron discharge power (60–80% [284, 285]) is transferred to the kinetic energy of ions that bombard the cathode, and this part is given to the water cooled cathode. The other part is partially transformed into heat, and we assume for definiteness that 20% of the discharge power is converted into heat, that is, under the conditions of the problem it is $P = 20\,\text{W}$. Assuming heat transport is due to the thermal conductivity of a buffer gas, we have the following heat balance equation for the buffer gas temperature:

$$\frac{1}{\rho}\frac{d}{d\rho}\left[\rho\kappa(T)\frac{dT}{d\rho}\right] + p(\rho) = 0\,, \tag{12.16}$$

Table 12.7 The boiling point T_b and the equilibrium temperatures T_{eq} according to formula (4.19) at which the rates of growth and evaporation of a cluster of a given size are equal. The temperatures T_{cr} correspond to the atom number densities $N_m = 1 \cdot 10^{11}\,\text{cm}^{-3}$ and $N_m = 1 \cdot 10^{12}\,\text{cm}^{-3}$ in parentheses.

Metal	T_b, K	ε_o, eV	A, eV	T_{cr}, K, $n = 10^4$	T_{eq}, 10^3 K, $n = 3 \cdot 10^4$
Ti	3560	4.89	3.2	1730(1860)	1720(1850)
Cu	2835	3.40	2.2	1260(1370)	1260(1360)
Ag	2435	2.87	2.0	1070(1160)	1060(1150)

where T is the gas temperature, ρ is the distance from the axis of the cylindrical magnetron chamber, $\kappa(T)$ is the thermal conductivity coefficient, and $p(\rho)$ is the specific power of heat release.

For estimations we take $p(\rho) = p_o$ to be independent of ρ, and we assume the temperature dependence of the thermal conductivity coefficient as

$$\kappa(T) = \kappa(T_o) * \left(\frac{T}{T_o}\right)^{\alpha} .$$

Under these conditions the solution of equation (12.16) has the form

$$\frac{T}{T_o} = \left[1 + \frac{Q(1+\alpha)}{4\pi\kappa(T_o)T_o}\right]^{1/\alpha+1} , \qquad (12.17)$$

where $Q = p_o * \pi\rho_o^2$ is the heat release power per unit chamber length, and T_o is the wall temperature.

We now estimate the maximum temperature of a buffer gas (argon), for which we have [148, 262] $\kappa(T_o) = 1.8 \cdot 10^{-4}\,\mathrm{W/cm^2 \cdot K}$, $T_o = 300\,\mathrm{K}$, and $\alpha = 0.75$. This gives

$$\frac{T_{max}}{T_o} \approx 1.8 Q^{0.57} ,$$

where Q is the power per unit length of the chamber, expressed in W/cm. Above we estimated that at a discharge power of 100 W approximately 20 W is released in the buffer gas as heat. Taking $Q = P/L$ and $L = \rho_o$, we find in this case $T_{max} \approx 1200\,\mathrm{K}$. This estimate corresponds roughly to the critical values in Table 12.7 above which the nucleation process ceases. Thus, the optimal power of magnetron discharge as a cluster source is on order of 100 W under given parameters of discharge.

Problem 12.21

Find the dependence of the conservation efficiency of atoms in clusters for thermal metal atoms on the number density of buffer gas atoms.

We take into account that new free atoms are added to the system continuously, and the clusters formed are carried away by the buffer gas flow. This character of cluster growth is similar to the case where metal clusters are formed from a metal-containing compound that is inserted into a buffer gas in the form of molecules and thermal metal-containing molecules decay in collisions with buffer gas atoms. As in this case, metal atoms are added to the system continuously, and the number density of free metal atoms is the parameter of the problem. We now give the spatial distribution of free metal atoms $N(x)$ in the form

$$N(x) = N_o \exp\left(-\frac{x}{\lambda}\right) ,$$

where x is the distance from the cathode and the parameter λ follows from the balance equations. According to scheme (9.20) for the nucleation process, its long stage is the formation of diatomic metal molecules, which are nuclei of condensation, and correspondingly the distribution function of clusters $f_n(x)$ consisting of n atoms and located at a distance x from the cathode corresponds to the number density of diatomic metal-containing molecules formed at a distance x' from the cathode. Therefore, from the second balance equation (9.22) we have for the size distribution function

$$f_n dn = K N_a N^2 dt = N_{cl} \exp\left(-2\frac{x'}{\lambda}\right) dx' , \tag{12.18}$$

where u is the velocity of the buffer gas flow, and the total number density of clusters N_{cl} at large distances from the cathode is

$$N_{cl} = \int f_n dn = \frac{K N_a N_o^2 \lambda}{2u} . \tag{12.19}$$

The size of the clusters at a point x that are formed at point x' follows from the balance equation (9.22), which gives

$$n^{1/3}(x, x') = \frac{k_o N_o}{3} \int\limits_{x'}^{x} \exp\left(-\frac{x}{\lambda}\right) dt = \frac{k_o N_o \lambda}{3u} \left[\exp\left(-\frac{x'}{\lambda}\right) - \exp\left(-\frac{x}{\lambda}\right)\right],$$

where u is the flow velocity. From this we have

$$n(x, x') = n_{max} \left[\exp\left(-\frac{x'}{\lambda}\right) - \exp\left(-\frac{x}{\lambda}\right)\right]^3 , \tag{12.20}$$

where the maximum cluster size n_{max} is given by

$$n_{max} = \left(\frac{k_o N_o \lambda}{3u}\right)^3 . \tag{12.21}$$

The size distribution function (12.18), with the size of the clusters (12.20) depending on the point of cluster formation and observation, gives all the information about the size distribution of clusters at each distance from the cathode. In particular, from this it follows that far from the cathode the average cluster size is

$$\overline{n} = \frac{1}{N_{cl}} \int\limits_{0}^{n_{max}} f_n n dn = \frac{2}{5} n_{max} . \tag{12.22}$$

We also determine the rate ν of atom attachment to clusters in accordance with the second process of (9.20):

$$\nu = \int k_o n^{2/3} f_n dn = \frac{k_o N_{cl} n_{max}^{2/3}}{2} \left[1 - \exp\left(-\frac{x}{\lambda}\right)\right]^3 \left[1 - \frac{1}{3}\exp\left(-\frac{x}{\lambda}\right)\right]^3 ,$$

which gives far from the cathode

$$\nu = \frac{5}{4} \frac{N_b k_o}{n_{max}^{1/3}} = 0.92 \frac{N_b k_o}{\overline{n}^{1/3}} , \tag{12.23}$$

where the number density of bound atoms is $N_b = N_{cl}\overline{n}$.

Let us apply this formula to typical experimental parameters. From experiments [283], for the titanium cathode we have $N_b \sim 10^{12}\,\text{cm}^{-3}$, $k_o = 3.2 \cdot 10^{-11}\,\text{cm}^3/\text{s}$, and $\overline{n} \sim 2 \cdot 10^4$, and we obtain on the basis of formula (12.23) for a typical cluster growth time

$$\tau_{cl} = \frac{1}{\nu} \sim 1\,\text{s}\,.$$

Problem 12.22

Ignoring nucleation processes and assuming the loss of metal atoms in the magnetron chamber to be due to the attachment of metal atoms to the walls, estimate the concentration and the number density of metal atoms in the magnetron chamber near the walls. Apply the results to typical conditions of cluster generation.

In analyzing the character of distribution of metal atoms in a magnetron chamber, we are based on equations of transport of metal atoms and heat near the walls of the magnetron chamber, which in the stationary regime of magnetron discharge have the forms

$$\mathbf{j} = -D_m \nabla N_m\,; \quad \mathbf{q} = -\kappa \nabla T\,, \tag{12.24}$$

where N_m is the number density of metal atoms, T is the gas temperature, D_m is the diffusion coefficient of metal atoms in a buffer gas, κ is the thermal conductivity coefficient, and $\mathbf{j}$ and $\mathbf{q}$ are the flux of metal atoms and the heat flux, respectively. This set of equations may be considered jointly with the boundary conditions, so that the number density of metal atoms is zero at the walls and the cathode, as well as far from the cathode, and the temperature T of the buffer gas is given at the walls and the cathode due to external cooling. In addition, this set of equations must be supplemented with terms for sources of metal atoms and heat, and equations (12.24) relate to a region where these sources are absent. Below we solve this set of equations in a region without sources of metal atoms and heat and extend these solutions over the entire magnetron chamber space for estimates.

Using the axial symmetry of the problem, we rewrite equations (12.24) in a region near the walls as

$$j = -D_m \frac{dN_m}{d\rho}\,; \quad q = -\kappa \frac{dT}{d\rho}\,, \tag{12.25}$$

with the boundary conditions $N_m(\rho_o) = 0$, $T(\rho_o) = T_o$, where ρ_o is the chamber radius and T_o is the wall temperature. We note almost identical temperature dependencies for the quantities D_m and κ, and below we take [148, 262] $\kappa(T) \backsim T^{1.75}$. Next, the diffusion coefficient D_m of metal atoms in a buffer gas can be expressed through the self-diffusion coefficient of buffer gas atoms if we assume the diffusion coefficients of atom scattering for both cases to be identical. In particular, for titanium atoms in argon we have $D_m = 1.1\,D_a$, where D_a is the self-diffusion coefficient for argon atoms. Note that the accuracy of this parameter is 10–20%.

Under the above assumptions we have from the solution of the heat balance equation

$$\rho_o - \rho = \frac{\kappa_o}{1.75 q T_o^{0.75}} \left(T^{1.75} - T_o^{1.75}\right) ,$$

with $\kappa_o = \kappa(T_o)$. One can rewrite this relation in the form

$$\frac{\rho_o - \rho}{\rho_o} = C \left[\left(\frac{T}{T_o}\right)^{1.75} - 1 \right] , \quad C = \frac{\kappa_o T_o}{1.75 q \rho_o} . \tag{12.26}$$

This relation corresponds to formula (12.17), which was obtained under similar assumptions.

Accounting for identical temperature dependencies for the parameters D_m and κ, one can transform equations (12.25) into the following form by dividing these equations:

$$\alpha \frac{dT}{dc} = \varepsilon_o , \quad \alpha = \frac{\kappa}{D_m N_a} , \quad \varepsilon_o = \frac{q}{j} . \tag{12.27}$$

Here $c = N_m/N_a$ is the concentration of metal atoms in a buffer gas and ε_o is the energy released in the magnetron chamber per metal atom formed. Note that this equation holds true in the region where there are no sources of heat or thermal metal atoms, that is, in regions near the walls. This equation has a simple solution:

$$c = \alpha \frac{T - T_o}{\varepsilon_o} . \tag{12.28}$$

From this solution it follows that the maximum concentration of metal atoms is in regions of heightened temperature.

We now apply this relation to experimental conditions [283], when titanium clusters are formed in the magnetron source with argon as the buffer gas. Taking $D_m = 1.1 D_a$, we obtain in this case $\alpha = 2.7$ with an accuracy of 10%. If we assume the heating of the buffer gas inside the magnetron chamber to be due to fast metal atoms that transfer their energy to gas atoms, we obtain the mean energy of sputtered metal atoms to be $\varepsilon_o \approx 10\,\text{eV}$. In this case we account for the fact that more than 70% of the discharge power is taken by ions [284, 285] that bombard the cathode, and only part of the rest of the power is scattered on the walls. We assume that approximately 20% of the discharge power is converted into heat and transport in the form of thermal flux to the walls. In our case of a power of 100 W, approximately 20 W is transformed into heat. On the other hand, the current of this discharge is 0.5 A, and the efficiency of atom formation is 0.3 [286, 287]. Hence, we have that the rate of generation of metal atoms under these conditions is approximately $10^{18}\,\text{s}^{-1}$, which gives for the ratio of the heat release power to the rate of atom generation $\varepsilon_o \approx 300\,\text{eV}$, and for the heat flux near walls we have $q \approx 0.1\text{W/cm}^2$.

Correspondingly, equation (12.26) under experimental parameters [283] contains the constant $C = 0.06$ for the chamber radius $\rho_o = 5\,\text{cm}$ if we reduce formula (12.26) to the initial temperature $T_o = 300\,\text{K}$. If we use this formula at the chamber center as an estimate, we obtain under experimental conditions [283] $T_{max} \approx 1500$ K, which corresponds to the maximum concentration of metal atoms $c_{max} \approx 4 \cdot 10^{-4}$ at the chamber center. This gives for the number density of metal atoms

$$N_m = N_{max}\left(1 - \frac{T_o}{T}\right) .$$

From this we find that the number density of free metal atoms is almost constant in the region $T \gg T_o$. Under experimental conditions [283] we obtain for the maximum number density of metal atoms $N_{max} \sim 10^{12}\,\mathrm{cm}^{-3}$.

Problem 12.23

Under typical experimental conditions of magnetron discharge as a cluster generator, determine a typical time of transport of metal atoms to walls.

The process of thermal conductivity has a diffusion character, and the analog of the diffusion coefficient in this case is the thermal diffusivity coefficient

$$\chi = \frac{\kappa}{C_p} = \frac{\kappa}{c_p N_a} ,$$

where $C_p = c_p N_a$ is the heat capacity per unit volume, $c_p = 5/2$ is the heat capacity per atom, and N_a is the number density of atoms. By analogy with the diffusion process we have the mean time for heat propagation from a point ρ to the walls if the distance from the walls is small, $\rho_o - \rho \ll \rho_o$:

$$\tau(\rho) = \frac{(\rho_o - \rho)^2}{2\chi} ,$$

where the thermal diffusivity coefficient χ assumes to be independent of the temperature in this region.

Accounting for the temperature dependence of χ, we generalize this formula to

$$\tau(\rho) = \int \frac{d(\rho_o - \rho)^2}{2\chi} .$$

Guided by argon as a buffer gas, we use formula (12.26), which connects the distance from the walls and the temperature. The temperature dependence of the argon thermal diffusivity coefficient has the form [148, 262] $\chi(T) = \chi_o(T/T_o)^{1.75}$, and $\chi_o = 0.21\,\mathrm{cm}^2/\mathrm{s}$ at a pressure of 1 atm. If we represent formula (12.26) in the form $\rho_o - \rho = Cz\rho_o$ and with $z = (T/T_o)^{1.75}$, we have for a typical time for metal atoms to reach the walls if they are initially located a point ρ

$$\tau(\rho) = \frac{C^2\rho_o^2}{2\chi_o}[z - \ln(z + 1)] . \tag{12.29}$$

In particular, reducing this formula to the chamber center, where it may be used as an estimate, we find under the experimental conditions [283] $z(0) = 1/C \approx 17$), that $\tau(0) \approx 1$ s. For this time under the action of the buffer gas drift, metal atoms propagate in a buffer gas at a distance that exceeds the chamber length. We can find from formula (12.29) the average time of drift of metal atoms to the walls $\overline{\tau}$ if we take the number density of metal atoms to be constant over the chamber cross section:

$$\overline{\tau} = \int \frac{d\rho^2}{d\rho_o^2} \tau(\rho) \approx \frac{\tau(0)}{3} \,,$$

which gives under the parameters of the experiment [283] $\overline{\tau} \approx 0.3$ s. This is less than a typical time of buffer gas drift through a magnetron chamber. Of course, we can use this result as an estimate.

Problem 12.24

On the basis of the experimental parameters of magnetron discharge [283] (a titanium cathode has a diameter of 5 cm, the argon pressure is $p = 10^{-4}$ atm, the current is 0.5 A, and the track width is $\Delta = 0.3$ cm), determine the efficiency of cluster formation as a function of the number density of buffer gas atoms, assuming the buffer gas temperature to be equal to room temperature.

Let us analyze the experimental results [283] from the standpoint of the possibility of cluster formation. In this experiment titanium clusters are formed in argon as a buffer gas under a pressure of 0.1 mbar and a wall temperature of 200 K, which corresponds to the number density $N_a = 4 \cdot 10^{15}\,\mathrm{cm}^{-3}$. At this temperature we have for the rate constants involving titanium clusters $k_o = 2.6 \cdot 10^{-11}\,\mathrm{cm}^3/\mathrm{s}$ and $K = 2.5 \cdot 10^{-33}\,\mathrm{cm}^6/\mathrm{s}$, which gives $G = 3 \cdot 10^6$ and $G^{1/4} = 40$. On the basis of formulas (9.25) and (9.27) we have for the average cluster size at the end of the nucleation process $\overline{n} = 2 \cdot 10^4$, which corresponds to experimental data under conditions where the subsequent coagulation of clusters is absent. This corresponds to a diffusion coefficient of clusters in argon of $D_n = 5\,\mathrm{cm}^2/\mathrm{s}$, and because the drift velocity of the argon flow is $u = 15$ cm/s, a typical dimension of a nucleation region is $l \sim D_n/u \sim 0.3$ cm if it is established owing to the motion of clusters. A typical time for a cluster to be located in this region is $\tau \sim D_n/u^2 \sim 0.02$ s. Since $l \ll \Delta x \sim 0.8$ cm, a size of the nucleation region is determined by free metal atoms traveling in a space, rather than by the diffusion of clusters.

We use other typical parameters of the experiment [283] under consideration. The voltage is 225 V, the current is 0.5 A, the cathode radius is 5 cm, the sputtering region occupies a small part of the cathode, and the metal atom yield is $\zeta \approx 0.4$ [286, 287]. From this we obtain for the flux of metal atoms from the cathode $j = 6 \cdot 10^{16}\,\mathrm{cm}^{-2}\mathrm{s}^{-1}$. We assume for simplicity that under these conditions half of the sputtered metal atoms return to the cathode and the other half are transformed into clusters or attach to walls. Then the flux of metal atoms that are transformed into clusters or attach to walls is $j_m = 3 \cdot 10^{16}\,\mathrm{cm}^{-2}\mathrm{s}^{-1}$.

Comparing a typical time for an atom to travel to the walls ($\rho_o^2/D_m \sim 0.01$ s) with a typical time of atom transport in a flux ($L/u \sim 1$ s, where $L \sim 10$ cm is a typical distance of atom transport and $u \sim 10$ cm/s is the flow velocity), we found a typical number density of free metal atoms to be $N_{\mathrm{m}} \sim 10^{14}\,\mathrm{cm}^{-3}$, which corresponds to a concentration of free atoms in a buffer gas of approximately 2%. Formula (12.28) gives for this concentration of metal atoms the temperature at the chamber center as approximately 3000 K. Cluster growth is absent at such temperatures according to the data in Table 12.7 and starts far from the cathode where the number density of metal atoms is below its typical value [281]. Note that formula (12.17) and (12.28) give in this case the maximum concentration of metal atoms: $c \sim 0.5\%$. A comparison of this and the above concentrations of metal atoms shows the accuracy of these estimations. Next, we have from this that a subsequent increase of the discharge power will lead to a decrease in the cluster yield.

This analysis of experimental conditions allows us to describe the general character of cluster formation in a magnetron plasma. Sputtered atoms remove to a large distance $\overline{\Delta x}$ from the cathode in comparison with the mean free path λ_T of thermalized atoms in a buffer gas under a given pressure and flow velocity. Sputtered atoms are thermalized at large distances from the cathode and then can return to the cathode or form clusters, so that this possibility depends on the number density N of free metal atoms. Let us denote by N_{o} such a number density of free atoms in a thermalization region at which the probabilities for a given atom to attach to clusters and to return to the cathode are equal. Then if the number density N of forming free metal atoms satisfies the condition $N \gg N_{\mathrm{o}}$, most of the atoms that are formed attach to clusters; in the opposite limiting case, free metal atoms being formed return to the cathode. Evidently, the number density of free atoms being formed is proportional to the discharge power and increases also with an increase in the discharge voltage. Therefore, the cluster regime of evolution of the magnetron plasma is realized at high powers and voltages of magnetron discharge.

12.5 Cluster Flow through an Exit Orifice

Problem 12.25

To conserve clusters in the flux of a buffer gas and to prevent them from attachment to walls near the exit orifice, a weak glow discharge is created in the flux and the voltage is applied near the orifice. Determine the electric potential near a round orifice that conserves clusters in a buffer gas flux.

The scheme of a buffer gas flow with clusters in the conic part of a chamber near the exit orifice is given in Figure 6.1. Then weak glow discharge is created, so that electrons and ions are formed. They attach to clusters and walls, and because electrons are more mobile particles, at a certain distance from the discharge they disappear. As a result, a buffer gas flow contains negatively charged clusters and positive ions. Because the mobility of positive ions, atomic or molecular, exceeds remarkably that of clusters, they also disappear in the next stage of the flow by attaching to walls. We take the optimal regime when clusters are singly charged and then their charge creates an electric field. Under the action of this field, clusters pass to the walls.

One can prevent the attachment of charged clusters to walls by applying a voltage to the walls [149], and this voltage φ must exceed that of the charged clusters, which gives

$$\varphi > 2\pi e N_{\rm cl}\rho_{\rm o}^2 , \tag{12.30}$$

where $N_{\rm cl}$ is the number density of clusters, which are assumed to be singly charged as is required under optimal conditions, and $\rho_{\rm o}$ is the orifice radius. The drift velocity $w_{\rm o}$ of clusters near the orifice is determined by formula (6.34), where the parameters of this formula ν and $\tau_{\rm or}$ are given by formulas (6.26) and (6.32). Let us assume the velocity of a buffer gas at the orifice to be equal to the sound speed $c_{\rm s}$. The ratio of the cluster flux to the buffer gas flux is conserved if clusters do not pass to the walls. Then the cluster number density far from the orifice $N_{\rm o}$, where cluster growth finishes, and the cluster number density near the orifice $N_{\rm cl}$ are connected by the relation with that $N_{\rm o}$ far from the orifice

$$N_{\rm cl} = \frac{0.37 N_{\rm o}}{(\nu\tau_{\rm or})^{2/3}} .$$

From this we obtain for the mass flow rate of clusters that may pass through the orifice

$$J = m_{\rm a} n \cdot \pi\rho_{\rm o}^2 N_{\rm cl} w_{\rm o} = 1.3 m_{\rm a} n \frac{\varphi}{e} c_{\rm s} (\nu\tau_{\rm or})^{2/3} ,$$

where $m_{\rm a}$ is the mass of a metal atom and n is the average number of cluster atoms. Since according to formula (6.26) $\nu \sim n^{-1/3}$, the rate of cluster generation depends on the cluster size as $J \sim n^{1/3}$.

Problem 12.26

Find the size dependence of the size distribution function of clusters moving in a buffer gas flow near the orifice.

We assume that clusters grow as a result of attachment of atoms to clusters in accordance with scheme (9.20), so that the size distribution function f_n depends on the number of cluster atoms n in the main range of sizes as $f_n \sim n^{-2/3}$ according to formula (9.24). The drift velocity of clusters near the orifice differs from that of a buffer gas flow and depends on the cluster size. This drift velocity is given by formula (6.34), where the rate of collisions between a cluster and buffer gas atoms according to formula (6.26) depends on the atom size as $\nu \sim n^{-1/3}$, so that the cluster drift velocity near the exit orifice $w_{\rm o}$ depends on the cluster size as $w_{\rm o} \sim n^{-4/9}$ according to formula (6.34). Because the flux of clusters of a given size is the same in different cross sections of the tube near the orifice, the size distribution function f_n, that is, the number density of clusters of this size, is given by

$$f_n \sim n^{-2/9} . \tag{12.31}$$

Problem 12.27

An argon flow with silver clusters passes through a round orifice, and the parameters of this flow are the same as in Problem 6.23. The electric potential $\varphi = 1000\,\text{V}$ is applied to the exit orifice with radius $\rho_o = 2\,\text{mm}$ to prevent clusters from attaching to walls. Silver clusters have an average size $n = 10^4$ and are singly charged. Determine the maximum rate of clusters passing through the orifice.

Negatively charged clusters move to the flow center until their electric potential does not exceed the external electric potential φ with respect to the flow. Assuming a uniform spatial distribution of clusters, and taking the orifice radius $\rho_o = 2\,\text{mm}$, we obtain from formula (12.30) the maximum number density of clusters N_{cl} that are kept in the flow by an external electric field:

$$N_{cl} = \frac{\varphi}{2\pi e \rho_o^2} = 2.8 \cdot 10^{10}\,\text{cm}^{-3} .$$

Taking the angle α of the conic tube part to be $\alpha = 30°$ (Figure 6.1) and the argon pressure to be $p = 0.1\,\text{Torr}$, we have the drift velocity of clusters near the orifice according to formula (6.34): $w_o = 3.8 \cdot 10^3\,\text{cm/s}$ (Problem 6.23). This gives for the flow rate of silver clusters through the orifice

$$J = m_a n \cdot \pi \rho_o^2 N_{cl} = 2 \cdot 10^{-4}\,\text{g/s} .$$

Of course, this is an upper limit for the flow rate of silver clusters.

Problem 12.28

Tungsten clusters are formed from $W F_6$ in a flow of an arc argon plasma according to the scheme in Figure 12.1, so that tungsten clusters remain in a narrow region near the flow axis. An exit consists of several holes, and the central exit through which the flow with clusters proceeds has a conic shape. Under the conditions in Table 9.2 and a typical cluster size $n = 10^3$, show that even at low orifice radii attachment of clusters to walls is weak.

We take the argon temperature at the exit to be $T = 1000\,\text{K}$, which corresponds to the sound speed at the orifice, $c_s = 5.9 \cdot 10^4\,\text{cm/s}$, and the number density of argon atoms at a pressure of $p = 1\,\text{atm}$ is $7.3 \cdot 10^{18}\,\text{cm}^{-3}$. A high gas pressure causes a strong interaction between clusters and argon atoms that quickly establishes an equilibrium between them. Therefore, formula (6.34) is not correct for the cluster drift velocity near the orifice, which is equal to the sound speed at the exit. Next, the diffusion coefficient of clusters in argon under given conditions is $D_{cl} = 3.9\,\text{cm}^2/\text{s}$ for $n = 10^3$, and the mean free path of clusters is $\lambda \approx 4\,\mu\text{m}$. A decrease in the cluster flow due to attachment to walls is characterized by the exponent $\exp(-\xi)$, where the parameter ξ is given by formula (6.40)

$$\xi = \frac{4 D_{cl}}{c_s \rho_o \tan\alpha} .$$

Taking an angle of the conic surface $\alpha = 30°$ and expressing the orifice radius ρ_o in 10 μm, we obtain under given conditions

$$\xi = \frac{0.46}{\rho_o} .$$

As is seen, the attachment of clusters to walls is weak owing to the high number density of argon atoms. In particular, for $\rho_o = 10$ μm approximately 63% of clusters pass through an orifice, and for $\rho_o = 100$ μm this value is 94%. From this it follows that cluster attachment to walls near the exit orifice is not important at high pressures, whereas if clusters move in a flow of a rare buffer gas, as takes place in a magnetron plasma, an external electric field is necessary to prevent clusters from attaching to walls.

12.6
Instability of Cluster Plasma

Problem 12.29

A cluster plasma consists of a buffer gas and an admixture of a metal vapor in the form of free atoms and large clusters. Consider an equilibrium between metal atoms and clusters in the case of a small temperature gradient if an equilibrium takes place between free atoms and clusters such that the temperature gradient creates the gradient of the number density of free atoms. This causes the diffusion flux of free atoms from a region of heightened temperature that is compensated by the diffusion of clusters in this region.

The cluster plasma under consideration is realized in an arc of high pressure in a buffer gas with a metal admixture. One can check the possibility of this equilibrium on the basis of the conservation of the total number of free and bound metal atoms:

$$N_m + N_b = \text{const} ,$$

where N_m, N_b are the number densities of free and bound metal atoms, respectively. If equilibrium between atoms and clusters is established quickly, this is given by formula (4.19), which leads to the following relation between their gradients:

$$\nabla N_m = N_m \frac{\varepsilon_o}{T^2} \nabla T ,$$

and the atom flux is

$$j_m = -D_m \nabla N_m = D_m N_m \frac{\varepsilon_o}{T^2} \nabla T , \tag{12.32}$$

where D_m is the diffusion coefficient of free metal atoms in a buffer gas.

The flux of atoms in clusters $j_{\rm cl}$ is

$$j_{\rm cl} = -\int D_n n dn \nabla f_n ,$$

where f_n is the size distribution function of clusters, and according to formula (6.5) the diffusion coefficient D_n of clusters consisting of n atoms is $D_n = D_{\rm o} n^{-2/3}$. From this one can estimate the flux of atoms in cluster diffusion:

$$j_{\rm cl} \sim D_n \nabla N_{\rm b} \sim \frac{j_m}{n^{2/3}} ,$$

which is small compared with the flux of free atoms ($D_m \sim D_{\rm o}$, $j_{\rm cl}/j_m \sim n^{-2/3}$).

This means the cluster is unstable and requires that the total number of atoms $N_{\rm m} + N_{\rm b}$ not be conserved in a space. This instability leads to another character of equilibrium. Atoms then propagate fast over a space, and $N_{\rm m} \approx$ const in that space. This creates another feature of the equilibrium in a weakly nonuniform cluster plasma, namely, the number density of free atoms is constant over a space, and the equilibrium consists in cluster growth owing to the attachment of atoms to clusters in a cold region and diffusion of clusters in a hot region, where they evaporate.

Problem 12.30

Show the instability of a weakly nonuniform cluster plasma consisting of a buffer gas with a metal admixture of an arc of high pressure. The equilibrium in this plasma results from cluster growth in a cold region and diffusion of clusters in a hot region.

This cluster plasma is realized in arc of high pressure when a flux of metal-containing molecules is injected into a buffer gas. Then metal clusters are formed in an intermediate region of the positive column of arc whose temperature satisfies criterion (9.54), $T_1 < T < T_2$. Assuming a typical cluster size to be large, one can estimate it by comparing a typical time of cluster growth and a time of diffusion transport of clusters from the region of their formation. In this case the kinetic equation for the size distribution function f_n of the clusters has the form

$$D_n \Delta f_n = I_{\rm col}(f_n) , \qquad (12.33)$$

where D_n is the diffusion coefficient in a buffer gas for clusters consisting of n atoms and $I_{\rm col}(f_n)$ is the collision integral for these clusters, which accounts for the processes of growth and evaporation of clusters and is given by formula (9.6).

Let us reduce equation (12.33) to the simpler form

$$\frac{f_n}{\tau_n} + I_{\rm col}(f_n) = 0 , \qquad (12.34)$$

where τ_n is a time of diffusion motion of a cluster through a cluster region, so that $l^2 = 2D_n\tau_n$, and l is a typical size of the cluster region. According to formula (6.5) the diffusion coefficient D_n of clusters consisting of n atoms in a buffer gas is $D_n = D_o n^{-2/3}$. Hence, we have $\tau_n = \tau n^{2/3}$, where the parameter τ is independent of n.

We consider such a regime of cluster growth when clusters reach a large size compared with the critical size during their lifetime, which corresponds to criterion (9.54). Let us find the asymptotic solution of the kinetic equation (12.34) in the limit of $n \gg n_c$, where n_c is the critical size, and if the evaporation of large clusters is not essential, that is, $N \gg N_{sat}(T)$ or $\Delta\varepsilon \gg T n_c^{1/3}$. The kinetic equation has the form

$$\frac{d}{dn}\left(n^{2/3} f_n\right) + \frac{f_n}{\beta n^{2/3}} = 0 \,.$$

From this equation it follows that large clusters accumulate in the cluster region. Indeed, the lifetime of large clusters due to their diffusion from the cluster region is $\tau_n \sim n^{2/3}$, and a typical rate of atom attachment is $k_n N \sim n^{2/3}$. Formally, under these conditions one can neglect the second term of the first equation (9.22). This means that the diffusion of clusters is not essential for the cluster balance, and the size distribution function of clusters has the form of formula (9.24)

$$f_n = \frac{C}{n^{2/3}} \,, \qquad n \gg n_{cr} \,.$$

One can see that normalization of the distribution function leads to divergence. This means instability the kinetic process, which is a nonstationary one because of the accumulation of large clusters in the cluster region [288].

Problem 12.31

Analyze the above cluster instability that develops in a buffer gas with an admixture of a metal vapor and a weak temperature variation in a space. As a result of equilibrium, clusters are formed in a cold region, and atoms move from a hot region and attach to clusters in a cold region.

One can expect an equilibrium in a buffer gas with an admixture of a metal vapor if the temperature varies in a direction that we denote as z. Then metal atoms travel to a cold region and attach to clusters there. In turn, clusters travel in a hot region and evaporate there. For the analysis of this equilibrium, we study the evolution of an individual cluster that travels to a hot region for evaporation. Growth of this cluster results from the attachment of free atoms and is described by the balance equation

$$\frac{dn}{dt} = k_o n^{2/3} N \,,$$

where n is the number of cluster atoms and N is the number density of metal atoms. Next, the cluster travels owing to its diffusion in a buffer gas, and its displacement is given by

$$\frac{d\overline{z^2}}{dt} = 2D_n \,,$$

where the dependence of the cluster diffusion coefficient D_n on its size is given by formula (6.15).

On the basis of these equations, we represent the evolution of the cluster size as the cluster is displaced from an initial point in the form

$$\frac{d\overline{z^2}}{dn} = \frac{2D_o}{k_o n^{4/3} N} .$$

From this equation it follows for the average distance squared $\overline{z^2}(t)$ from an initial point that at the end of the process

$$\overline{z^2}(\infty) = \frac{\Delta^2}{n^{1/3}} , \quad \Delta^2 = \frac{6D_o}{k_o N} . \tag{12.35}$$

One can see that the larger a cluster is, the less distance it can go. As a result, an instability occurs, and, owing to the motion of atoms, all the metal is collected in a cold region in the form of clusters.

Table 12.8 [36, 288] gives values of the reduced parameter $\Delta\sqrt{N_o N}$, which does not depend on the density of metal atoms or that of a buffer gas. Under typical

Table 12.8 The reduced diffusion coefficient according to formula (6.15) and Table 6.1 for metal clusters in argon at a temperature of $T = 1000$ K and the normal number density of argon atoms $N_a = 2.69 \cdot 10^{19}$ cm^{-3}. The reduced displacement of clusters $\Delta\sqrt{N_a N}$ is given by formula (12.35) [36, 288].

Element	D_o, cm^2/s	$\Delta\sqrt{N_o N}$, 10^{15} cm^{-2}
Ti	0.91	1.59
V	1.05	3.51
Fe	1.17	2.13
Co	1.20	2.22
Ni	1.22	2.31
Zr	0.74	1.52
Nb	0.90	1.85
Mo	0.98	2.05
Rh	1.05	2.23
Pd	1.01	2.16
Ta	0.90	2.19
W	0.98	2.42
Re	1.01	2.49
Os	1.05	2.60
Ir	1.01	2.51
Pt	0.98	2.45
Au	0.93	2.32
U	0.81	2.11

parameters of a dense cluster plasma $N_a \sim 10^{19}\,\mathrm{cm}^{-3}$ and $N \sim 10^{13}-10^{15}\,\mathrm{cm}^{-3}$, we have $\Delta \sim 0.01$–0.1 cm, that is, the cluster instability is typically realized under real laboratory conditions.

Problem 12.32

Estimate the depth of penetration of a flux of metal atoms in a cluster plasma consisting of a dense buffer gas of weakly varied temperature and an admixture of an atomic metal vapor and metal clusters.

Under this equilibrium, atoms attach to clusters and quickly are formed as a result of cluster evaporation, and the atom flux is connected with the temperature gradient for a buffer gas by formula (12.32). Ignoring for simplicity cluster evaporation in a cold region, we take the rate of cluster growth due to atom attachment in the standard form

$$\frac{dn}{dt} = k_o n^{2/3} N_m \,.$$

Then the depth of cluster penetration l follows from the balance of the rates of atom attachment to clusters and atom transport, that is,

$$k_o n^{2/3} N_m N_{cl} l \sim D_m N_m \frac{\varepsilon_o}{T^2} \nabla T \,,$$

This gives for the depth l of penetration of free atoms into a cold region of the cluster plasma where clusters are located

$$l = \frac{j_m}{k_o n^{2/3} N_{cl} N_m} \sim \frac{D_m}{k_o N_b} \frac{\varepsilon_n}{T} \frac{\nabla T}{T} n^{1/3} \,. \tag{12.36}$$

Here $N_b = n N_{cl}$ is the total number density of bound atoms in clusters and n is a typical number of cluster atoms. In this regime, which provides accumulation of bound metal atoms in a narrow region of the plasma at the end of the process, most of the metal atoms are bound in clusters.

13
Conclusion

In conclusion we discuss cluster applications and also the role of cluster processes analyzed in the previous chapters in various phenomena both in the laboratory and in nature. Because of the high reactivity of clusters, their contact leads to joining, and specific properties of incident clusters are lost in a joined cluster. Therefore, clusters must be used soon after their generation, and clusters in various applications are found in the form of cluster beams or are located in a buffer gas where the interaction of clusters with gas atoms does not violate the cluster stability. Therefore, generators of cluster beams are the basis of cluster applications.

The history of cluster beam generation began half a century ago [289–291], and we now have a wide arsenal of cluster generators. Cluster generators are based on the expansion of a gas or vapor that has passed through a nozzle [290, 292], laser evaporation for the generation of heat proof metals [263], magnetron generation of metal clusters [276–280], and so on. Each of these methods corresponds to certain objects and conditions. One can observe an expansion of the types of cluster generators for contemporary applications. Some of these methods were analyzed in the problems discussed in the previous chapter.

The main cluster application is in the manufacture of new materials as a result of deposition of cluster beams on a substratum, and this aspect may have various versions. The production of thin films by cluster deposition is a branch of nanotechnology [293–298] that consists in manufacturing materials and devices with nanosized elements, and cluster beams are a convenient tool for this goal [299–302]. Cluster beams allow one to produce so-called cluster-assembled materials [303–306] that may have specific properties and are nanostructures.

If a cluster beam includes liquid clusters, their deposition on a surface leads to the formation of a fine film. For example, one can prepare in this manner mirrors of high accuracy [276]. Of course, they can be made by other methods, for example, by the deposition of an atomic beam, but the cluster beam has certain advantages. First, the intensity of cluster beams is stronger than that of atomic beams. Second, there is the possibility to govern by a cluster beam. The high mass of clusters provides better divergence of a cluster beam in comparison to an atomic beam. Charging some clusters by a crossed electron beam, one can operate by a cluster beam. Third, the deposition of clusters provides a softer regime for the process of film formation. Indeed, the heat release as a result of cluster deposition proceeds

Cluster Processes in Gases and Plasmas. Boris M. Smirnov

ISBN: 978-3-527-40943-3

owing to the surface cluster energy, whereas in the case of an atomic beam, the binding energy of each atom forming a bond is converted into heat, and strong heating in the course of deposition can lead to the production of a nonuniform film. Fourth, charging of clusters and the possibility of accelerating them allows one to optimize the velocity of a cluster beam for the production of a qualitative film.

The simplest method of film production by deposition of cluster beams, the Ion Cluster Beam Method [271], uses a beam of charged clusters that is operated by electric optics and deposited on a target. This method is of interest for microelectronics, where cluster beams allow one to fabricate thin uniform films of various materials, that is, metallic, dielectric, semiconductor, and organic films (e.g., [271, 307–310]). The maximum rate of deposition by this method when clusters are formed by free jet expansion of an evaporated metal is determined by the saturation vapor pressure of this metal at the melting point. A small value of this saturation vapor pressure compared with the atmospheric gas pressure restricts the deposition rates for some metals. The maximum deposition rate related to silver clusters is 74 nm/s [274, 275], and for deposition of zinc clusters the deposition rate reaches 100 nm/s [311, 312]. Clusters deposited by this method are found in the liquid state [260]. Note that the specific heat release in the course of this deposition by cluster beams is as low as 0.1–1 W/cm^2. Because of the relatively low intensity of cluster beams, the ion cluster beam method is only applied to the fabrication of small elements in microelectronics.

Another method of application of cluster beams is based on the low-energy cluster beam deposition technique [313–315] and involves a low-energy beam of neutral solid clusters. In this case the deposition of a cluster beam is accompanied by the growth of a target. As a result, a forming film consists of a deposited uniform matrix with embedded clusters. In contrast to the deposition of fine uniform films, in this case the manufacturing material is a uniform matrix with embedded clusters – grains. Because magic numbers of cluster atoms are preferable for the formation of solid clusters, a cluster beam consists of solid clusters of almost identical sizes in this case. Thus, this method allows one to create nanometer films deposited from a vapor with embedded clusters of almost identical sizes deposited from cluster beams. It is impossible to produce such structures by other methods, and clusters of various materials and sizes can be used for this purpose (e.g., [316–319]). These materials are uniform films with embedded clusters of similar sizes. Films with embedded clusters can be used as filters because clusters are absorbers in a certain spectral range. The spectral characteristics of these filters can be controlled by the sort, size, and density of the embedded clusters. Along with filters, films consisting of a transparent matrix with embedded clusters can be used as elements of optoelectronics. Some transitions of clusters as atomic systems can be saturated, so that these films can be used as optical locks owing to their nonlinear transparency.

Films with embedded clusters of magnetic materials (Fe, Co, Ni) are magnetic nanostructures and are like multidomain magnetic systems. In this context, the advantages of such films are as follows. First, the size of individual grains of these films, which coincides with the cluster size, is several times less than for normal

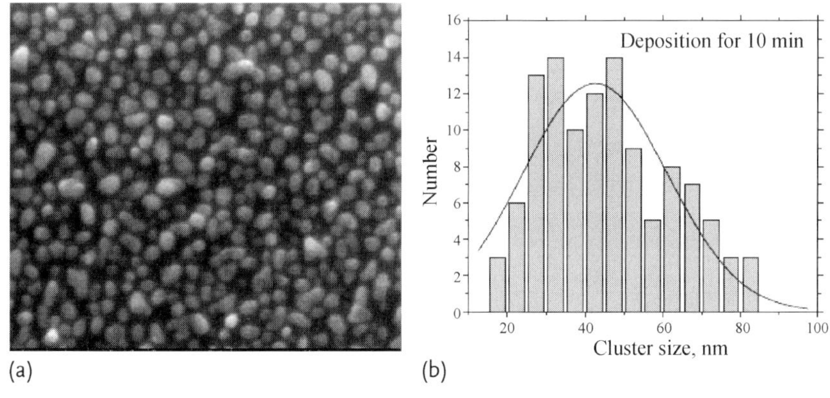

Fig. 13.1 Surface structure as a result of deposition of 5 nm silver clusters onto a silicon substratum (a) and the size distribution function of surface clusters (b) [320].

magnetic films. This fact reduces the saturated magnetic field for this magnetic material. Second, nearby sizes of embedded clusters, magnetic grains, provide improved precision and selectivity for devices on the basis on such magnetic materials. Third, the possibility to vary the type and size of embedded clusters allows one to adjust the properties of magnetic films. Therefore, films with embedded clusters such as cluster-assembled materials are a new potential material for precise devices.

It should be noted that these films, which resulted from the deposition of solid magnetic clusters onto substrates, may be of interest as a magnetic material. Indeed, in such films clusters partially conserve their individuality and can be individual domains in magnetic materials. Because the size distribution function of deposited clusters can be narrow, one can obtain in this way magnetic materials with resonant parameters that depend on cluster sizes in a cluster beam. For natural magnetic materials with a wide size distribution function of domains this is impossible.

The deposition of solid metal clusters leads to the formation of amorphous porous metal films, as shown in Figure 13.1 [320]. To illustrate the peculiarities of such films, we describe below one of applications of such films. A silver covering is an effective antibacterial remedy [321–323] that is used to kill microbes. In this process silver is a catalyst, and its action on microbes depends on the sizes of the nonuniformities of the silver surface and the sizes of microbial elements that are decomposed when in contact with the silver surface. A special study [324] demonstrates that the strongest action on bacteria proceeds from the silver surface with nonuniformities of 1–10 nm, that corresponds to the cluster method of preparation of porous metal films. This example shows that cluster methods are of interest not only for nanotechnology and microelectronics, but also for medicine.

Note that the fabrication of films by deposition may be realized by chemical methods (e.g., [325, 326]). Comparing cluster and chemical methods of film formation, when both methods may be used, one can conclude that the chemical methods are characterized by higher intensity, and cluster methods are more expensive. The main advantage of cluster methods is the quality of the films, so that cluster methods may be used under specific conditions where high-quality materials are required.

Another group of cluster applications relates to the excitation of a cluster beam by laser radiation. A pecularity of a cluster beam is that atoms are found in clusters in a condensed state, whereas the average density is that of a gaseous one. Therefore, on the other hand, this laser radiation interacts with clusters as with a condensed system, and, on the other hand this interaction causes a larger specific excitation of atoms than that in the case of an atomic gas. Because of a small total mass of clusters, the action of strong laser radiation on a cluster beam leads to maximum excitation of a cluster matter that is possible on the basis of nonexplosive laboratory methods. In the first stage of this interaction, cluster atoms are ionized owing to an overbarrier ionization under the action of strong electric fields of laser radiation [327, 328], and the subsequent interaction of a laser pulse with a plasma that is formed has a complex character and depends on various conditions of this process [327–331]. In the context of this discussion, of interest are the application properties of a nonequilibrium plasma that results from the excitation of a cluster beam. This plasma is a source of X-ray radiation [332], and the efficiency of conversion of laser pulse energy into the energy of X-rays attains 10% [333, 334]. A high degree of excitation allows one to use an excited clusters for neutron generation [335–337].

The cluster applications described above represent only one aspect to applying the understanding of cluster processes that constitutes the main task of this book. These processes are of importance for phenomena in which they partake. Let us consider this from the standpoint of the action of the electric machine of the Earth's atmosphere. It is based on the negative charge of the Earth that is supported by atmospheric processes, and this leads to secondary electric processes in the atmosphere such as lightning, and this view has a long history. In particular, Feynman [226] stated in his lectures that lightning led to charging of the Earth. The basis of this conclusion is that according to measurements some clouds have a high electric potential, and the lower part of clouds is mostly negatively charged. Then a negative charge is transferred to the Earth's surface by lightning. It is clear that the charging of the Earth starts from ionization processes under the action of cosmic radiation and radiation resulting from the radioactive decay of elements on the Earth's surface. One can expect that the charge of atomic and molecular ions is transferred to aerosols, and the of falling of negatively charged aerosol particles leads to negative charging of the Earth.

This simple explanation is true in principle, but cannot be supported by the numerical parameters of atmospheric processes. Indeed, the formation of negatively charged aerosols may result from the higher mobility of negative ions that attach to aerosols, and the falling of these aerosols leads to charge separation and charging of the Earth's surface. But the numerical parameters of the processes cannot explain the observable parameters of the charging of the Earth. The effect of aerosol charging is amplified as a result of the collision of needle-shaped ice particles with liquid drops if this process proceeds in an external electric field. Then an induced charge of a needlelike ice particle is transferred to a liquid drop from the ends of an ice particle. This process increases the rate of formation of charged aerosol particles, and therefore electric atmospheric fields are formed in clouds at altitudes of several kilometers, where ice and water aerosols exist simultaneously. But this solves only part of the problem. This mechanism leads to an equal probability of positive and negative charging of water particles, and subsequent attachment of positive and negative ions to charged aerosols will determine their charge sign. In principle, both situations are possible depending on the ion chemistry when this process leads to positive and negative charging of the Earth. Observations show that in approximately 90% of cases negative charging occurs.

From this one can conclude that charging of the Earth is not a global process that takes place over the entire surface of the Earth uniformly. It is a local process that depends on the local chemistry of atmospheric ions. Hence, one can consider that the charging of the Earth results from specific instabilities at local points of the Earth's atmosphere, and these instabilities depend on the local chemistry of atmospheric ions. This complicates the problem and is responsible for the many views on how the atmospheric electric field is created. Nevertheless, the analysis of processes involving charged aerosols creates a basis for certain explanations.

We consider one more problem involving clusters, which relates to the combustion of solid fuels and coal. As a result, solid particles are present in combustion products. For example, at thermoelectric power stations solid inorganic particles are charged in a gas stream, and then these charged particles may be removed from the gas stream by electric fields. This method cannot act on small soot particles, so their decomposition may be realized by the creation of an additional plasma. Thus, the positive solution of this problem depends on the parameters of the combustion products and the conditions of their flow with a gas. The choice of optimal conditions of gas purification may be based on a detailed understanding of the accompanying processes involving soot particles. Thus, we conclude again that a detailed description of cluster processes is necessary for the optimal purification of combustion products.

Appendix A
Mechanical and Electrical Parameters of Particles with Ellipsoidal and Similar Shapes

Below we give analytical expressions for some parameters of symmetric particles that have circular cross sections. The main geometric figure of these particles is an ellipsoid whose surface satisfies the equation

$$\frac{z^2}{a^2} + \frac{\rho^2}{b^2} = 1 \,, \tag{A1}$$

where ρ, z are cylindrical coordinates with the origin at the figure's center and a, b, b are the lengths of the principal axes of the figure. The case $b > a$ corresponds to a flattened (prolate) ellipsoid, while $b < a$ corresponds to a stretched (elongated) ellipsoid, and for the sphere we have $a = b$.

A.1
The Effective Hydrodynamic Radius [340]

The effective radius [340] of a particle of any form R_{ef} is introduced on the basis of the Stokes formula (6.7) such that in the case of a sphere the effective radius coincides with the radius of the sphere

$$\mathbf{F} = 6\pi\eta\mathbf{v}R_{\mathrm{ef}} \,. \tag{A2}$$

Here $\mathbf{F}$ is the resistive force that acts on a particle moving in a gas, η is the gas viscosity, $\mathbf{v}$ is the velocity of the particle, and this equation is valid for small Reynolds numbers and if R_{ef} greatly exceeds the mean free path of gas atoms or molecules. Values of R_{ef} are as follows.

Flattened Ellipsoid, Motion along Its Axis (z-axis)

$$R_{\mathrm{ef}} = \frac{8b}{3}\left[\frac{2\varphi}{1-\varphi^2} + \frac{2(1-2\varphi^2)}{(1-\varphi^2)^{3/2}} \arctan\frac{\sqrt{1-\varphi^2}}{\varphi}\right]^{-1} , \quad \varphi = \frac{a}{b} \le 1 \,. \tag{A3}$$

The limiting cases:

Sphere ($a = b, \varphi = 1$)

$$R_{\mathrm{ef}} = a$$

Cluster Processes in Gases and Plasmas. Boris M. Smirnov

ISBN: 978-3-527-40943-3

Disk ($a = 0, \varphi = 0$)

$$R_{\text{ef}} = \frac{8b}{3\pi}\,.$$

Flattened Ellipsoid, Motion Directed Perpendicular to Its Axis (*x*-axis)

$$R_{\text{ef}} = \frac{8b}{3}\left[-\frac{\varphi}{1-\varphi^2} - \frac{2\varphi^2-3}{(1-\varphi^2)^{3/2}}\arcsin\sqrt{1-\varphi^2}\right]^{-1}, \quad \varphi = \frac{a}{b} \le 1\,. \tag{A4}$$

The limiting cases:

Sphere ($a = b, \varphi = 1$)

$$R_{\text{ef}} = a$$

Disk ($a = 0, \varphi = 0$)

$$R_{\text{ef}} = \frac{16b}{9\pi}\,.$$

Stretched Ellipsoid, Motion along Its Axis (*z*-axis)

$$R_{\text{ef}} = \frac{8b}{3}\left[-\frac{2\varphi}{\varphi^2-1} + \frac{2\varphi^2-1}{(\varphi^2-1)^{3/2}}\ln\frac{\left(\varphi+\sqrt{\varphi^2-1}\right)}{\left(\varphi-\sqrt{\varphi^2-1}\right)}\right]^{-1}, \quad \varphi = \frac{a}{b} \ge 1\,. \tag{A5}$$

The limiting cases:

Sphere ($a = b, \varphi = 1$)

$$R_{\text{ef}} = a\,.$$

Rod ($a \gg b, \varphi \to \infty$)

$$R_{\text{ef}} = \frac{2a}{3\ln[(2a/b) - 1/2]}\,.$$

Stretched Ellipsoid, Motion Directed Perpendicular to Its Axis (*x*-axis)

$$R_{\text{ef}} = \frac{8b}{3}\left[\frac{\varphi}{\varphi^2-1} + \frac{2\varphi^2-3}{(\varphi^2-1)^{3/2}}\ln\left(\varphi+\sqrt{\varphi^2-1}\right)\right]^{-1}, \quad \varphi = \frac{a}{b} \ge 1\,. \tag{A6}$$

The limiting cases:

Sphere ($a = b, \varphi = 1$)

$$R_{\text{ef}} = a\,.$$

Rod ($a \gg b, \varphi \to \infty$)

$$R_{\text{ef}} = \frac{4a}{3\ln(2a/b + 1/2)}\,.$$

Torus (Large Radius Is R, Small Radius Is a, $R \gg a$)

Motion directed perpendicular to the plane of the torus

$$R_{\text{ef}} = \frac{4\pi R}{3\ln[2\pi R/a - 0.75]}\,.$$

Motion is in the plane of the torus

$$R_{\text{ef}} = \frac{\pi R}{3\ln[2\pi R/a - 2.09]}\,.$$

A.2
Capacity (C) [48, 159]

The Ellipsoid's Capacity:

$$\frac{1}{C} = \frac{1}{2}\int_0^\infty \frac{d\xi}{\sqrt{\xi^2 + a^2}(\xi^2 + b^2)}\,. \tag{A7}$$

Flattened Ellipsoid ($a < b$):

$$C = \frac{\sqrt{b^2 - a^2}}{\arccos(a/b)} \tag{A8}$$

The limiting cases:

Sphere ($a = b$)

$$C = a\,.$$

Disk ($b \gg a$)

$$C = \frac{2b}{\pi}\,.$$

Stretched Ellipsoid $(a > b)$:

$$C = \frac{\sqrt{a^2 - b^2}}{\ln \frac{a+\sqrt{a^2-b^2}}{b}} . \tag{A9}$$

The limiting cases:

Sphere $(a = b)$

$$C = a .$$

Rod $(l = 2a, l \gg b)$

$$C = \frac{l}{\left(2 \ln \frac{l}{b}\right)} .$$

Torus (Large Radius Is R, Small Radius Is a, $R \gg a$):

$$C = \frac{\pi R}{\ln\left(\frac{8R}{a}\right)} . \tag{A10}$$

Spherical Segment (R Is the Radius, θ Is the Angle between the Polar Axis and a Conic Boundary:)

$$C = \frac{R}{\pi}(\sin\theta + \theta) . \tag{A11}$$

The limiting cases:

Sphere $(\theta = \pi)$

$$C = R .$$

Hemisphere

$$C = \left(\frac{1}{2} + \frac{1}{\pi}\right) R = 0.818 R .$$

A.3
Polarizability [48, 339]

Polarizability of a Metallic Ellipsoid

Stretched Ellipsoid $(a > b)$:

$$\alpha_{\parallel} = \frac{ab^2}{3n_z} , \quad \alpha_{\perp} = \frac{ab^2}{3n_\rho} ,$$

where $\alpha_{\|}, \alpha_{\perp}$ are the components of the polarizability tensor ($\alpha_{\|} = \alpha_{zz}, \alpha_{\perp} = \alpha_{xx} = \alpha_{yy}$), $n_z + 2n_\rho = 1$, and

$$n_z = \frac{1-\epsilon^2}{2\epsilon^2}\left(\ln\frac{1+\epsilon}{1-\epsilon} - 2\epsilon\right), \quad \epsilon = \sqrt{1-\frac{b^2}{a^2}}.$$

The limiting cases:

Ball ($a = b$)

$$\alpha_{\|} = \alpha_{\perp} = a^3 .$$

Almost a Ball ($a \approx b$)

$$\alpha_{\|} = ab^2\left(1 - \frac{2}{5}\epsilon^2\right)^{-1}, \quad \alpha_{\perp} = ab^2(1+\epsilon^2/5)^{-1} .$$

Rod ($a \gg b$)

$$\alpha_{\|} = \frac{a^3}{3\ln[(2a/b) - 1]}, \quad \alpha_{\perp} = \frac{2ab^2}{3} .$$

Flattened Ellipsoid ($a > b$):

$$\alpha_{\|} = \frac{ab^2}{3n_z}, \quad \alpha_{\perp} = \frac{ab^2}{3n_\rho}, \quad n_z = \frac{1+\epsilon^2}{\epsilon^3}(\epsilon - \arctan\epsilon), \quad \epsilon = \sqrt{1-\frac{b^2}{a^2}};$$

The limiting cases:

Ball ($a = b$)

$$\alpha_{\|} = \alpha_{\perp} = a^3 .$$

Disk ($a \ll b$)

$$\alpha_{\|} = \frac{4b^3}{3\pi} .$$

Polarizability of a Dielectric Ellipsoid

The relation between the polarizability α of a dielectric particle that occupies a volume V and consists of a material of a dielectric constant ε, and the polarizability of the same metallic particle α_{m} is given by formula (5.13):

$$\frac{1}{\alpha} = \frac{1}{\alpha_{\mathrm{m}}} + \frac{4\pi}{V(\varepsilon - 1)} .$$

This leads to the following values of the polarizability of a particle of the corresponding form:

Ball (radius is *r*)

$$\alpha_{\|} = \alpha_{\perp} = r^3 \frac{\varepsilon - 1}{\varepsilon + 1} .$$

Rod (length is *l*, radius is *r*)

$$\alpha_{\|} = r^2 l \frac{\varepsilon - 1}{2} , \quad \alpha_{\perp} = r^2 l \frac{\varepsilon - 1}{\varepsilon + 1} .$$

Ellipsoid

$$\alpha_{\|} = r^2 l/(3n_x) , \quad \alpha_{\perp} = r^2 l/(3n_y) , \tag{A12}$$

where

$$\begin{aligned} & n_x + 2n_y = 1 , \\ & n_x = \frac{1 - \epsilon^2}{2\epsilon^2} \left(\ln \frac{1 + \epsilon}{1 - \epsilon} - 2\epsilon \right) , \\ & \epsilon = \sqrt{1 - \frac{r^2}{l^2}} . \end{aligned} \tag{A13}$$

Appendix B
Conversion Factors of Cluster Physics

Number	Equation	Proportionality factor C	Units used
1.	$v = C\sqrt{\varepsilon/m}$	$5.931 \cdot 10^7$ cm/s	ε in eV, m in emu*)
		$1.389 \cdot 10^6$ cm/s	ε in eV, m in amu*)
		$5.506 \cdot 10^5$ cm/s	ε in K, m in emu
		$1.289 \cdot 10^4$ cm/s	ε in K, m in amu
2.	$v = C\sqrt{T/m}$	$1.567 \cdot 10^6$ cm/s	T in eV, m in amu
		$1.455 \cdot 10^4$ cm/s	ε in K, m in amu
3.	$\varepsilon = Cv^2$	$3.299 \cdot 10^{-12}$ K	v in cm/s, m in emu
		$6.014 \cdot 10^{-9}$ K	v in cm/s, m in amu
		$2.843 \cdot 10^{-16}$ eV	v in cm/s, m in emu
		$5.182 \cdot 10^{-13}$ eV	v in cm/s, m in amu
4.	$r_W = Cm/\rho^{1/3}$	0.7346 Å	m in m_a, ρ in g/cm^3
	$n = C(r_o/r_W)^3$	4.189	r_o and r_W in Å
5.	$k_o = Cr_W^2\sqrt{T/m}$	$4.5714 \cdot 10^{-12}$ cm^3/s	r_W in Å, T in K, m in amu
6.	$v = C\rho r_o^2/\eta$	0.2179 cm/s	r_o in µm, ρ in g/cm^3, η in 10^{-5} g/(cm · s)
		0.01178 cm/s	r_o in µm, ρ in g/cm^3, η for air at $p = 1$ atm, $T = 300$ K
7.	$D_o = C\sqrt{T/m}/(Nr_W^2)$	$1.469 \cdot 10^{21}$ cm^2/s	r_W in Å, N in cm^{-3}, T in K, m in amu
		0.508 cm^2/s	r_W in Å, $N = 2.687 \cdot 10^{19}$ cm^{-3}, T in K, m in amu
		54.69 cm^2/s	r_W in Å, $N = 2.687 \cdot 10^{19}$ cm^{-3}, T in eV, m in amu
8.	$K_o = C(\sqrt{Tm}Nr_W^2)^{-1}$	$1.364 \cdot 10^{19}$ cm^2/(V · s)	r_W in Å, N in cm^{-3}, T in K, m in aumu
		0.508 cm^2/(V · s)	r_W in Å, $N = 2.687 \cdot 10^{19}$ cm^{-3}, T in K, m in amu
		54.69 cm^2/(V · s)	r_W in Å, $N = 2.687 \cdot 10^{19}$ cm^{-3}, T in K, m in amu

Cluster Processes in Gases and Plasmas. Boris M. Smirnov

ISBN: 978-3-527-40943-3

Key to table

1. The particle velocity is $v = \sqrt{2\varepsilon/m}$, where ε is the energy and m is the particle mass.
2. The average particle velocity is $v = \sqrt{8T/(\pi m)}$ with the Maxwell velocity distribution function of particles, where T is the temperature expressed in energy units and m is the particle mass.
3. The particle energy is $\varepsilon = mv^2/2$, where m is the particle mass and v is the particle velocity.
4. The Wigner–Seitz radius according to formula (2.6).
5. The reduced rate constant for atom attachment to a cluster according to formula (2.7).
6. The free-fall velocity for a cluster with radius r_o that is given by formula (6.13), $v = 2\rho g r_o^2/(9\eta)$, where g is the free-fall acceleration, ρ is the density of the cluster material, and η is the viscosity of the gas through which the cluster moves.
7. The diffusion coefficient for a spherical cluster is $D_n = D_o/n^{2/3}$, where n is the number of cluster atoms, and according to formula (6.15) is $D_o = 3\sqrt{2T/\pi m}/(16Nr_W^2)$, where T is the gas temperature, N is the number density of gas atoms, m is the mass of a gas atom, and r_W is the Wigner–Seitz radius.
8. The zero-field mobility for a spherical cluster according to formula (6.20) is $K_n = K_o/n^{2/3}$, where n is the number of cluster atoms and $K_o = 3e/(8Nr_W^2\sqrt{2\pi mT})$, where the notation is given above.

Appendix C
Thermodynamic Parameters of Elements

Thermodynamic parameters for elements.

Period \ Group	I	II	III	IV	V	VI	VII	VIII		
1	1.0079 218.0 $_{1}$H mol 115 Hydrogen 0.46	0 4.003 126 gas $_{2}$He 0.082 Helium								
2	6.941 159 $_{3}$Li sol 139 Lithium 135	9.012 324 $_{4}$Be sol 136 Beryllium 309	565 10.81 153 sol $_{5}$B 539 Boron	717 12.011 158 sol $_{6}$C 711 Carbon	472.7 14.007 153 mol $_{7}$N 5.6 Nitrogen	249.2 15.999 161 mol $_{8}$O 6.8 Oxygen	79.4 18.998 159 mol $_{9}$F 6.5 Fluorine	0 20.180 146 gas $_{10}$Ne 1.7 Neon		
3	22.990 108 $_{11}$Na sol 154 Sodium 89	24.305 147 $_{12}$Mg sol 149 Magnesium 129	330 26.982 165 sol $_{13}$Al 294 Aluminum	450 28.086 168 sol $_{14}$Si 383 Silicon	316 30.974 163 sol $_{15}$P 52 Phosphorus	279 32.065 168 sol $_{16}$S 9.6 Sulfur	121.3 35.453 165 mol $_{17}$Cl 20.4 Chlorine	0 39.948 155 gas $_{18}$Ar 6.5 Argon		
4	39.098 89 $_{19}$K sol 160 Potassium 78	40.08 178 $_{20}$Ca sol 155 Calcium 150	44.956 378 $_{21}$Sc sol 175 Scandium 305	47.867 473 $_{22}$Ti sol 180 Titanium 429	50.942 514 $_{23}$V sol 183 Vanadium 459	51.996 397 $_{24}$Cr sol 174 Chromium 349	54.938 281 $_{25}$Mn sol 174 Manganese 220	55.845 416 $_{26}$Fe sol 180 Iron 351	58.933 425 $_{27}$Co sol 180 Cobalt 382	58.69 430 $_{28}$Ni sol 182 Nickel 372
	337 63.546 166 sol $_{29}$Cu 305 Copper	130 65.409 161 sol $_{30}$Zn 115 Zinc	277 69.723 169 sol $_{31}$Ga 256 Gallium	377 72.64 168 sol $_{32}$Ge 334 Germanium	302 74.922 174 sol $_{33}$As 32 Arsenic	227 78.96 177 sol $_{34}$Se 26 Selenium	111.9 79.904 175 liq $_{35}$Br 30 Bromine	0 83.798 164 gas $_{36}$Kr 9.0 Krypton		
5	85.468 81 $_{37}$Rb sol 170 Rubidium 69	87.62 164 $_{38}$Sr sol 165 Strontium 139	88.906 421 $_{39}$Y sol 179 Yttrium 393	91.224 609 $_{40}$Zr sol 181 Zirconium 582	92.906 726 $_{41}$Nb sol 186 Niobium 697	95.94 658 $_{42}$Mo sol 182 Molybdenum 594	[97.907] 678 $_{43}$Tc sol 181 Technetium 585	101.07 643 $_{44}$Ru sol 186 Ruthenium 568	102.91 557 $_{45}$Rh sol 185 Rhodium 495	106.42 379 $_{46}$Pd sol 167 Palladium 393
	285 107.87 173 sol $_{47}$Ag 255 Silver	111.8 112.41 168 sol $_{48}$Cd 100 Cadmium	243 114.82 174 sol $_{49}$In 226 Indium	302 118.70 168 sol $_{50}$Sn 290 Tin	262 121.76 180 sol $_{51}$Sb 68 Antimony	197 127.60 183 sol $_{52}$Te 51 Tellurium	106.8 126.90 181 sol $_{53}$I 42 Iodine	0 131.29 170 gas $_{54}$Xe 12.6 Xenon		
6	132.90 76.5 $_{55}$Cs sol 175 Cesium 65.9	137.33 180 $_{56}$Ba sol 170 Barium 151	138.90 431 $_{57}$La sol 182 Lanthanum 400	178.49 619 $_{72}$Hf sol 187 Hafnium 661	180.95 $_{73}$Ta sol Tantalum	183.84 849 $_{74}$W sol 174 Tungsten 799	186.21 770 $_{75}$Re sol 189 Rhenium 707	190.23 791 $_{76}$Os sol 193 Osmium 628	192.22 665 $_{77}$Ir sol 194 Iridium 564	195.08 520 $_{78}$Pt sol 192 Platinum 510
	336 196.97 180 sol $_{79}$Au 324 Gold	61.4 200.59 174 liq $_{80}$Hg 59 Mercury	182 204.38 181 sol $_{81}$Tl 162 Thallium	195 207.2 175 sol $_{82}$Pb 179 Lead	207 208.98 187 sol $_{83}$Bi 179 Bismuth	[208.98] sol $_{84}$Po 101 Polonium	[209.99] sol $_{85}$At Astatine	0 [222.02] 176 gas $_{86}$Rn Radon		

Key (example cell): 190.23 791 / $_{76}$Os sol 193 / Osmium 628 — Atomic weight; Symbol; Atomic number; Element; Aggregate state; Enthalpy of transition to atomic vapor, kJ/mol; Entropy of atomic vapor state, kJ/(mol·K); Enthalpy of evaporation, kJ/mol

Lantanides

140.12 423 $_{58}$Ce sol 192 Cerium 314	140.91 356 $_{59}$Pr sol 190 333 Praseodymium	144.24 328 $_{60}$Nd sol 189 Neodymium 284	[144.91] $_{61}$Pm sol 188 Promethium	150.36 207 $_{62}$Sm sol 183 Samarium 192	151.96 175 $_{63}$Eu sol 189 Europium 175	157.25 398 $_{64}$Gd sol 193 Gadolinium 312
158.92 389 $_{65}$Tb sol 204 Terbium 391	162.50 290 $_{66}$Dy sol 197 Dysprosium 293	164.93 301 $_{67}$Ho sol 196 Holmium 251	167.26 317 $_{68}$Er sol 196 Erbium 293	168.93 233 $_{69}$Tm sol 190 Thulium 247	173.04 152 $_{70}$Yb sol 173 Ytterbium 159	174.97 428 $_{71}$Lu sol 185 Lutetium 428

Actinides

232.04 190 602 $_{90}$Th sol Thorium 544	231.04 198 607 $_{91}$Pa sol Protactinium 481	238.03 200 533 $_{92}$U sol Uranium 423

Aggregate state at room temperature : **gas** - atomic gas, **mol** - gas of diatomic molecules, **liq** - liquid state, **sol** - solid state.

Fig. C.1 Thermodynamic parameters of elements.

Cluster Processes in Gases and Plasmas. Boris M. Smirnov

ISBN: 978-3-527-40943-3

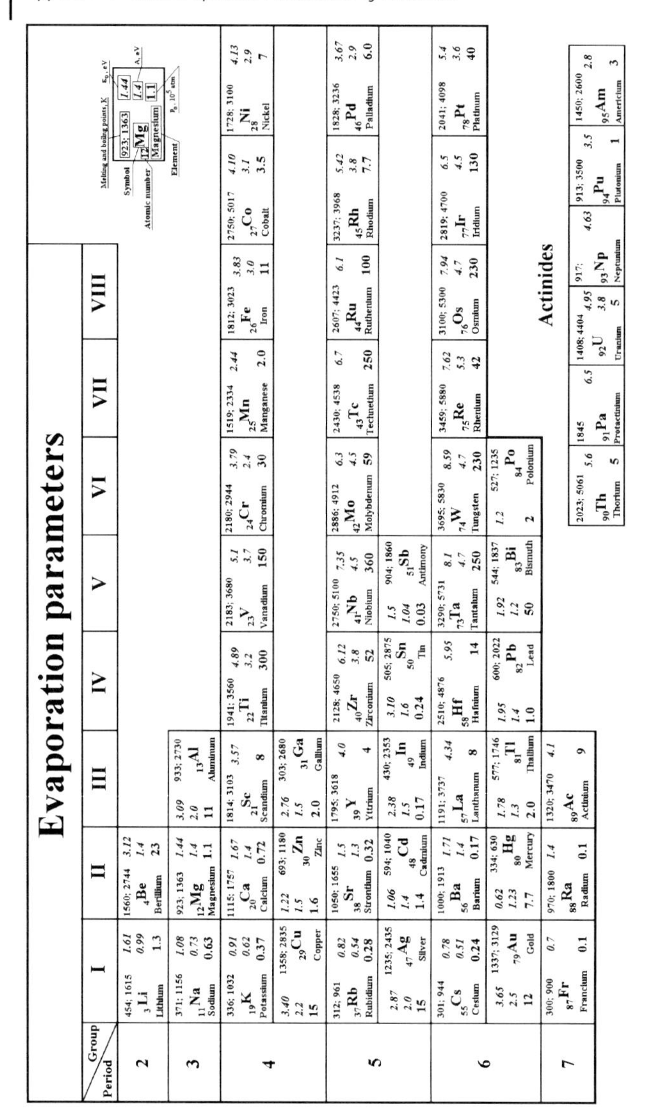

Fig. C.2 Evaporation parameters.

References

1 Faraday, M. (1957) *The Chemical History of a Candle*, Crowell, New York.
2 Faraday, M. (1988) *Faraday's Chemical History of a Candle*, Chicago Review Press, Chicago.
3 Faraday, M. (1857) *Philos. Trans. R. Soc.*, **147**, 145.
4 Ostwald, W. (1897) *Z. Phys. Chem.*, **22**, 289.
5 Ostwald, W. (1900) *Z. Phys. Chem.*, **34**, 495.
6 Aitken, J. (1881) *Nature*, **23**, 195.
7 Aitken, J. (1892) *Proc. R. Soc.*, **A51**, 408.
8 Johnson, B.F.G. (1980) *Transition Metal Clusters*, John Wiley & Sons, Ltd., Chichester.
9 Cotton, F.A. (1985) *Clusters: Structure and Bonding*, Springer, Berlin.
10 Moskovits, M. (1986) *Metal Clusters*, John Wiley & Sons, Inc., New York.
11 Sugano, S. (1991) *Microcluster Physics*, Springer, Berlin.
12 Smirnov, B.M. (1992) *Ion Clusters and van der Waals Molecules*, Gordon, Philadelphia.
13 Gonzalez-Moraga, G. (1993) *Cluster Chemistry: Introduction to the Chemistry of Transition Metal and Main Group Element Molecular Clusters*, Springer, Berlin.
14 Haberland, H. (Ed.) (1994) *Clusters of Atoms and Molecules I. (Theory, Experiment, and Clusters of Atoms)*, Springer, Berlin.
15 Haberland, H. (Ed.) (1994) *Clusters of Atoms and Molecules II (Solvation and Chemistry of Free Clusters, and Embedded, Supported and Compressed Clusters)*, Springer, New York.
16 Kreibig, U. and Vollmer, M. (1995) *Optical Properties of Metal Clusters*, Springer, Berlin.
17 Smirnov, B.M. (1999) *Clusters and Small Particles in Gases and Plasmas*, Springer, New York.
18 Lakhno, V.D. (2001) *Clusters in Physics, Chemistry and Biology*, R&C Dynamics, Moscow-Izhevsk.
19 Reinhard, P.G. and Suraud, E. (2003) *Introduction to Cluster Dynamics*, John Wiley & Sons, Inc., New York.
20 Khanna, S.N. and Castleman, A.W. (2003) *Quantum Phenomena in Clusters and Nanostructures*, Springer, Heidelberg.
21 Connerade, J.P. and Solov'yov, A. (2005) *Latest Advances in Atomic Cluster Collisions*, World Scientific, Singapore.
22 Alonso, J.A. (2005) *Structure and Properties of Atomic Nanoclusters*, World Scientific, Singapore.
23 Echt, O. Sattler, K. and Recknagel, E. (1981) *Phys. Rev. Lett.*, **94**, 54.
24 Echt, O. *et al.* (1982) *Ber. Bunsenges. Phys. Chem.*, **86**, 860.
25 Ding, A. and Hesslich, J. (1983) *Chem. Phys. Lett.*, **94**, 54.
26 Knight, W.D. *et al.* (1984) *Phys. Rev. Lett.*, **52**, 2141.
27 Harris, I.A., Kidwell, R.S. and Northby, J.A. (1984) *Phys. Rev. Lett.*, **53**, 2390.
28 Phillips, J.C. (1986) *Chem. Rev.*, **86**, 619.
29 Harris, I.A., Norman, K.A., Mulkern, R.V. and Northby, J.A. (1986) *Chem. Phys. Lett.*, **130**, 316.
30 Miehle, W., Kandler, O., Leisner, T. and Echt, O. (1989) *J. Chem. Phys.*, **91**, 5940.
31 Smirnov, B.M. (1992) *Sov. Phys. Usp.*, **35**, 1052.

Cluster Processes in Gases and Plasmas. Boris M. Smirnov

ISBN: 978-3-527-40943-3

32 Wigner, E.P. and Seitz, F. (1934) *Phys. Rev.*, **46**, 509.
33 Wigner, E.P. (1934) *Phys. Rev.*, **46**, 1002.
34 Smirnov, B.M. (2008) *Reference Data on Atomic Physics and Atomic Processes*, Springer, Heidelberg.
35 Smirnov, B.M. (1994) *Phys. Usp.*, **37**, 621.
36 Smirnov, B.M. (2000) *Phys. Usp.*, **43**, 453.
37 Ino, S. (1969) *J. Phys. Soc. Japan*, **27**, 941.
38 Ter Haar, D. (1966) *Elements of Thermostatics*, Addison-Wesley, New York.
39 Landau, L.D. and Lifshitz, E.M. (1980) *Statistical Physics*, vol. 1, Pergamon Press, Oxford.
40 Smirnov, B.M. (1993) *Phys. Usp.*, **36**, 933.
41 Gibbs, J.W. (1875) *Trans. Conn. Acad. Arts. Sci.*, **3**, 108; (1878) **3**, 343.
42 Gibbs, J.W. (1928) *The Collected Works*, Longmans and Green, New York.
43 Curie, P. (1885) *Bull. Soc. Miner. Fr.*, **8**, 145.
44 Bjørnholm, S. *et al.* (1990) *Phys. Rev. Lett.*, **65**, 1627.
45 Berry, R.S. and Smirnov, B.M. (2007) *Phase Transitions of Simple Systems*, Springer, Berlin.
46 Smirnov, B.M. (2001) *Phys. Usp.*, **44**, 1229.
47 Smirnov, B.M. (1994) *Phys. Usp.*, **37**, 1079.
48 Landau, L.D. and Lifshitz, E.M. (1984) *Electrodynamics of Continuous Media*, Pergamon Press, Oxford.
49 Rayleigh, L. (1882) *Philos. Mag.*, **14**, 184.
50 Bohr, N. and Wheeler, J.A. (1939) *Phys. Rev.*, **56**, 426.
51 Meitner, L. and Frish, O.R. (1939) *Nature*, **143**, 239.
52 Last, I. and Jortner, J. (2000) *Phys. Rev. A*, **62**, 013201.
53 Lyalin, A., Semenov, S.K., Solov'ev, A.V. and Greiner, W. (2002) *Phys. Rev. A*, **65**, 023201, 043202.
54 Schreier, P. and Märk, T.D. (1987) *J. Chem. Phys.*, **87**, 1456.
55 Echt, O. *et al.* (1988) *Phys. Rev. A*, **38**, 3286.
56 Schreier, P. Stamatovic, A. and Märk, T.D. (1988) *J. Chem. Phys.*, **89**, 2956.
57 Lezius, M., Schreier, P., Stamatovic, A. and Märk, T.D. (1989) *J. Chem. Phys.*, **91**, 3240.
58 Martin, T.P., Näher, U., Göhlich, H. and Lange, T. (1984) *Chem. Phys. Lett.*, **196**, 113.
59 Heinebrodt, M. *et al.* (1997) *Z. Phys.*, **D40**, 334.
60 Näher, U., Frank, S., Malinowski, N., Zimmermann, U. and Martin, T.P. (1994) *Z. Phys. D*, **31**, 191.
61 Näher, U., Bjørnholm, S. and Frauendorf, S. (1997) *Phys. Rep.*, **285**, 245.
62 Lyalin, A., Semenov, S.K., Solov'ev, A.V. and Greiner, W. (2002) *Phys. Rev. A* **65**, 023201.
63 Eletskii, A.V. and Smirnov, B.M. (1989) *Sov. Phys. Usp.*, **32**, 763.
64 Ashcroft, N.M. and Mermin, N.D. (1976) *Solid State Physics*, Hort, Rinehart and Wilson, New York.
65 Wang, S.W., Falicov, L.M. and Searcy, W. (1984) *Surf. Sci.*, **143**, 609.
66 Lennard-Jones, J.E. and Ingham, A.E. (1924) *Proc. R. Soc. Lond. Ser. A*, **107**, 463.
67 Lennard-Jones, J.E. (1925) *Proc. R. Soc. Lond. Ser. A*, **106**, 636.
68 Smirnov, B.M. (1992) *Sov. Phys. Usp.*, **35**, 37.
69 Kittel, C. (1986) *Introduction to Solid State Physics*, John Wiley & Sons, Inc., New York.
70 Doye, P.K., Wales, D.J. and Berry, R.S. (1995) *J. Chem. Phys.*, **103**, 4234.
71 Morse, P.M. (1929) *Phys. Rev.*, **34**, 57.
72 Berry, R.S., Smirnov, B.M. and Strizhev, A.Yu. (1997) *JETF*, **85**, 588.
73 Smirnov, B.M., Strizhev, A.Yu. and Berry, R.S. (1999) *J. Chem. Phys.*, **110**, 7412.
74 Wulff, G. (1901) *Z. Kristallogr.*, **34**, 449.
75 Smirnov, B.M. (2006) *Principles of Statistical Physics*, Wiley-VCH Verlag GmbH, Berlin.
76 Leontovich, M.A. (1983) *Introduction to Thermodynamics. Stat. Phys.*, Nauka, Moscow.
77 Smirnov, B.M. (1993) *Phys. Usp.*, **36**, 933.
78 Smirnov, B.M. (1995) *High Temp.*, **33**, 700.
79 Smirnov, B.M. (1995) *Phys. Scr.*, **51**, 402.
80 Mackay, A.L. (1962) *Acta Crystallogr.*, **15**, 916.
81 Smirnov, B.M. (1995) *Chem. Phys. Lett.*, **232**, 395.
82 Farges, J., de Feraudy, M.F., Raoult, B. and Torchet, G. (1983) *J. Chem. Phys.*, **78**, 5067.
83 Farges, J., de Feraudy, M.F., Raoult, B. and Torchet, G. (1986) *J. Chem. Phys.*, **84**, 3491.

84 Landau, L.D. and Lifshitz, E.M. (1980) *Mechanics*, Pergamon Press, Oxford.
85 Landau, L.D. and Lifshits, E.M. (1980) *Quantum Mechanics*, Pergamon Press, Oxford.
86 Smirnov, B.M. (1997) *Phys. Usp.* **40**, 1117.
87 Langevin, P. (1905) *Ann. Chem. Phys.*, **8**, 245.
88 Smirnov, B.M. (1993) *Plasma Chem. Plasma Proc.*, **13**, 673.
89 Smoluchowski, M.V. (1916) *Z. Phys.*, **17**, 585.
90 Smith I.M. and Tyler, R.J. (1972) *Combust. Flame*, **51**, 312.
91 Smith I.M. and Tyler, R.J. (1974) *Combust. Sci. Technol.*, **9**, 87.
92 Grigor'ev, G.Yu., Dorofeev, S.B. and Smirnov, B.M. (1984) *Khim. Fiz.*, **3**, 603.
93 Grigor'ev, G.Yu., Dorofeev, S.B., Kuvshinov, B.N. and Smirnov, B.M. (1984) *Fiz. Goren. vzryva*, **N5**, 3.
94 Hubbard, J. (1963) *Proc. R. Soc. Lond. Ser. A*, **276**, 238.
95 Hubbard, J. (1964) *Proc. R. Soc. Lond. Ser. A*, **277**, 237.
96 Hubbard, J. (1964) *Proc. R. Soc. Lond. Ser. A*, **281**, 401.
97 Mott, N.F. (1974) *Metal-Insulator Transitions*, Taylor & Francis, London.
98 Mott, N.F. (1990) *Metal-Insulator Transitions*, Taylor & Francis, London.
99 Chesnovski, O., Yang, S.H., Pettiette, C.L. *et al.* (1987) *Chem. Phys. Lett.*, **138**, 119.
100 Zolkin, A.S. (1992) *The Sources of Metal Vapors*, Preprint of Thermophysics Institute of SBRAS, Thermophysics Institute, Novosibirsk.
101 Planck, M. (1901) *Ann. Phys.*, **4**, 453.
102 Reif, F. (1965) *Statistical and Thermal Physics*, McGraw-Hill, Boston.
103 Busani, R., Folkers, M. and Chesnovski, O. (1998) *Phys. Rev. Lett.*, **81**, 3836.
104 Rayleigh, J. (1900) *Philos. Mag.*, **49**, 539.
105 Lorentz, H.A. (1903) *Proc. Akad. Amst.*, 666.
106 Jeans, J.H. (1909) *Philos. Mag.*, **Feb**, 229.
107 Wien, W. (1896) *Wied. Ann.*, **58**, 662.
108 Baltes, H.P. and Geist, J. (1981) *J. Quant. Spectrosc. Radiat. Transf.*, **26**, 535.
109 Weber, B. and Scholl, R. (1993) *J. Appl. Phys.*, **74**, 607.
110 Luizova, L.A., Smirnov, B.M. and Khakhaev, A.D. (1989) *Dokl. Akad. Nauk SSSR*, **309**, 1359.
111 Luizova, L.A., Smirnov, B.M., Khakhaev, A.D. and Chugin, V.P. (1991) *High Temp.*, **28**, 897.
112 Smirnov, B.M. (1993) *Phys. Usp.*, **36**, 592.
113 Kresin, V.V. (1992) *Phys. Rep.*, **220**, 1.
114 Langmuir, I. and Tonks, L. (1927) *Phys. Rev.*, **34**, 376.
115 Bréchignac, C., Cahuzac, P., Leygnier, J. and Sarfati, A. (1993) $\sigma(\mathrm{Li}_n^+)$. *Phys. Rev. Lett.*, **70**, 2036.
116 Bréchignac, C., Cahuzac, P., Carlier, F. and Leygnier, J. (1989) $\sigma(\mathrm{K}_n^+)$. *Chem. Phys. Lett.*, **164**, 433.
117 Bréchignac, C., Cahuzac, P., Kebaili, N., Leygnier, J. and Sarfati, A. (1992) *Phys. Rev. Lett.*, **68**, 3916.
118 Alexander, M.L., Johnson, M.A., Levinger, N.E. and Lindinger, W.C. (1986) *Phys. Rev. Lett.*, **57**, 976.
119 Tiggesburker, J., Köller, L., Lutz, H.O. and Meiwes-Broer, K.-H. (1992) *Chem. Phys. Lett.*, **190**, 42.
120 Guet, C. and Jonston, W.R. (1992) *Phys. Rev.*, **45B**, 11283.
121 Haberland, H. and von Issendorff, B. (1996) *Phys.Rev.Lett.*, **76**, 1445.
122 Bréchignac, C., Cahuzac, P., Carlier, F., de Frutos, M. and Leygnier, J. (1992) *Chem. Phys. Lett.*, **189**, 28.
123 Pedersen, J., Borggreen, J., Chowdhury, P. *et al.* (1993) *Z. Phys. D*, **26**, 281.
124 Frenzel, U., Kalmbach, U., Kreisle, D. and Recknagel, E. (1996) *Surf. Rev. Lett.*, **3**, 505.
125 Frenzel, U. (1996) *Schwarzkörperstrahlung von Metallclustern*, Hartung-Gorre Verlag, Constance.
126 Frenzel, U., Hammer, U., Westje, H. and Kreisle, D. (1997) *Z. Phys.*, **D40**, 108.
127 Smirnov, B.M. and Weidele, H. (1999) *JETP Lett.*, **69**, 490.
128 Smirnov, B.M. and Weidele, H. (1999) *JETP*, **89**, 1030.
129 Smirnov, B.M. (1993) *Int. J. Theor. Phys.*, **32**, 1453.
130 Mandelbrott, B.B. (1982) *The Fractal Geometry of Nature*, Freeman, San Francisco.
131 Jullien, R. and Botet, R. (1987) *Aggregation and Fractal Aggregates*, World Scientific, Singapore.

132 Feder, J. (1988) *Fractals*, Plenum Press, New York.

133 Avnir, D. (1989) *The Fractal Approach to Heterogeneous Chemistry: Surfaces, Colloids, Polymers*, Wiley, Chichester.

134 Lifshits, E.M. and Pitaevskii, L.P. (1981) *Physical Kinetics*, Pergamon Press, Oxford.

135 Chapman, S. and Cowling, T.G. (1952) *The Mathematical Theory of Non-uniform Gases*, Cambridge Univ. Press, Cambridge.

136 Ferziger, J.H. and Kaper, H.G. (1972) *Mathematical Theory of Transport Processes in Gases*, North Holland, Amsterdam.

137 Einstein, A. (1905) *Ann. Phys.*, **17**, 549.

138 Einstein, A. (1906) *Ann. Phys.*, **19**, 371.

139 Einstein, A. (1908) *Z. Electrochem.*, **14**, 235.

140 Nernst, W. (1988) *Z. Phys. Chem.*, **2**, 613.

141 Townsend, J.S. and Bailey, V.A. (1899) *Philos. Trans. A*, **193**, 129.

142 Townsend, J.S. and Bailey, V.A. (1900) *Philos. Trans. A*, **195**, 259.

143 Huxley, L.G.H. and Crompton, R.W. (1973) *The Diffusion and Drift of Electrons in Gases*, John Wiley & Sons, Inc., New York.

144 McDaniel, E.W. and Mason, E.A. (1973) *The Mobility and Diffusion of Ions in Gases* John Wiley & Sons, Inc., New York.

145 Stokes, G.G. (1951) *Trans. Camb. Philos. Soc.*, **9**, **II**, 8.

146 Smirnov, B.M. (2001) *Physics of Ionized Gases*, John Wiley & Sons, Inc., New York.

147 Landau, L.D. and Lifshitz, E.M. (1986) *Fluid Dynamics*, Pergamon Press, Oxford.

148 Vargaftik, N.B. (1975) *Tables of Thermophysical Properties of Liquids and Gases*, Halsted Press, New York.

149 Smirnov, B.M., Shyjumon, I. and Hippler, R. (2007) *Phys. Rev. E*, **77**, 066402.

150 Rao, A.K. and Whitby, K.T. (1978) *J. Aerosol Sci.*, **9**, 77.

151 Reist, P.C. (1984) *Introduction to Aerosol Science*, Macmillan Publishing Company, New York.

152 Green, H.L. and Lane, W.R. (1964) *Particulate Clouds: Dust, Smokes and Mists*, Van Nostrand, Princeton.

153 Licht, W. (1980) *Air Pollution Control Engineering*, Marcel Dekker, New York.

154 Fuks, N.A. (1964) *Mechanics of Aerosols*, Macmillan, New York.

155 Debye, P. and Hückel, E. (1923) *Phys. Z.*, **24**, 185.

156 Jackson, J.D. (1998) *Classical Electrodynamics*, John Wiley & Sons, Inc., New York.

157 Sena, L.A. (1946) *ZhETF*, **16**, 734.

158 Sena, L.A. (1948) *Collisions of Electrons and Ions with Atoms*, Gostekhizdat, Leningrad.

159 Smythe, W.R. (1950) *Static and Dynamic Electricity*, McGraw Hill, New York.

160 Zaslavskii, G.M. and Sagdeev, R.Z. (1988) *Introduction to Nonlinear Physics. From Pendulum to Turbulence and Chaos*, Nauka, Moscow.

161 Bernstein, I.B. and Rabinovitz, I.N. (1959) *Phys. Fluids*, **2**, 112.

162 Zobnin, A.V., Nefedov, A.P., Sinel'shchikov, V.A. and Fortov, V.E. (2000) *JETP*, **91**, 483.

163 Bystrenko, O. and Zagorodny, A. (2003) *Phys. Rev. E*, **67**, 066403.

164 Sukhinin, G.I. and Fedoseev, A.V. (2007) *Plasma Phys. Rep.*, **33**, 1023.

165 Lampe, M., Gavrishchaka, V. Ganguli, G. and Joyce, G. (2001) *Phys. Rev. Lett.*, **86**, 5278.

166 Lampe, M., Goswami, R., Sternovsky, Z. *et al.* (2003) *Phys. Plasmas*, **10**, 1500.

167 Sternovsky, Z., Lampe, M. and Robertson, S. (2004) *IEEE Trans. Plasma Sci.*, **32**, 632.

168 Cini, M. (1975) *J. Catal.*, **37**, 187.

169 Beck, D.E. (1984) *Solid State Commun.*, **49**, 381.

170 Perdew, J.P. (1988) *Phys.Rev. B*, **37**, 6175.

171 Makov, G., Nitzan, A. and Brus, L.E. (1988) *J. Chem. Phys.*, **88**, 5076.

172 Smith, J.M. (1965) *Am. Inst. Aeronaut. Astronaut. J.*, **3**, 648.

173 Wood, D.M. (1981) *Phys. Rev. Lett.*, **46**, 749.

174 Van Staveren, M.P.J., Brom, H.B., de Jong, L.J. and Ishii, Y. (1987) *Phys. Rev. B*, **35**, 7749.

175 Bréchignac, C., Cahuzac, P., Carlier, F., de Frutos, M. and Leygnier, J. (1990) *J. Chem. Soc. Faraday Trans.*, **86**, 2525.

176 Leopold, D.G., Ho, J.H. amd Lineberger, W.C. (1987) *J. Chem. Phys.*, **86**, 1715.

177 Ganteför, G., Gausa, M., Meiwes-Broer, K.H. and Lutz, H.O. (1988) *Faraday Discuss. Chem. Soc.*, **86**, 197.

178 Gausa, M., Ganteför, G., Lutz, H.O. and Meiwes-Broer, K.H. (1990) *Int. J. Mass Spectrom. Ion Process.*, **102**, 227.
179 Seidl, M., Meiwes-Broer, K.H. and Brack, M. (1991) *J. Chem. Phys.*, **95**, 1295.
180 Weidele, H., Kreisle, D., Recknagel, E. *et al.* (1995) *Chem. Phys. Lett.*, **237**, 425.
181 Cobine, J.D. (1958) *Gaseous Conductors*, Dover, New York.
182 Neuman, W. (1987) *The Mechanism of the Thermoemitting Arc Cathode*, Akademie-Verlag, Berlin.
183 Lide, D.R. (ed.) (2003–2004) *Handbook of Chemistry and Physics*, 86th edn, CRC Press, London.
184 Illenberger, E. and Smirnov, B.M. (1998) *Phys. Usp.*, **41**, 651.
185 Kittel, C. (1970) *Thermal Physics*, John Wiley & Sons, Inc., New York.
186 Smirnov, B.M. (2003) *Phys. Usp.*, **46**, 589.
187 Voloshchuk, V.M. (1984) *Kinetic Theory of Coagulation*, Hidrometeoizdat, Leningrad.
188 Smoluchowski, M.V. (1918) *Z. Phys. Chem.*, **92**, 129.
189 Fokker, A.D. (1914) *Ann. Phys. (Leipzig)*, **43**, 810.
190 Planck, M. (1917) *Preuss. Acad. Wiss. Phys. Mat. Kl.*, 324.
191 Gurevich, L.E. (1940) *Introduction to Physical Kinetics*, GIITL, Leningrad.
192 Martin, T.P., Näher, U., Schaber, H. and Zimmermann, U. (1994) *J. Chem. Phys.*, **100**, 2322.
193 Martin, T.P. (1996) *Phys. Rep.*, **273**, 199.
194 Zel'dovich, J.B. (1942) *ZhETF*, **12**, 525.
195 Frenkel, J.I. (1946) *Kinetic Theory of Liquids*, Oxford Univ. Press, Oxford.
196 Abraham, F.F. (1974) *Homogeneous Nucleation Theory*, Academic Press, New York.
197 Landau, L.D. and Lifshitz, E.M. (1980) *Statistical Physics*, vol. 2., Pergamon Press, Oxford.
198 Gutzow, I. and Schmelzer, J. (1995) *The Vitreous State*, Springer, Berlin.
199 Smirnov, B.M. and Strizhev, A.Yu. (1994) *Phys. Scr.*, **49**, 615.
200 Rao, B.K. and Smirnov, B.M. (1997) *Phys. Scr.*, **56**, 588.
201 Rao, B.K. and Smirnov, B.M. (2002) *Mater. Phys. Mech.*, **5**, 1.
202 Lifshitz, I.M. and Slezov, V.V. (1958) *Sov. Phys. JETP*, **35**, 331.
203 Lifshitz, I.M. and Slezov, V.V. (1959) *Fiz. Tver. Tela*, **1**, 1401.
204 Lifshitz, I.M. and Slezov, V.V. (1961) *J. Phys. Chem. Sol.*, **19**, 35.
205 Slezov, V.V. and Sagalovich, V.V. (1987) *Sov. Phys. Usp.*, **30**, 23.
206 Zhukhovizkii, D.I., Khrapak, A.G. and Jakubov, I.T. (1983) *High Temp.*, **21**, 1197.
207 Zhukhovizkii, D.I. (1989) *High Temp.*, **27**, 515.
208 Smirnov, B.M., Shyjumon, I. and Hippler, R. (2006) *Phys. Scr.*, **73**, 288.
209 Smirnov, B.M. (2000) *Phys. Scr.*, **62**, 148.
210 Smirnov, B.M. (1995) *Phys. Scr.*, **51**, 380.
211 Fortov, V.E., Petrov, O.F. and Nefedov, A.P. (1997) *Phys. Usp.*, **40**, 1163.
212 Shukla, P.K. and Mamun, A.A. (2001) *Introduction to Dusty Plasma Physics*, IOP Publishing, Bristol.
213 Fortov, V.E. *et al.* (2004) *Phys. Usp.*, **47**, 447.
214 Fortov, V.E., Khrapak, A.G. and Jakubov, I.T. (2004) *Physics of Non-Ideal Plasma*, Fizmatlit, Moscow.
215 Melzer, A. and Goree, J. (2008) Fundamentals of Dusty Plasmas, in *Low Temperature Plasmas*, vol. 1 (eds R. Hippler, H. Kersten, M. Schmidt and K.H. Schoenbach), Wiley-VCH Verlag GmbH, Berlin, p. 129.
216 Hippler, R. and Kersten, H. (2008) Applications of Dusty Plasmas, in *Low Temperature Plasmas*, vol. 2 (eds R. Hippler, H. Kersten, M. Schmidt and K.H. Schoenbach), Wiley-VCH Verlag GmbH, Berlin, p. 787.
217 Dutton, J. (1975) *J. Phys. Chem. Ref. Data*, **4**, 577.
218 Harris, S.H. (2005) in *Encyclopedia of Physics* (eds R.G. Lerner and G.L. Trigg) VCH Publishers, New York, 1990, p. 30; Wiley-VCH Verlag GmbH, Weinheim, p. 61.
219 Fleagle, R.G. and Businger J.A. (1980) *An Introduction to Atmospheric Physics*, Academic Press, San Diego.
220 Salby, M.L. (1996) *Fundamentals of Atmospheric Physics*, Academic Press, San Diego.
221 Friendlander, S.K. (1977) *Smoke, Dust and Haze: Fundamentals of Aerosol Behavior*, Elsevier, Amsterdam.

222 Uman, M.A. (1986) *About Lightning*, Dover, New York.
223 Hinds, W.C. (1999) *Aerosol Technology: Properties, Behavior and Measurement of Airborne Particles*, John Wiley & Sons, Inc., New York.
224 Israel, H. (1973) *Atmospheric Electricity*, Keter Press Binding, Jerusalem.
225 Moore, C.B. and Vonnegut, B. (1977) in *Lightning* (ed. R.H. Golde), Academic Press, London, p. 51.
226 Feynman, R.P., Leighton, R.B. and Sands, M. (1964) *The Feynman Lectures of Physics*, vol. 2, Addison-Wesley, Reading.
227 Berger, K. (1977) in *Lightning* (ed. R.H. Golde), Academic Press, London, p. 119.
228 Fowler, R.G. (1982) in *Applied Atomic Collision Physics*, vol. 5 (eds H.S.W. Massey, E.W. McDaniel and B. Bederson), Academic Press, New York, p. 35.
229 Latham, J. and Stromberg, I.M. (1977) in *Lightning* (ed. R.H. Golde), Academic Press, London, p. 99.
230 Seinfeld, J.H. and Pandis, S.N. (1998) *Atmospheric Chemistry and Physics*, John Wiley & Sons, Inc., New York.
231 Ellis, H.W., Pai, R.Y., McDaniel, E.W., Mason and Viehland, L.A. (1976) *At. Data Nucl. Data Tables*, **17**, 177.
232 Ellis, H.W., McDaniel, E.W., Albritton, D.L., Viehland, L.A., Lin, S.L. and Mason, E.A. (1978) *At. Data Nucl. Data Tables* **22**, 179.
233 Ellis, H.W., Trackston, M.G., McDaniel, E.W. and Mason, E.A. (1984) *At. Data Nucl. Data Tables* **31**, 113.
234 Viehland, L.A. and Mason, E.A. (1995) *At. Data Nucl. Data Tables*, **60**, 37.
235 Byers, H.R. (1965) *Elements of Cloud Physics*, University of Chicago Press, Chicago.
236 Fletcher, N.H. (1969) *The Physics of Rainclouds*, Cambridge Univ. Press, London.
237 Mason, B.J. (1971) *The Physics of Clouds*, Clarendon Press, Oxford.
238 Twomey, S. (1977) *Atmospheric Aerosols*, Elsevier, Amsterdam.
239 Proppacher, H. and Klett, J. (1978) *Microphysics of Clouds and Precipitation*, Reidel, London.
240 Jayaratne, E.R., Saunders, C.P.R. and Hallett, J. (1983) *Q. J. R. Meteorol. Soc.*, **109**, 609.
241 Williams, E.R., Zhang, R. and Rydock, J. (1991) *J. Atmos. Sci.*, **48**, 2195.
242 Dong, Y. and Hallett, J. (1992) *J. Geophys. Res.*, **97**, 20361.
243 Mason, B.L. and Dash, J.G. (2000) *J. Geophys. Res.*, **105**, 10185.
244 Dash, J.G., Mason, B.L. and Wettlaufer, J.S. (2001) *J. Geophys. Res.*, **106**, 20395.
245 Berdeklis, P. and List, R. (2001) *J. Atmos. Sci.*, **58**, 2751.
246 Nelson, J. and Baker, M. (2003) *Atmos. Chem. Phys. Discuss.*, **3**, 41.
247 Brinker, C.J. and Scherer, G.W. (1990) *Sol-Gel Science. The Physics and Chemistry of Sol-Gel Processing*, Academic Press, New York.
248 Smirnov, B.M. (1990) *Physics of Fractal Clusters*, Nauka, Moscow.
249 Witten, T.A. and Sander, L.M. (1981) *Phys. Rev. Lett.* **47**, 1400.
250 Meakin, P. (1983) *Phys. Rev. Lett.* **51**, 1119.
251 Lushnikov, A.A., Negin, A.E. and Pakhomov, A.V. (1990) *Chem. Phys. Lett.*, **175**, 138.
252 Lushnikov, A.A., Negin, A.E., Pakhomov, A.V. and Smirnov, B.M. (1991) *Sov. Phys. Usp.*, **34**, 160.
253 Smirnov, B.M. (1991) *Sov. Phys. Usp.*, **34**, 711.
254 Smirnov, B.M. (1993) *Phys. Rep.*, **224**, 151.
255 Smirnov, B.M. (2004) *Contrib. Plasma Phys.*, **44**, 558.
256 Smirnov, B.M. (1998) *JETP Lett.*, **68**, 779.
257 Smirnov, B.M. (2000) *J. Phys. D*, **33**, 115.
258 Smirnov, B.M. (2000) *JETP Lett.*, **71**, 588.
259 Hagena, O.F. (1981) *Surf. Sci.*, **106**, 101.
260 Gspann, J. (1986) *Z. Phys. D*, **3**, 143.
261 Hagena, O.F. (1987) *Z. Phys. D*, **4**, 291(1987); **17**, 157(1990); **20**, 425(1991).
262 Vargaftic, N.B., Filipov, L.N., Tarismanov, A.A. and Totzkii, E.E. (1990) *Reference Data for Thermal Conductivitites of Liquids and Gases*, Energoatomizdat, Moscow.
263 Smalley, R.E. (1983) *Laser Chem.*, **2**, 167.
264 Hopkins, J.B., Langridge-Smith, P.R.R., Morse, M.D. and Smalley, R.E. (1983) *J. Chem. Phys.*, **78**, 1627.
265 Cheshnovsky, O., Yang, S.H., Pettiette, C.L. *et al.* (1987) *Chem. Phys. Lett.*,

139, 233(1987); *Rev. Sci. Instrum.* **58**, 2131(1987).
266 Milany, P. and de Heer, W.A. (1990) *Sci. Instrum.*, **61**, 1835.
267 Vorob'ev, V.S. (1993) *Phys. Usp.*, **36**, 1129.
268 Anisimov, S.I., Imas, Ya.A., and Khodyko, Yu.V. (1970) *Action of High-Power Radiation on Metals*, Nauka, Moscow.
269 Bronin, S.Ya. and Polishchuk, V.P. (1984) *High Temp.*, **22**, 755.
270 Brykin, M.V., Vorob'ev, V.S. and Shelukhaev, B.P. (1987) *High Temp.*, **25**, 468.
271 Takagi, T. (1988) *Ionized Cluster Beam Depositon and Epitaxy*, Noyes Publications, Park Ridge.
272 Smirnov, B.M. (1992) *Plasma Chem. Plasma Process.*, **12**, 177.
273 Smirnov, B.M. (2001) *Phys. Scr.*, **64**, 152.
274 Hagena, O.F. (1991) *Z. Phys. D*, **20**, 425.
275 Hagena, O.F. (1992) *Rev. Sci. Instrum.*, **3**, 2374.
276 Haberland, H. *et al.* (1992) *J. Vac. Sci. Technol. A*, **10**, 3266.
277 Haberland, H., von Issendorff, B., Yufeng, J., and Kolar, T. (1992) *Phys. Rev. Lett.*, **69**, 3212.
278 Haberland, H. *et al.* (1993) *Mater. Sci. Eng. B*, **19**, 31.
279 Haberland, H. *et al.* (1993) *Z. Phys. D*, **26**, 8.
280 Haberland, H., Mall, M., Moseler, M., Qiang, Y., Reiners, T. and Thurner, Y. (1994) *J. Vac. Sci. Technol. A*, **12**, 2925.
281 Wolter, M. *et al.* (2005) *J. Phys. D*, **38**, 2390.
282 Kashtanov, P.V., Smirnov, B.M. and Hippler, R. (2007) *Phys. Usp.*, **50**, 455.
283 Shyjumon, I., Gopinadhan, M., Helm, C.A., Smirnov, B.M. and Hippler, R. (2006) *Thin Solid Films*, **500**, 41.
284 Rossnagel, S.M. and Kaufman, H.R. (1988) *J. Vac. Sci. Technol. A*, **10**, 223.
285 Goeckner, M.J., Goree, J.A. and Sheridan, T.E. (1991) *IEEE Trans. Plasma Sci.*, **PS-19**, 301.
286 Samsonov, D. and Goree, J. (1999) *J. Vac. Sci. Technol. A*, **17**, 2835.
287 Samsonov, D. and Goree, J. (1999) *Phys. Rev. E*, **59**, 1047.
288 Smirnov, B.M. (1998) *Phys. Scr.*, **58**, 363.
289 Becker, E.W. Bier, K. and Henkes, W. (1956) *Z. Phys.*, **146**, 333.
290 Henkes, W. (1961) *Z. Naturforsh. A*, **16**, 842.
291 Henkes, W. (1962) *Z. Naturforsh. A*, **17**, 786.
292 Hagena, O.F. and Obert, W. (1972) *J. Chem. Phys.*, **56**, 1793.
293 Korvink, J.G. and Greiner, A. (2002) *Semiconductors for Micro- and Nanotechnology*, Wiley-VCH Verlag GmbH, Berlin.
294 Poole, C.P. and Owens, F.J. (2003) *Introduction to Nanotechnology*, Wiley-VCH Verlag GmbH, Berlin.
295 Wolf, E.L. (2004) *Nanophysics and Nanotechnology*, Wiley-VCH Verlag GmbH, Berlin.
296 Köhler, M. and Fritzsche, W. (2004) *Nanotechnology*, Wiley-VCH Verlag GmbH, Berlin.
297 Borisenko, E.W. and Ossicini, S. (2004) *What Is What in Nanoscience*, Wiley-VCH Verlag GmbH, Berlin.
298 Kelsall, R. and Hamley, I.W. (2005) *Nanoscale Science and Technology*, Wiley-VCH Verlag GmbH, Berlin.
299 Gleiter, H. (1992) *Nanostruct. Mater.*, **1**, 1.
300 Gleiter, H. (1995) *Nanostruct. Mater.*, **6**, 3.
301 Edelstein, A.S., Cammaratra, R.C. (eds) (1996) *Nanomaterials: Synthesis, Properties and Applications*, IOP, Bristol.
302 Jena, P. and Khanna, S.N. (1996) *Mater. Sci. Eng. A*, **217/218**, 218.
303 de Heer, W.A., Milani, P. and Chatelain, A. (1990) *Phys. Rev. Lett.*, **65**, 488.
304 Khanna, S.N. and Linderoth, S. (1991) *Phys. Rev. Lett.*, **67**, 742.
305 Khanna, S.N. and Jena, P. (1992) *Phys. Rev. Lett.*, **69**, 1664.
306 Wiel, R. (1993) *Z. Phys. D*, **27**, 89.
307 Sosnowski, M. and Yamada, I. (1990) *Nucl. Instrum. Methods. Phys. Res. B*, **46**, 397.
308 Huq, S.E., McMahon, R.A. and Ahmed, H. (1990) *Semicond. Sci. Technol.* **5**, 771.
309 Takaoka, G.H., Ishikawa, J. and Takagi, T. (1990) *J. Vac. Sci. Technol. A*, **8**, 840.
310 Sosnowski, M., Usui, H. and Yamada, I. (1990) *J. Vac. Sci. Technol. A*, **8**, 1470.
311 Gspann, J. (1993) *Nucl. Instrum. Methods B*, **80/81**, 1336.
312 Gspann, J. (1993) *Z. Phys. D*, **26**, S174.
313 Melinon, P., Paillard, V., Dupuis, V., Perez, A. *et al.* (1995) *Int. J. Mod. Phys. B* **9**, 339.

314 Perez, A., Melinon, P., Dupuis, V., Jensen, P. *et al.* (1997) *J. Phys. D*, **30**, 709.
315 Jensen, P. (1999) *Rev. Mod. Phys.*, **71**, 1695.
316 Perlarin, M. *et al.* (1997) *Chem. Phys. Lett.*, **277**, 96.
317 Melinon, P. *et al.* (1997) *J. Chem. Phys.*, **107**, 10278.
318 Palpant, B. *et al.* (1998) *Phys. Rev. B*, **57**, 1963.
319 Ray, C. *et al.* (1998) *Phys. Rev. Lett.*, **80**, 5365.
320 Bhattacharyya, S.R., Chini, T.K., Datta, D. *et al.* (2009) *J. Phys. D*, **42**, 81.
321 Liau, S.Y., Read, D.C., Pugh, W.J., Furr, J.R. and Russell, A.D. (1997) *Lett. Appl. Microbiol.*, **25**, 279.
322 Gupta, A. and Silver, S. (1998) *Nat. Biotechnol.*, **16**, 888.
323 Nomiya, K., Yoshizawa, A., Tsukagoshi, K., Kasuga, N.C., Hirakawa, S. and Watanabe, J. (2004) *J. Inorg. Biochem.* **98**, 46.
324 Morones J.R., Elechiguerra, J.L., Camacho, A., Holt, K., Kouri, J.B., Ramirez, J.P. and Yacaman, M.J. (2005) *Nanotechnology*, **16**, 2346.
325 Kukushkin, S.A. and Slezov, V.V. (1996) *Disperse Systems on Solid Surfaces*, Nauka, St. Petersburg; in Russian
326 Kukushkin, S.A. and Osipov, A.V. (1998) *Phys. Usp.*, **168**, 1083.
327 Krainov, V.P. and Smirnov, M.B. (2000) *Phys. Usp.*, **43**, 901.
328 Krainov, V.P. and Smirnov, M.B. (2002) *Phys. Rep.*, **370**, 237.
329 Rül, E. (2003) *Int. J. Mass Spectrom. Ion Process.*, **229**, 117.
330 Saalmann, U., Siedschlag, C. and Rost, J.M. (2006) *J. Phys. B*, **39**, R39.
331 Krainov, V.P., Smirnov, B.M. and Smirnov, M.B. (2007) *Phys. Rep.*, **370**, 237.
332 Ditmire, T. Donnelly, T., Rubenchik, A.M., Falcone, R.W. and Perry, M.D. (1996) *Phys. Rev. A*, **53**, 3379.
333 Ter-Avetisyan, S., Schnürer, M., Stiel, H., Vogt, U., Radloff, W., Karpov, W., Sandner, W. and Nickles, P.V. (2001) *Phys. Rev. E*, **64**, 036404.
334 Schnürer, M., Ter-Avetisyan, S., Stiel, H. *et al.* (2001) *Eur. Phys. J. D*, **14**, 331.
335 Ditmire, T. *et al.* (1999) *Nature*, **398**, 489.
336 Ditmire, T., Zweiback, J., Yanovsky, V.P. *et al.* (2000) *Phys. Plasmas*, **7**, 1993.
337 Zweiback, J., Smith, R.A., Cowan, T.E. *et al.* (2000) *Phys. Rev. Lett.*, **84**, 2634.
338 Bhattacharyya, S.R. *et al.* (2008) *ZhETF*, **134**, 1181(2008); *J. Phys. D*, **42**, 035306(2009).
339 Polya, G., Szebo, G. (1945) *Amer. J. Math.*, **67**, 1.
340 Happel, J. and Brenner, H. (1983) *Low Reynolds Number Hydrodynamics*, Prentice Hall, New York.

Index

Cluster Processes in Gases and Plasmas. Boris M. Smirnov

ISBN: 978-3-527-40943-3